Designing Embedded Systems with 32-Bit PIC Microcontrollers and MikroC

Designing Embedded Systems with 32-Bit PIC Microcontrollers and MikroC

Dogan Ibrahim

AMSTERDAM • BOSTON • HEIDELBERG • LONDON
NEW YORK • OXFORD • PARIS • SAN DIEGO
SAN FRANCISCO • SINGAPORE • SYDNEY • TOKYO

Newnes is an imprint of Elsevier

ELSEVIER

Newnes

Newnes is an imprint of Elsevier
The Boulevard, Langford Lane, Kidlington, Oxford OX5 1GB, UK
225 Wyman Street, Waltham, MA 02451, USA

First edition 2014

Notice
No responsibility is assumed by the publisher for any injury and/or damage to persons or property as a matter of products liability, negligence or otherwise, or from any use or operation of any methods, products, instructions or ideas contained in the material herein. Because of rapid advances in the medical sciences, in particular, independent verification of diagnoses and drug dosages should be made.

British Library Cataloguing in Publication Data
A catalogue record for this book is available from the British Library

Library of Congress Cataloging-in-Publication Data
A catalog record for this book is available from the Library of Congress

ISBN–13: 978-0-08-097786-7

For information on all Newnes publications
visit our website at www.newnespress.com

Printed and bound by CPI Group (UK) Ltd, Croydon, CR0 4YY

Working together
to grow libraries in
developing countries

www.elsevier.com • www.bookaid.org

Contents

Preface

A microcontroller is a single-chip microprocessor system which contains data and program memory, serial and parallel I/O, timers, external and internal interrupts, all integrated into a single chip that can be purchased for as little as $2.00. About 40% of microcontroller applications are in office automation, such as PCs, laser printers, fax machines, intelligent telephones, and so forth. About one-third of microcontrollers are found in consumer electronic goods. Products like CD players, hi-fi equipment, video games, washing machines, cookers and so on fall into this category. The communications market, automotive market, and the military share the rest of the application areas.

This book is written for advanced students, for practicing engineers, and for hobbyists who want to learn more about the programming and applications of 32-bit series of microcontrollers. The book has been written with the assumption that the reader has taken a course on digital logic design, and has been exposed to writing programs using at least one high-level programming language. Knowledge of the C programming language will be useful. Also, familiarity with at least one member of the PIC series of microcontrollers (e.g. PIC16 or PIC18) will be an advantage. The knowledge of assembly language programming is not required because all the projects in the book are based on using the C language.

Chapter 1 presents the basic features of microcontrollers and the important topic of numbering systems, and also describes how to convert between different number bases.

Chapter 2 provides a review of the PIC32 series of 32-bit microcontrollers. Various features of these microcontrollers are described in detail.

Chapter 3 provides a short tutorial on the C language and then examines the features of the mikroC Pro for PIC32 compiler used extensively in the book.

Chapter 4 is about the advanced features of the mikroC Pro for PIC32 language. Topics such as built-in functions and libraries are discussed in this chapter with examples.

Chapter 5 explores the various software and hardware development tools for the PIC32 series of 32-bit microcontrollers. Examples of various commercially available development kits are given in this chapter. Also, development tools such as simulators, emulators, and in-circuit debuggers are described with simple example projects.

Chapter 6 describes the use of program development tools such as flowcharts and the program description language.

Chapter 7 provides some simple projects using the PIC32 series of 32-bit microcontrollers and the mikroC Pro for PIC32 C language compiler. All the projects in this chapter are based on the PIC32MX460F512L 32-bit microcontroller, and all the projects have been tested and working. This chapter should be useful for those who are new to using the 32-bit microcontrollers, and for those who want to extend their knowledge of programming the PIC32 series of microcontrollers using the mikroC Pro for PIC32 language.

Chapter 8 covers more advanced projects such as digital filtering, using the PIC32 series of 32-bit microcontrollers. All the projects given in the book have been tested and are fully working. The block diagram, circuit diagram, full program listing, and operation of all the projects are described in detail.

Finally, the Appendix describes basic features of the popular MPLAB PIC32 C compiler, developed by Microchip Inc. for their 32-bit series of microcontrollers. An example application is also given to show how to use this compiler.

Dogan Ibrahim
London, 2012

Acknowledgments

The following material is reproduced in this book with the kind permission of the respective copyright holders and may not be reprinted, or reproduced in any way, without their prior consent.

Figures 2.2, 2.3, 2.5, 2.11, 2.15—2.38 are taken from Microchip Technology Inc. Data Sheet PIC32MX3XX/4XX (DS61143E). Figures 5.3—5.5 are taken from the website of Microchip Technology Inc.

Figures 5.6—5.11 are taken from the website of Digilent Inc.

Figures 5.12—5.16, 5.19, and 5.20 are taken from the website of mikroElektronica.

Figures 5.17 and 5.18 are taken from the website of Olimex.

PIC®, PICSTART®, and MPLAB® are all the trademarks of Microchip Technology Inc.

CHAPTER 1

Microcomputer Systems

Chapter Outline

1.1 Introduction

The term microcomputer is used to describe a system that includes a minimum of a microprocessor, program memory, data memory, and input—output (I/O). Some microcomputer systems include additional components such as timers, counters, analog-to-digital (A/D) converters, and so on. Thus, a microcomputer system can be anything from a large computer having hard disks, floppy disks, and printers to a single-chip embedded controller.

In this book, we are going to consider only the type of microcomputers that consist of a single silicon chip. Such microcomputer systems are also called microcontrollers and they are used in many household goods such as microwave ovens, TV remote control units (CUs), cookers, hi-fi equipment, compact disc players, personal computers, fridges, and so on. There are a large number of microcontrollers available in the market. In this book, we shall be looking at the programming and system design using the 32-bit *programmable interface controller* (PIC) series of microcontrollers manufactured by Microchip Technology Inc.

1.2 Microcontroller Systems

A microcontroller is a single-chip computer. *Micro* suggests that the device is small, and *controller* suggests that the device can be used in control applications. Another term used for microcontrollers is *embedded controller* since most of the microcontrollers are built into (or embedded in) the devices they control.

A microprocessor differs from a microcontroller in many ways. The main difference is that a microprocessor requires several other components for its operation, such as program memory and data memory, I/O devices, and external clock circuit. A microcontroller on the other hand has all the support chips incorporated inside the same chip. All microcontrollers operate on a set of instructions (or the user program) stored in their memory. A microcontroller fetches the instructions from its program memory one by one, decodes these instructions, and then carries out the required operations.

Microcontrollers have traditionally been programmed using the assembly language of the target device. Although the assembly language is fast, it has several disadvantages. An assembly program consists of mnemonics and it is difficult to learn and maintain a program written using the assembly language. Also, microcontrollers manufactured by different firms have different assembly languages and the user is required to learn a new language every time a new microcontroller is to be used. Microcontrollers can also be programmed using a high-level language, such as BASIC, PASCAL, and C. High-level languages have the advantage that it is much easier to learn a high-level language than an assembler. Also, very large and complex programs can easily be developed using a high-level language. In this book, we shall be learning the programming of 32-bit PIC microcontrollers using the popular mikroC Pro for PIC C language, developed by mikroElektronika.

In general, a single chip is all that is required to have a running microcontroller system. In practical applications, additional components may be required to allow a microcomputer to interface to its environment. With the advent of the PIC family of microcontrollers, the development time of an electronic project has reduced to several hours.

Basically, a microcomputer (or microcontroller) executes a user program which is loaded in its program memory. Under the control of this program, data are received from external devices (inputs), manipulated and then sent to external devices (outputs). For example, in a microcontroller-based fluid-level control system, the fluid level is read by the microcomputer via a level-sensor device and the microcontroller attempts to control the fluid level at the required value. If the fluid level is low, the microcomputer operates a pump to draw more fluid from the reservoir in order to keep the fluid at the required level. Figure 1.1 shows the block diagram of our simple fluid-level control system.

The system shown in Figure 1.1 is a very simplified fluid-level control system. In a more sophisticated system, we may have a keypad to set the required fluid level, and a liquid-crystal display (LCD) to display the current level in the tank. Figure 1.2 shows the block diagram of this more sophisticated fluid-level control system.

We can make our design even more sophisticated (Figure 1.3) by adding an audible alarm to inform us if the fluid level is outside the required value. Also, the actual level at any time can be sent to a PC every second for archiving and further processing. For example, a graph of the daily fluid-level changes can be plotted on the PC. As you can see, because the microcontrollers are programmable, it is very easy to make the final system as simple or as complicated as we like.

A microcontroller is a very powerful tool that allows a designer to create sophisticated I/O data manipulation under program control. Microcontrollers are classified by the number of bits they process. Eight-bit devices are the most popular ones and are used in most low-cost

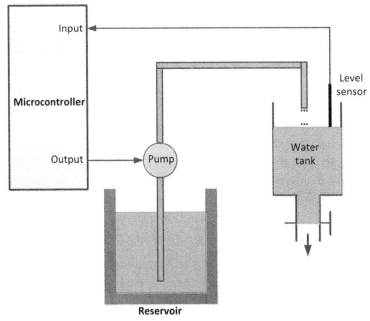

Figure 1.1
Microcontroller-based fluid-level control system.

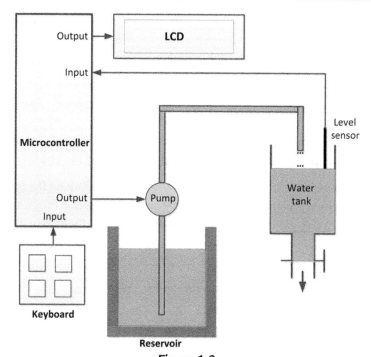

Figure 1.2
Fluid-level control system with a keypad and an LCD.

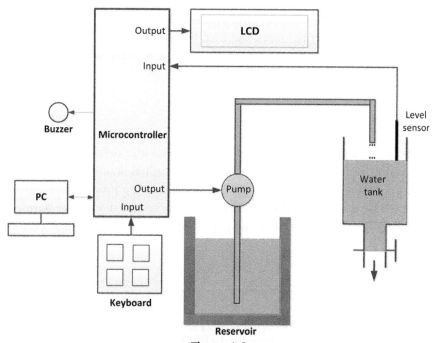

Figure 1.3
More sophisticated fluid-level controller.

low-speed microcontroller-based applications. Microcontrollers of 16 and 32 bits are much more powerful, but usually more expensive and their use may not be justified in many small- to medium-size general-purpose applications.

The simplest microcontroller architecture consists of a microprocessor, memory, and I/O. The microprocessor consists of a central processing unit (CPU) and the CU. The CPU is the brain of the microcontroller and this is where all the arithmetic and logic operations are performed. The CU controls the internal operations of the microprocessor and sends out control signals to other parts of the microcontroller to carry out the required instructions.

Memory is an important part of a microcontroller system. Depending on the type used, we can classify memories into two groups: program memory and data memory. Program memory stores the programs written by the programmer and this memory is usually nonvolatile, i.e. data are not lost after the removal of power. Data memory is where the temporary data used in a program are stored and this memory is usually volatile, i.e. data are lost after the removal of power.

There are basically six types of memories, summarized as follows:

1.2.1 Random Access Memory

Random access memory (RAM) is a general-purpose memory which usually stores the user data in a program. RAM memory is volatile in the sense that it cannot retain data in the absence of power, i.e. data are lost after the removal of power. Most microcontrollers have some amount of internal RAM. Several kilobytes is a common amount, although some microcontrollers have much more, and some less. For example, the PIC32MX460F512L 32-bit microcontroller has 512 kB of RAM. In general, it is possible to extend the memory by adding external memory chips.

1.2.2 Read Only Memory

Read only memory (ROM) usually holds program or fixed user data. ROM is nonvolatile. If power is removed from ROM and then reapplied, the original data will still be there. ROM memories are programmed at factory during the manufacturing process and their contents cannot be changed by the user. ROM memories are only useful if we have developed a program and wish to order several thousand copies of it, or if we wish to store some configuration data.

1.2.3 Programmable Read Only Memory

Programmable read only memory (PROM) is a type of ROM that can be programmed in the field, often by the end user, using a device called a PROM programmer. Once a PROM has been programmed, its contents cannot be changed. PROMs are usually used in low-production applications where only several such memories are required.

1.2.4 Erasable Programmable Read Only Memory

Erasable programmable read only memory (EPROM) is similar to ROM, but the EPROM can be programmed using a suitable programming device. EPROM memories have a small, clear glass window on the top of the chip where the data can be erased under strong ultraviolet light. Once the memory is programmed, the window can be covered with dark tape to prevent accidental erasure of the data. An EPROM memory must be erased before it can be reprogrammed. Many development versions of microcontrollers are manufactured with EPROM memories where the user program can be stored. These memories are erased and reprogrammed until the user is satisfied with the program. Some versions of EPROMs, known as one-time programmable (OTP), can be programmed using a suitable programmer device but these memories cannot be erased. OTP memories cost much less than the EPROMs. OTP is useful after a project has been developed completely and it is required to make many copies of the program memory.

1.2.5 Electrically Erasable Programmable Read Only Memory

Electrically erasable programmable read only memory (EEPROM) is a nonvolatile memory. These memories can be erased and also be reprogrammed using suitable programming devices. EEPROMs are used to save configuration information, maximum and minimum values, identification data, etc.

1.2.6 Flash EEPROM

This is another version of EEPROM-type memory. This memory has become popular in microcontroller applications and is generally used to store the user program. Flash EEPROM is nonvolatile and is usually very fast. The data can be erased and then reprogrammed using a suitable programming device. These memories can also be programmed without removing them from their circuits. Some microcontrollers have only 1 K flash EEPROM while some others have 32 K or more.

1.3 Microcontroller Features

Microcontrollers from different manufacturers have different architectures and different capabilities. Some may suit a particular application while others may be totally unsuitable for the same application. The hardware features of microcontrollers, in general, are described in this section.

1.3.1 Supply Voltage

Most microcontrollers operate with the standard logic voltage of +5 V. Some microcontrollers can operate at as low as +2.7 V and some will tolerate +6 V without any problems. You should check the manufacturers' data sheets about the allowed limits of the power supply voltage. For example, PIC32MX460F512L 32-bit microcontrollers can operate with a power supply of +2.3 to +3.6 V.

A voltage regulator circuit is usually used to obtain the required power supply voltage when the device is to be operated from a mains adaptor or batteries. For example, a 5 V regulator may be required if the microcontroller and peripheral devices operate from a +5 or +3.3 V supply and a 9 V battery is to be used as the power supply.

1.3.2 The Clock

All microcontrollers require a clock (or an oscillator) to operate. The clock is usually provided by connecting external timing devices to the microcontroller. Most microcontrollers will generate clock signals when a crystal and two small capacitors are connected. Some will operate with resonators or external resistor–capacitor pair. Some microcontrollers have built-in timing circuits and they do not require any external timing components. If your application is not time sensitive, you should use external or internal (if available) resistor–capacitor timing components for simplicity and low cost.

An instruction is executed by fetching it from the memory and then decoding it. This usually takes several clock cycles and is known as the *instruction cycle*. PIC32 series of microcontrollers can operate with clock frequencies up to 80 MHz.

1.3.3 Timers

Timers are important parts of any microcontroller. A timer is basically a counter which is driven either from an external clock pulse or from the internal oscillator of the microcontroller. A PIC32 microcontroller can have 16- or 32-bit-wide timers (two 16-bit timers are combined to create a 32-bit timer). Data can be loaded into a timer under program control and the timer can be stopped or started by program control. Most timers can be configured to generate an interrupt when they reach a certain count (usually when they overflow). The interrupt can be used by the user program to carry out accurate timing-related operations inside the microcontroller.

Some microcontrollers offer capture and compare facilities where a timer value can be read when an external event occurs, or the timer value can be compared to a preset value and an interrupt can be generated when this value is reached. PIC32 microcontrollers can have up to five capture inputs.

1.3.4 Watchdog

Most microcontrollers have at least one watchdog facility. The watchdog is basically a timer which is refreshed by the user program and a reset occurs if the program fails to refresh the watchdog. The watchdog timer is used to detect a system problem, such as the program being in an endless loop. A watchdog is a safety feature that prevents runaway software and stops the microcontroller from executing meaningless and unwanted code. Watchdog facilities are commonly used in real-time systems for safety, where it may be required to regularly check the successful termination of one or more activities.

1.3.5 Reset Input

A reset input is used to reset a microcontroller externally. Resetting puts the microcontroller into a known state such that the program execution starts from a known address. An external reset action is usually achieved by connecting a push-button switch to the reset input such that the microcontroller can be reset when the switch is pressed.

1.3.6 Interrupts

Interrupts are very important concepts in microcontrollers. An interrupt causes the microcontroller to respond to external and internal (e.g. a timer) events very quickly. When an interrupt occurs, the microcontroller leaves its normal flow of program execution and jumps to a special part of the program, known as the *interrupt service routine* (ISR). The program code inside the ISR is executed and upon return from the ISR, the program resumes its normal flow of execution.

The ISR starts from a fixed address of the program memory. This address is also known as the *interrupt vector address*. Some microcontrollers with multi-interrupt features have just one interrupt vector address, while some others have unique interrupt vector addresses, one for each interrupt source. Interrupts can be nested such that a new interrupt can suspend the execution of another interrupt. Another important feature of a microcontroller with multi-interrupt capability is that different interrupt sources can be given different levels of priority.

1.3.7 Brown-out Detector

Brown-out detectors are also common in many microcontrollers and they reset a microcontroller if the supply voltage falls below a nominal value. Brown-out detectors are safety features and they can be employed to prevent unpredictable operation at low voltages, especially to protect the contents of EEPROM-type memories.

1.3.8 A/D Converter

An A/D converter is used to convert an analog signal such as voltage to a digital form, so that it can be read and processed by a microcontroller. Some microcontrollers have built-in A/D converters. It is also possible to connect an external A/D converter to any type of microcontroller. PIC32 microcontroller A/D converters are usually 10-bits wide, having 1024 quantization levels. Most PIC microcontrollers with A/D features have multiplexed A/D converters where more than one analog-input channel is provided. For example, PIC32MX460F512L microcontroller has 16 channels of A/D converters, each 10 bits wide.

The A/D conversion process must be started by the user program and it may take several tens of microseconds for a conversion to complete. A/D converters usually generate interrupts when a conversion is complete, so that the user program can read the converted data quickly and efficiently.

A/D converters are very useful in control and monitoring applications since most sensors (e.g. temperature sensor, pressure sensor, force sensor, etc.) produce analog-output voltages.

1.3.9 Serial I/O

Serial communication (also called RS232 communication) enables a microcontroller to be connected to another microcontroller or to a PC using a serial cable. Some microcontrollers have built-in hardware called universal asynchronous receiver transmitter (UART) to implement a serial communication interface. The baud rate (bits per second) and the data format can usually be selected by the user program. If any serial I/O hardware is not provided, it is easy to develop software to implement serial data communication using any I/O pin of a microcontroller. PIC32MX460F512L microcontroller provides two UART modules, each with RS232, RS485, LIN, or IrDA compatible interfaces.

In addition, two serial peripheral interface and two Integrated Inter Connect hardware bus interfaces are available on the PIC32MX460F512L microcontroller. As we shall see in later chapters, mikroC Pro for PIC32 language provides special instructions for reading and writing to other compatible serial devices.

1.3.10 EEPROM Data Memory

EEPROM-type data memory is also very common in many microcontrollers. The advantage of an EEPROM memory is that the programmer can store nonvolatile data in such a memory, and can also change these data whenever required. For example, in a temperature monitoring application, the maximum and the minimum temperature readings can be stored in an EEPROM memory. Then, if the power supply is removed for whatever reason, the values of the latest readings will still be available in the EEPROM memory.

mikroC Pro for PIC32 language provides special instructions for reading and writing to the EEPROM memory of a PIC32 microcontroller.

1.3.11 LCD Drivers

LCD drivers enable a microcontroller to be connected to an external LCD display directly. These drivers are not common since most of the functions provided by them can easily be implemented in software.

1.3.12 Analog Comparator

Analog comparators are used where it is required to compare two analog voltages. These modules are not available in low-end PIC microcontrollers. PIC32MX460F512L microcontrollers have two built-in analog comparators.

1.3.13 Real-Time Clock

Real-time clock enables a microcontroller to have absolute date and time information continuously. Built-in real-time clocks are not common in most microcontrollers since they can easily be implemented by either using an external dedicated real-time clock chip or by writing a program. For example, PIC32MX460F512L microcontroller has built-in real-time clock and calendar module.

1.3.14 Sleep Mode

Some microcontrollers offer built-in sleep modes where executing this instruction puts the microcontroller into a mode where the internal oscillator is stopped and the power consumption is reduced to an extremely low level. The main reason of using the sleep mode is to conserve the battery power when the microcontroller is not doing anything useful. The microcontroller usually wakes up from the sleep mode by external reset or by a watchdog time-out.

1.3.15 Power-on Reset

Some microcontrollers have built-in power-on reset circuits which keep the microcontroller in reset state until all the internal circuitry has been initialized. This feature is very useful as it starts the microcontroller from a known state on power-up. An external reset can also be provided where the microcontroller can be reset when an external button is pressed.

1.3.16 Low-Power Operation

Low-power operation is especially important in portable applications where the microcontroller-based equipment is operated from batteries, and very long battery life is a main requirement. Some microcontrollers can operate with <2 mA with 5 V supply, and around 15 µA at 3 V supply.

1.3.17 Current Sink/Source Capability

This is important if the microcontroller is to be connected to an external device which may draw large current for its operation. PIC32 microcontrollers can source and sink 18 mA of current from each output port pin. This current is usually sufficient to drive LEDs, small lamps, buzzers, small relays, etc. The current capability can be increased by connecting external transistor switching circuits or relays to the output port pins.

1.3.18 USB Interface

USB is currently a very popular computer interface specification used to connect various peripheral devices to computers and microcontrollers. PIC32 microcontrollers normally provide built-in USB modules.

1.3.19 Motor Control Interface

Some PIC microcontrollers, for example, PIC18F2x31 provide motor control interface.

1.3.20 CAN Interface

CAN bus is a very popular bus system used mainly in automation applications. Some PIC18F series of microcontrollers (e.g. PIC18F4680) provide CAN interface capabilities.

1.3.21 Ethernet Interface

Some PIC microcontrollers (e.g. PIC18F97J60) provide Ethernet interface capabilities. Such microcontrollers can easily be used in network-based applications.

1.3.22 ZigBee Interface

ZigBee is an interface similar to Bluetooth and is used in low-cost wireless home automation applications. Some PIC series of microcontrollers provide ZigBee interface capabilities making the design of such wireless systems very easy.

1.3.23 Multiply and Divide Hardware

PIC32 microcontrollers have built-in hardware for fast multiplication and division operations.

1.3.24 Operating Temperature

It is important to know the operating temperature range of a microcontroller chip before a project is developed. PIC32 microcontrollers can operate in the temperature range −40 to +105 °C.

1.3.25 Pulse Width Modulated Outputs

Most microcontrollers provide pulse width modulated (PWM) outputs for driving analog devices such as motors, lamps, and so on. The PWM is usually a separate module and runs in hardware, independent of the CPU.

1.3.26 JTAG Interface

Some high-end microcontrollers (e.g. PIC32MX460F512L) have built-in JTAG interfaces for enhanced debugging interface.

1.3.27 Package Size

It is sometimes important to know the package size of a microcontroller chip before a microcontroller is chosen for a particular project. Low-end microcontrollers are usually packaged in 18, 28 or 40-pin DIL packages. High-end microcontrollers, for example, PIC32MX460F512L is housed in a 100-pin TQFP-type package.

1.3.28 Direct Memory Access

Some high-end microcontrollers have built-in direct memory access (DMA) channels that can be used to transfer large amounts of data between different devices without the intervention of the CPU. For example, PIC32MX460F512L microcontroller has four DMA channels.

1.4 Microcontroller Architectures

Usually two types of architectures are used in microcontrollers (Figure 1.4): *Von Neumann* architecture and *Harvard* architecture. Von Neumann architecture is used by a large percentage of microcontrollers and here all memory space is on the same bus and instruction and data use the same bus. In the Harvard architecture (used by most PIC microcontrollers), code and data are on separate busses and this allows the code and data to be fetched simultaneously, resulting in an improved performance.

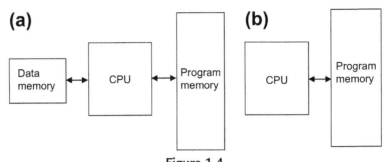

Figure 1.4
(a) Harvard Architecture, (b) Von Neumann Architecture.

1.4.1 Reduced Instruction Set Computer and Complex Instruction Computer

Reduced instruction set computer (RISC) and complex instruction computer (CISC) refer to the instruction set of a microcontroller. In an eight-bit RISC microcontroller, data are 8 bits wide but the instruction words are more than 8-bits wide (usually 12, 14 or 16 bits) and the instructions occupy one word in the program memory. Thus, the instructions are fetched and executed in one cycle, resulting in an improved performance.

In a CISC microcontroller, both data and instructions are 8 bits wide. CISC microcontrollers usually have over 200 instructions. Data and code are on the same bus and cannot be fetched simultaneously.

1.5 8, 16, or 32 Bits?

People are usually confused for making a decision between 8, 16, or 32 bits of microcontrollers. It is important to realize that the number of bits just refers to the width of the data handled by the processor. This number actually limits the precision of mathematical operations carried out by the CPU (although it is possible to emulate higher order mathematics in software or by using special hardware).

In general, 8-bit microcontrollers have been around since the first days of the microcontroller development. They are cheap, easy to use (only small package size), low speed, and can be used in most general-purpose control and data manipulation operations. For example, it is still very efficient to design low- to medium-speed control systems (e.g. temperature control, fluid-level control, or robotics applications) using 8-bit microcontrollers. In such applications, low cost is more important than high speed. Many commercial and industrial applications fall into this category and can easily be designed using standard 8-bit microcontrollers.

Microcontrollers of 16 and 32 bit on the other hand usually cost more, but they offer much higher speeds, and much higher precision in mathematical operations. These microcontrollers are usually housed in larger packages (e.g. 64 or 100 pins) and offer much more features, such as larger data and program memories, more timer/counter modules, more and faster A/D channels, more I/O ports, and so on. Microcontrollers of 32 bit are usually used in high-speed, real-time digital signal processing applications, where also high precision is a requirement, such as digital image processing, digital audio processing, and so on. Most consumer products, such as electronic games and mobile phones, are based on 32-bit processors as they demand high-speed real-time operation with color graphical displays and with touch-screen panels. Other high-speed applications such as video capturing, image filtering, video editing, video streaming, speech recognition, and speech processing all require very fast 32-bit processors with lots of data and program memories, and very high precision while implementing the digital signal processing algorithms.

This book is about 32-bit PIC microcontrollers. We shall be seeing the basic architectures and features of these microcontrollers. In addition, many working projects will be given in the book to illustrate how these microcontrollers can be programmed and used in real applications.

The remainder of this chapter is about the important topic of number systems. The readers should be familiar with the number systems before attempting to develop programs for microcontrollers.

1.6 Number Systems

The efficient use of a microprocessor or microcontroller requires a working knowledge of binary, decimal, and hexadecimal numbering systems. This section provides a background for those who are unfamiliar with these numbering systems and who do not know how to convert from one number system to another one.

Number systems are classified according to their bases. The numbering system used in everyday life is base 10 or the decimal number system. The most commonly used numbering system in microprocessor and microcontroller applications is base 16, or hexadecimal. In addition, base 2 (binary), or base 8 (or octal) number systems are also used.

1.6.1 Decimal Number System

As you all know, the numbers in this system are 0, 1, 2, 3, 4, 5, 6, 7, 8, and 9. We can use the subscript 10 to indicate that a number is in decimal format. For example, we can show decimal number 235 as 235_{10}.

In general, a decimal number is represented as follows:

$$a_n \times 10^n + a_{n-1} \times 10^{n-1} + a_{n-2} \times 10^{n-2} + \cdots\cdots + a_0 \times 10^0$$

For example, decimal number 825_{10} can be shown as follows:

$$825_{10} = 8 \times 10^2 + 2 \times 10^1 + 5 \times 10^0$$

Similarly, decimal number 26_{10} can be shown as follows:

$$26_{10} = 2 \times 10^1 + 6 \times 10^0$$

or

$$3359_{10} = 3 \times 10^3 + 3 \times 10^2 + 5 \times 10^1 + 9 \times 10^0$$

1.6.2 Binary Number System

In binary number system, there are two numbers: 0 and 1. We can use the subscript 2 to indicate that a number is in binary format. For example, we can show binary number 1011 as 1011_2.

In general, a decimal number is represented as follows:

$$a_n \times 2^n + a_{n-1} \times 2^{n-1} + a_{n-2} \times 2^{n-2} + \cdots\cdots + a_0 \times 2^0$$

For example, binary number 1110_2 can be shown as follows:

$$1110_2 = 1 \times 2^3 + 1 \times 2^2 + 1 \times 2^1 + 0 \times 2^0$$

Similarly, binary number 10001110_2 can be shown as follows:

$$10001110_2 = 1 \times 2^7 + 0 \times 2^6 + 0 \times 2^5 + 0 \times 2^4 + 1 \times 2^3 + 1 \times 2^2 + 1 \times 2^1 + 0 \times 2^0$$

1.6.3 Octal Number System

In octal number system, the valid numbers are 0, 1, 2, 3, 4, 5, 6, and 7. We can use the subscript 8 to indicate that a number is in octal format. For example, we can show octal number 23 as 23_8.

In general, an octal number is represented as follows:

$$a_n \times 8^n + a_{n-1} \times 8^{n-1} + a_{n-2} \times 8^{n-2} + \cdots\cdots\cdots + a_0 \times 8^0$$

For example, octal number 237_8 can be shown as follows:

$$237_8 = 2 \times 8^2 + 3 \times 8^1 + 7 \times 8^0$$

Similarly, octal number 1777_8 can be shown as follows:

$$1777_8 = 1 \times 8^3 + 7 \times 8^2 + 7 \times 8^1 + 7 \times 8^0$$

1.6.4 Hexadecimal Number System

In hexadecimal number system, the valid numbers are 0, 1, 2, 3, 4, 5, 6, 7, 8, 9, A, B, C, D, E, and F. We can use the subscript 16 of H to indicate that a number is in hexadecimal format. For example, we can show hexadecimal number 1F as $1F_{16}$ or as $1F_H$.

In general, a hexadecimal number is represented as follows:

$$a_n \times 16^n + a_{n-1} \times 16^{n-1} + a_{n-2} \times 16^{n-2} + \cdots\cdots + a_0 \times 16^0$$

For example, hexadecimal number $2AC_{16}$ can be shown as follows:

$$2AC_{16} = 2 \times 16^2 + 10 \times 16^1 + 12 \times 16^0$$

Similarly, hexadecimal number $3FFE_{16}$ can be shown as follows:

$$3FFE_{16} = 3 \times 16^3 + 15 \times 16^2 + 15 \times 16^1 + 14 \times 16^0$$

1.7 Converting Binary Numbers into Decimal

To convert a binary number into decimal, write the number as the sum of the powers of 2.

Example 1.1

Convert binary number 1011_2 into decimal.

Solution 1.1
Write the number as the sum of the powers of 2:

$$\begin{aligned} 1011_2 &= 1 \times 2^3 + 0 \times 2^2 + 1 \times 2^1 + 1 \times 2^0 \\ &= 8 + 0 + 2 + 1 \\ &= 11 \end{aligned}$$

or, $1011_2 = 11_{10}$

Example 1.2

Convert binary number 11001110_2 into decimal.

Solution 1.2
Write the number as the sum of the powers of 2 as follows:

$$\begin{aligned} 11001110_2 &= 1 \times 2^7 + 1 \times 2^6 + 0 \times 2^5 + 0 \times 2^4 + 1 \times 2^3 + 1 \times 2^2 + 1 \times 2^1 + 0 \times 2^0 \\ &= 128 + 64 + 0 + 0 + 8 + 4 + 2 + 0 \\ &= 206 \end{aligned}$$

or, $11001110_2 = 206_{10}$

Table 1.1: Decimal Equivalent of Binary Numbers

Binary	Decimal	Binary	Decimal
00000000	0	00010000	16
00000001	1	00010001	17
00000010	2	00010010	18
00000011	3	00010011	19
00000100	4	00010100	20
00000101	5	00010101	21
00000110	6	00010110	22
00000111	7	00010111	23
00001000	8	00011000	24
00001001	9	00011001	25
00001010	10	00011010	26
00001011	11	00011011	27
00001100	12	00011100	28
00001101	13	00011101	29
00001110	14	00011110	30
00001111	15	00011111	31

Table 1.1 shows the decimal equivalent of numbers from 0 to 31.

1.8 Converting Decimal Numbers into Binary

To convert a decimal number into binary, divide the number repeatedly by two and take the remainders. The first remainder is the least significant digit (LSD), and the last remainder is the most significant digit (MSD).

Example 1.3

Convert decimal number 28_{10} into binary.

Solution 1.3
Divide the number by two repeatedly and take the remainders:

```
28/2  →  14   Remainder 0 (LSD)
14/2  →  7    Remainder 0
7/2   →  3    Remainder 1
3/2   →  1    Remainder 1
1/2   →  0    Remainder 1 (MSD)
```

The required binary number is 11100_2.

Example 1.4

Convert decimal number 65_{10} into binary.

Solution 1.4

Divide the number by two repeatedly and take the remainders:

```
65/2  →  32   Remainder 1 (LSD)
32/2  →  16   Remainder 0
16/2  →   8   Remainder 0
8/2   →   4   Remainder 0
4/2   →   2   Remainder 0
2/2   →   1   Remainder 0
1/2   →   0   Remainder 1 (MSD)
```

The required binary number is 1000001_2.

Example 1.5

Convert decimal number 122_{10} into binary.

Solution 1.5

Divide the number by two repeatedly and take the remainders:

```
122/2  →  61   Remainder 0 (LSD)
61/2   →  30   Remainder 1
30/2   →  15   Remainder 0
15/2   →   7   Remainder 1
7/2    →   3   Remainder 1
3/2    →   1   Remainder 1
1/2    →   0   Remainder 1 (MSD)
```

The required binary number is 1111010_2.

1.9 Converting Binary Numbers into Hexadecimal

To convert a binary number into hexadecimal, arrange the number in groups of four and find the hexadecimal equivalent of each group. If the number cannot be divided exactly into groups of four, insert zeroes to the left-hand side of the number.

Example 1.6

Convert binary number 10011111_2 into hexadecimal.

Solution 1.6

First, divide the number into groups of four and then find the hexadecimal equivalent of each group:

```
10011111 = 1001 1111
              9    F
```

The required hexadecimal number is $9F_{16}$.

Example 1.7

Convert binary number 11101111000011110_2 into hexadecimal.

Solution 1.7

First, divide the number into groups of four and then find the equivalent of each group:

```
1110111100001110 = 1110 1111 0000 1110
                     E    F    0    E
```

The required hexadecimal number is $EF0E_{16}$.

Example 1.8

Convert binary number 111110_2 into hexadecimal.

Solution 1.8

Since the number cannot be divided exactly into groups of four, we have to insert zeroes to the left of the number:

```
111110 = 0011 1110
          3    E
```

The required hexadecimal number is $3E_{16}$.

Table 1.2 shows the hexadecimal equivalent of numbers 0–31.

Table 1.2: Hexadecimal Equivalents of Decimal Numbers

Decimal	Hexadecimal	Decimal	Hexadecimal
0	0	16	10
1	1	17	11
2	2	18	12
3	3	19	13
4	4	20	14
5	5	21	15
6	6	22	16
7	7	23	17
8	8	24	18
9	9	25	19
10	A	26	1A
11	B	27	1B
12	C	28	1C
13	D	29	1D
14	E	30	1E
15	F	31	1F

1.10 Converting Hexadecimal Numbers into Binary

To convert a hexadecimal number into binary, write the 4-bit binary equivalent of each hexadecimal digit.

Example 1.9

Convert hexadecimal number $A9_{16}$ into binary.

Solution 1.9
Writing the binary equivalent of each hexadecimal digit:

$$A = 1010_2 \quad 9 = 1001_2$$

The required binary number is 10101001_2.

Example 1.10

Convert hexadecimal number $FE3C_{16}$ into binary.

Solution 1.10
Writing the binary equivalent of each hexadecimal digit:

$$F = 1111_2 \quad E = 1110_2 \quad 3 = 0011_2 \quad C = 1100_2$$

The required binary number is 1111111000111100_2.

1.11 Converting Hexadecimal Numbers into Decimal

To convert a hexadecimal number into decimal, we have to calculate the sum of the powers of 16 of the number.

Example 1.11

Convert hexadecimal number $2AC_{16}$ into decimal.

Solution 1.11
Calculating the sum of the powers of 16 of the number:

$$
\begin{aligned}
2AC_{16} &= 2 \times 16^2 + 10 \times 16^1 + 12 \times 16^0 \\
&= 512 + 160 + 12 \\
&= 684
\end{aligned}
$$

The required decimal number is 684_{10}.

Example 1.12

Convert hexadecimal number EE_{16} into decimal.

Solution 1.12
Calculating the sum of the powers of 16 of the number:

$$EE_{16} = 14 \times 16^1 + 14 \times 16^0$$
$$= 224 + 14$$
$$= 238$$

The required decimal number is 238_{10}.

1.12 Converting Decimal Numbers into Hexadecimal

To convert a decimal number into hexadecimal, divide the number repeatedly into 16 and take the remainders. The first remainder is the LSD, and the last remainder is the MSD.

Example 1.13

Convert decimal number 238_{10} into hexadecimal.

Solution 1.13
Dividing the number repeatedly by 16:

```
238/16  →  14   Remainder 14 (E) (LSD)
14/16   →   0   Remainder 14 (E) (MSD)
```

The required hexadecimal number is EE_{16}.

Example 1.14

Convert decimal number 684_{10} into hexadecimal.

Solution 1.14
Dividing the number repeatedly into 16:

```
684/16  →  42   Remainder 12 (C) (LSD)
42/16   →   2   Remainder 10 (A)
2/16    →   0   Remainder 2  (MSD)
```

The required hexadecimal number is $2AC_{16}$.

1.13 Converting Octal Numbers into Decimal

To convert an octal number into decimal, calculate the sum of the powers of 8 of the number.

Example 1.15

Convert octal number 15_8 into decimal.

Solution 1.15
Calculating the sum of the powers of 8 of the number:

$$15_8 = 1 \times 8^1 + 5 \times 8^0$$
$$= 8 + 5$$
$$= 13$$

The required decimal number is 13_{10}.

Example 1.16

Convert octal number 237_8 into decimal.

Solution 1.16
Calculating the sum of the powers of 8 of the number:

$$237_8 = 2 \times 8^2 + 3 \times 8^1 + 7 \times 8^0$$
$$= 128 + 24 + 7$$
$$= 159$$

The required decimal number is 159_{10}.

1.14 Converting Decimal Numbers into Octal

To convert a decimal number into octal, divide the number repeatedly by 8 and take the remainders. The first remainder is the LSD, and the last remainder is the MSD.

Example 1.17

Convert decimal number 159_{10} into octal.

Solution 1.17
Dividing the number repeatedly by 8:

```
159/8  →  19  Remainder 7 (LSD)
19/8   →   2  Remainder 3
2/8    →   0  Remainder 2 (MSD)
```

The required octal number is 237_8.

Example 1.18

Convert decimal number 460_{10} into octal.

Solution 1.18
Dividing the number repeatedly by 8:

```
460/8  →  57 Remainder 4 (LSD)
57/8   →  7 Remainder 1
7/8    →  0 Remainder 7 (MSD)
```

The required octal number is 714_8.

Table 1.3 shows the octal equivalent of decimal numbers 0–31.

1.15 Converting Octal Numbers into Binary

To convert an octal number into binary, write the 3-bit binary equivalent of each octal digit.

Example 1.19

Convert octal number 177_8 into binary.

Solution 1.19
Write the binary equivalent of each octal digit:

$1 = 001_2 \quad 7 = 111_2 \quad 7 = 111_2$

The required binary number is 001111111_2.

Table 1.3: Octal Equivalents of Decimal Numbers

Decimal	Octal	Decimal	Octal
0	0	16	20
1	1	17	21
2	2	18	22
3	3	19	23
4	4	20	24
5	5	21	25
6	6	22	26
7	7	23	27
8	10	24	30
9	11	25	31
10	12	26	32
11	13	27	33
12	14	28	34
13	15	29	35
14	16	30	36
15	17	31	37

Example 1.20

Convert octal number 75_8 into binary.

Solution 1.20
Write the binary equivalent of each octal digit:

$7 = 111_2 \quad 5 = 101_2$

The required binary number is 111101_2.

1.16 Converting Binary Numbers into Octal

To convert a binary number into octal, arrange the number in groups of three and write the octal equivalent of each digit.

Example 1.21

Convert the binary number 110111001_2 into octal.

Solution 1.21
Arranging in groups of three:

$110111001 = 110 \ 111 \ 001$
$\qquad\qquad\quad 6 \quad 7 \quad 1$

The required octal number is 671_8.

1.17 Negative Numbers

The most significant bit of a binary number is usually used as the sign bit. By convention, for positive numbers, this bit is 0, and for negative numbers, this bit is 1. Figure 1.5 shows the

Binary number	Decimal equivalent
0111	+7
0110	+6
0101	+5
0100	+4
0011	+3
0010	+2
0001	+1
0000	0
1111	−1
1110	−2
1101	−3
1100	−4
1011	−5
1010	−6
1001	−7
1000	−8

Figure 1.5
Four-bit positive and negative numbers.

4-bit positive and negative numbers. The largest positive and negative numbers are $+7$ and -8, respectively.

To convert a positive number into negative, take the complement of the number and add 1. This process is also called the 2's complement of the number.

Example 1.22

Write decimal number -6 as a 4-bit number.

Solution 1.22
First, write the number as a positive number, then find the complement and add 1:

```
0110  +6

1001  compliment
   1  add 1
-------
1010  which is -6
```

Example 1.23

Write decimal number -25 as an 8-bit number.

Solution 1.23
First, write the number as a positive number, then find the complement and add 1:

```
00011001  +25

11100110  compliment
       1  add 1
--------------
11100111  which is -25
```

1.18 Adding Binary Numbers

The addition of binary numbers is similar to the addition of decimal numbers. Numbers in each column are added together with a possible carry from a previous column. The primitive addition operations are the following:

```
0 + 0 = 0
0 + 1 = 1
1 + 0 = 1
1 + 1 = 10        generate a carry bit
1 + 1 + 1 = 11    generate a carry bit
```

some examples are given below.

Example 1.24

Find the sum of binary numbers 011 and 110.

Solution 1.24
We can add these numbers as in the addition of decimal numbers:

```
   011      First column:    1 + 0 = 1
 + 110      Second column:   1 + 1 = 0 and a carry bit
 ------      Third column:    1 + 1 = 10
  1001
```

Example 1.25

Find the sum of binary numbers 01000011 and 00100010.

Solution 1.25
We can add these numbers as in the addition of decimal numbers:

```
   01000011    First column:    1 + 0 = 1
 + 00100010    Second column:   1 + 1 = 10
 ----------    Third column:    0 + carry = 1
   01100101    Fourth column:   0 + 0 = 0
               Fifth column:    0 + 0 = 0
               Sixth column:    0 + 1 = 1
               Seventh column:  1 + 0 = 1
               Eighth column:   0 + 0 = 0
```

1.19 Subtracting Binary Numbers

To subtract two numbers, convert the number to be subtracted into negative and then add the two numbers.

Example 1.26

Subtract binary number 0010 from 0110.

Solution 1.26
First, let us convert the number to be subtracted into negative:

```
  0010   number to be subtracted

  1101   compliment
     1   add 1
 -----
  1110
```

Now, add the two numbers:

```
   0110
 + 1110
 - - - - - - -
   0100
```

Since we are using 4 bits only we cannot show the carry bit.

1.20 Multiplication of Binary Numbers

Multiplication of two binary numbers is the same as the multiplication of two decimal numbers. The four possibilities are:

```
0 × 0 = 0
0 × 1 = 0
1 × 0 = 0
1 × 1 = 1
```

Some examples are given below.

Example 1.27

Multiply the two binary numbers 0110 and 0010.

Solution 1.27
Multiplying the numbers:

```
                0110
                0010
                - - - - - - -
                0000
               0110
              0000
             0000
             - - - - - - - - - -
             001100 or 1100
```

In this example, 4 bits are needed to show the final result.

Example 1.28

Multiply binary numbers 1001 and 1010.

Solution 1.28
Multiplying the numbers:

```
        1001
        1010
       -------
        0000
       1001
      0000
      1001
   -----------
     1011010
```

In this example, 7 bits are required to show the final result.

1.21 Division of Binary Numbers

The division of binary numbers is similar to the division of decimal numbers. An example is given below.

Example 1.29

Divide binary number 1110 into binary number 10.

Solution 1.29
Dividing the numbers:

```
              111
      10 | 1110
           10
           ---
           11
           10
           ---
           10
           10
           ---
           00
```

giving the result 111_2.

1.22 Floating Point Numbers

Floating point numbers are used to represent noninteger fractional numbers. For example, 3.256, 2.1, 0.0036 and so forth. Floating point numbers are used in most engineering and technical calculations. The most commonly used floating point standard is the IEEE standard. According to this standard, floating point numbers are represented with 32 bit (single precision) or 64 bit (double precision).

In this section, we shall be looking at the format of 32-bit floating point numbers only and see how mathematical operations can be performed with such numbers.

According to the IEEE standard, 32-bit floating point numbers are represented as follows:

```
31 30    23 22                    0
X  XXXXXXX XXXXXXXXXXXXXXXXXXXXXXX
↑    ↑                 ↑
sign exponent      mantissa
```

The most significant bit indicates sign of the number where 0 indicates positive, and 1 indicates that the number is negative.

The 8-bit exponent shows the power of the number. To make the calculations easy, the sign of the exponent is not shown, but instead excess 128 numbering system is used. Thus, to find the real exponent, we have to subtract 127 from the given exponent. For example, if the mantiss is "10000000", the real value of the mantissa is $128 - 127 = 1$.

The mantissa is 23-bits wide and represents the increasing negative powers of 2. For example, if we assume that the mantissa is: "11100000000000000000000", the value of this mantissa is calculated as: $2^{-1} + 2^{-2} + 2^{-3} = 7/8$.

The decimal equivalent of a floating point number can be calculated using the formula:

$$\text{Number} = (-1)^{s}\, 2^{e-127} 1 \cdot f$$

where,

s = 0 for positive numbers, 1 for negative numbers
e = exponent (between 0 and 255)
f = mantissa.

As shown in the above formula, there is a hidden "1" in-front of the mantissa, i.e. mantissa is shown as "1·f".

The largest and the smallest numbers in 32-bit floating point format are as follows:

1.22.1 The Largest Number

```
0 11111110 11111111111111111111111
```

This number is $(2 - 2^{-23})\, 2^{127}$ or decimal 3.403×10^{38}. The numbers keep their precision up to six digits after the decimal point.

1.22.2 The Smallest Number

```
0 00000001 00000000000000000000000
```

This number is: 2^{-126} or decimal 1.175×10^{-38}.

1.23 Converting a Floating Point Number into Decimal

To convert a given floating point number into decimal, we have to find the mantissa and the exponent of the number and then convert into decimal as shown above.

Some examples are given here.

Example 1.30

Find the decimal equivalent of the floating point number given below:

```
0 10000001 10000000000000000000000
```

Solution 1.30
Here,

```
sign = positive
exponent = 129 − 127 = 2
mantissa = 2⁻¹ = 0.5
```

The decimal equivalent of this number is $+1.5 \times 2^2 = +6.0$.

Example 1.31

Find the decimal equivalent of the floating point number given below:

```
0 10000010 11000000000000000000000
```

Solution 1.31
In this example,

```
sign = positive
exponent = 130 − 127 = 3
mantissa = 2⁻¹ + 2⁻² = 0.75
```

The decimal equivalent of the number is $+1.75 \times 2^3 = 14.0$.

1.23.1 Normalizing the Floating Point Numbers

Floating point numbers are usually shown in normalized form. A normalized number has only one digit before the decimal point (a hidden number 1 is assumed before the decimal point).

To normalize a given floating point number, we have to move the decimal point repetitively one digit to the left and then increase the exponent after each move.

Some examples are given below.

Example 1.32

Normalize the floating point number 123.56.

Solution 1.32
If we write the number with a single digit before the decimal point, we get:

$$1.2356 \times 10^2$$

Example 1.33

Normalize the binary number 1011.1_2

Solution 1.33
If we write the number with a single digit before the decimal point, we get:

$$1.0111 \times 2^3$$

1.23.2 Converting a Decimal Number into Floating Point

To convert a given decimal number into floating point, we have to carry out the following steps:

- Write the number in binary.
- Normalize the number.
- Find the mantissa and the exponent.
- Write the number as a floating point number.

Some examples are given below:

Example 1.34

Convert decimal number 2.25_{10} into floating point.

Solution 1.34
Writing the number in binary:

$$2.25_{10} = 10.01_2$$

Normalizing the number,

$$10.01_2 = 1.001_2 \times 2_1$$

Here, $s = 0$, $e - 127 = 1$ or $e = 128$, and $f = 00100000000000000000000$.

(Remember that a number 1 is assumed on the left-hand side, even though it is not shown in the calculation). We can now write the required floating point number as follows:

```
s    e              f
0 10000000 (1)001 0000 0000 0000 0000 0000
```

or, the required 32-bit floating point number is as follows:

```
01000000000100000000000000000000
```

Example 1.35

Convert the decimal number 134.0625_{10} into floating point.

Solution 1.35
Writing the number in binary:

$$134.0625_{10} = 10000110.0001$$

Normalizing the number,

$$10000110.0001 = 1.00001100001 \times 2^7$$

Here, $s = 0$, $e - 127 = 7$ or $e = 134$, and $f = 00001100001000000000000$.

We can now write the required floating point number as follows:

```
s    e              f
0   10000110 (1)00001100001000000000000
```

or, the required 32 bit floating point number is as follows:

```
01000011000001100001000000000000
```

1.23.3 Multiplication and Division of Floating Point Numbers

The multiplication and division of floating point numbers is rather easy and the steps are given below:

- Add (or subtract) the exponents of the numbers.
- Multiply (or divide) the mantissa of the numbers.
- Correct the exponent.
- Normalize the number.
- The sign of the result is the EXOR of the signs of the two numbers.

Since the exponent is processed twice in the calculations, we have to subtract 127 from the exponent.

An example is given below to show the multiplication of two floating point numbers.

Example 1.36

Show the decimal numbers 0.5_{10} and 0.75_{10} in floating point and then calculate the multiplication of these numbers.

Solution 1.36
We can convert the numbers into floating point as follows:

$0.5_{10} = 1.0000 \times 2^{-1}$

here, s = 0, e − 127 = −1 or e = 126 and f = 0000.

or,

$0.5_{10} = 0\ 01110110\ (1)000\ 0000\ 0000\ 0000\ 0000\ 0000$

Similarly,

$0.75_{10} = 1.1000 \times 2^{-1}$

here, s = 0, e = 126 and f = 1000

or,

$0.75_{10} = 0\ 01110110\ (1)100\ 0000\ 0000\ 0000\ 0000\ 0000$

Multiplying the mantissas we get "(1)100 0000 0000 0000 0000 0000". The sum of the exponents is $126 + 126 = 252$. Subtracting 127 from the mantissa we obtain, $252 - 127 = 125$. The EXOR of the signs of the numbers is 0. Thus, the result can be shown in floating point as follows:

$0\ 01111101\ (1)100\ 0000\ 0000\ 0000\ 0000\ 0000$

the above number is equivalent to decimal 0.375 ($0.5 \times 0.75 = 0.375$) which is the correct result.

1.23.4 Addition and Subtraction of Floating Point Numbers

The exponents of floating point numbers must be the same before they can be added or subtracted. The steps to add or subtract floating point numbers is as follows:

• Shift the smaller number to the right until the exponents of both numbers are the same. Increment the exponent of the smaller number after each shift.
• Add (or subtract) the mantissa of each number as an integer calculation, without considering the decimal points.
• Normalize the obtained result.

An example is given below.

Example 1.37

Show decimal numbers 0.5_{10} and 0.75_{10} in floating point and then calculate the sum of these numbers.

Solution 1.37

As shown in Example 1.36, we can convert the numbers into floating point as follows:

$0.5_{10} = 0\ 01110110\ (1)000\ 0000\ 0000\ 0000\ 0000\ 0000$

similarly,

$0.75_{10} = 0\ 01110110\ (1)100\ 0000\ 0000\ 0000\ 0000\ 0000$

Since the exponents of both numbers are the same, there is no need to shift the smaller number. If we add the mantissa of the numbers without considering the decimal points, we get:

$$(1)000\ 0000\ 0000\ 0000\ 0000\ 0000$$
$$(1)100\ 0000\ 0000\ 0000\ 0000\ 0000$$
$$\text{-----------------------------------} +$$
$$(10)100\ 0000\ 0000\ 0000\ 0000\ 0000$$

To normalize the number, we can shift it right by one digit and then increment its exponent. The resulting number is as follows:

$0\ 01111111\ (1)010\ 0000\ 0000\ 0000\ 0000\ 0000$

The above floating point number is equal to decimal number 1.25, which is the sum of decimal numbers 0.5 and 0.75.

To convert floating point numbers into decimal and decimal numbers into floating point, the freely available program given in the following website can be used: http://babbage.cs.qc.edu/courses/cs341/IEEE-754.html

1.24 Binary Coded Decimal Numbers

Binary coded decimal (BCD) numbers are usually used in display systems such as LCDs and seven-segment displays to show numeric values. In BCD, each digit is a 4-bit number from 0 to 9. As an example, Table 1.4 shows the BCD numbers between 0 and 20.

Example 1.38

Write the decimal number 295 as a BCD number.

Solution 1.38

Writing the 4-bit binary equivalent of each digit,

$2 = 0010_2 \quad 9 = 1001_2 \quad 5 = 0101_2$

The required BCD number is $0010\ 1001\ 0101_2$.

Table 1.4: BCD Numbers between 0 and 20

Decimal	BCD	Binary
0	0000	0000
1	0001	0001
2	0010	0010
3	0011	0011
4	0100	0100
5	0101	0101
6	0110	0110
7	0111	0111
8	1000	1000
9	1001	1001
10	0001 0000	1010
11	0001 0001	1011
12	0001 0010	1100
13	0001 0011	1101
14	0001 0100	1110
15	0001 0101	1111
16	0001 0110	1 0000
17	0001 0111	1 0001
18	0001 1000	1 0010
19	0001 1001	1 0011
20	0010 0000	1 0100

Example 1.39

Write the decimal equivalent of BCD number 1001 1001 0110 0001_2.

Solution 1.39

Writing the decimal equivalent of each group of 4 bit, we get the required decimal number:

9961

1.25 The American Standard Code for Information Interchange Table

American Standard Code for Information Interchange (ASCII) is the most commonly used format for text files in computers. ASCII is currently in two formats: standard ASCII and extended ASCII. In standard ASCII format, there are 128 symbols and each alphabetic, numeric, or special character is represented with a 7-bit binary number. In extended ASCII format, there are 256 characters represented by 8 bits, and these additional characters are used to represent additional symbols (e.g. foreign letters, currency symbols, drawing symbols, etc.).

First 32 symbols (0−31) in the ASCII table are nonprintable and are known as the *control characters*. For example, carriage-return, line-feed, form-feed, bell, etc. are all control

*	0	1	2	3	4	5	6	7	8	9	A	B	C	D	E	F
0	NUL	SOH	STX	ETX	EOT	ENQ	ACK	BEL	BS	TAB	LF	VT	FF	CR	SO	SI
1	DLE	DC1	DC2	DC3	DC4	NAK	SYN	ETB	CAN	EM	SUB	ESC	FS	GS	RS	US
2		!	"	#	$	%	&	'	(	)	*	+	,	-	.	/
3	0	1	2	3	4	5	6	7	8	9	:	;	<	=	>	?
4	@	A	B	C	D	E	F	G	H	I	J	K	L	M	N	O
5	P	Q	R	S	T	U	V	W	X	Y	Z	[	\	]	^	_
6	`	a	b	c	d	e	f	g	h	i	j	k	l	m	n	o
7	p	q	r	s	t	u	v	w	x	y	z	{	|	}	~	

Figure 1.6
Standard ASCII characters.

	0	1	2	3	4	5	6	7	8	9	A	B	C	D	E	F
8	□	□	,	ƒ	„	…	†	‡	ˆ	‰	Š	‹	Œ	□	□	□
9	□	`	'	̏	"	•	–	—	~	™	š	›	œ	□	□	Ÿ
A		¡	¢	£	¤	¥	¦	§	¨	©	ª	«	¬	-	®	¯
B	°	±	²	³	´	µ	¶	·	¸	¹	º	»	¼	½	¾	¿
C	À	Á	Â	Ã	Ä	Å	Æ	Ç	È	É	Ê	Ë	Ì	Í	Î	Ï
D	Ð	Ñ	Ò	Ó	Ô	Õ	Ö	×	Ø	Ù	Ú	Û	Ü	Ý	Þ	ß
E	à	á	â	ã	ä	å	æ	ç	è	é	ê	ë	ì	í	î	ï
F	ð	ñ	ò	ó	ô	õ	ö	÷	ø	ù	ú	û	ü	ý	þ	ÿ

Figure 1.7
Extended ASCII characters.

symbols. Symbols 32–127 are printable characters. Figures 1.6 and 1.7 show a list of the standard and extended ASCII characters, respectively.

1.26 Summary

Chapter 1 has given an introduction to the microprocessor and microcontroller systems. The basic building blocks of microcontrollers have been described briefly. The chapter has also provided an introduction to various number systems, and described how to convert a given number from one base into another one. The important topic of floating point numbers and floating point arithmetic has also been described with examples.

1.27 Exercises

1. What is a microcontroller? What is a microprocessor? Explain the main differences between a microprocessor and a microcontroller.

2. Give some example applications of microcontrollers around you.

3. Where would you use an EPROM memory?

4. Where would you use a RAM memory?

5. Explain what types of memories are usually used in microcontrollers.

6. What is an I/O port?

7. What is an A/D converter? Give an example use for this converter.

8. Explain why a watchdog timer could be useful in a real-time system.

9. What is serial I/O? Where would you use serial communication?

10. Why is the current sinking/sourcing important in the specification of an output port pin?

11. What is an interrupt? Explain what happens when an interrupt is recognized by a microcontroller?

12. Why is brown-out detection important in real-time systems?

13. Explain the differences between an RISC-based microcontroller and a CISC-based microcontroller. What type of microcontroller is PIC?

14. Convert the following decimal numbers into binary:
    a) 23   b) 128   c) 255 d) 1023
    e) 120 f) 32000 g) 160 h) 250

15. Convert the following binary numbers into decimal:
    a) 1111   b) 0110   c) 11110000
    d) 00001111 e) 10101010 f) 10000000

16. Convert the following octal numbers into decimal:
    a) 177   b) 762 c) 777   d) 123
    e) 1777 f) 655 g) 177777 h) 207

17. Convert the following decimal numbers into octal:
    a) 255   b) 1024 c) 129 d) 2450
    e) 4096 f) 256 g) 180 h) 4096

18. Convert the following hexadecimal numbers into decimal:
    a) AA   b) EF   c) 1FF d) FFFF
    e) 1AA f) FEF g) F0   h) CC

19. Convert the following binary numbers into hexadecimal:
    a) 0101 b) 11111111 c) 1111 d) 1010
    e) 1110 f) 10011111 g) 1001 h) 1100

20. Convert the following binary numbers into octal:

     a) 111000 b) 000111   c) 1111111 d) 010111
     e) 110001 f) 11111111 g) 1000001 h) 110000

21. Convert the following octal numbers into binary:

     a) 177   b) 7777 c) 555 d) 111
     e) 1777777 f) 55571 g) 171 h) 1777

22. Convert the following hexadecimal numbers into octal:

     a) AA b) FF c) FFFF d) 1AC
     e) CC f) EE g) EEFF h) AB

23. Convert the following octal numbers into hexadecimal:

     a) 177   b) 777   c) 123 d) 23
     e) 1111 f) 17777777 g) 349 h) 17

24. Convert the following decimal numbers into floating point:

     a) 23.45 b) 1.25 c) 45.86 d) 0.56

25. Convert the following decimal numbers into floating point and then calculate their sum:

     0.255 and 1.75

26. Convert the following decimal numbers into floating point and then calculate their product:

     2.125 and 3.75

27. Convert the following decimal numbers into BCD:

     a) 128 b) 970 c) 900 d) 125

28. Find the binary values of the following ASCII characters:

     a) "C" b) "z" c) "8" d) "I"

29. Find the hexadecimal values of the following ASCII characters:

     a) "1" b) "A" c) "z" d) "m"

30. Find the decimal values of the following ASCII characters:

     a) "3" b) "M" c) "p" d) "G"

CHAPTER 2

PIC32 Microcontroller Series

Chapter Outline

2.2 Summary   89
2.3 Exercises   90

PIC32 is a 32-bit family of general-purpose, high-performance microcontrollers manufactured by Microchip Technology Inc. Looking at the PIC microcontroller development history, based on their performance, we can divide the PIC microcontroller families into low-performance, low- to medium-performance, medium-performance, and high-performance devices.

Low-performance PIC microcontrollers consist of the basic 8-bit PIC10 and PIC16 series of devices which have been around for over a decade. These devices are excellent general-purpose low-speed microcontrollers which have been used successfully in thousands of applications worldwide.

The PIC18 series of microcontrollers were then introduced by Microchip Technology Inc. as low- to medium-performance devices for use in high-pin count, high-density, complex applications requiring large number of I/O ports, large program and data memories, and supporting complex communication protocols such as CAN, USB, TCP/IP, or ZigBee. Although these devices are also based on 8-bit architecture, they offer higher speeds, from DC to 40 MHz, with a performance rating of up to 10 MIPS (Million Instructions Per Second).

The PIC24 series of microcontrollers are based on 16-bit architecture and have been introduced as medium-performance devices to be used in applications requiring high compatibility with lower performance PIC microcontroller families, and at the same time offering higher throughput and complex instruction sets. These microcontrollers have been used in many real-time applications such as digital signal processing, automatic control, speech and image processing, and so on, where higher accuracy than 8-bits is required, and at the same time higher speed is the main requirement.

PIC32 microcontroller family has been developed for high-performance general-purpose microcontroller applications. The family offers 80 MIPS performance with a wide range of on-chip peripherals, large data and program memories, large number of I/O ports, and an architecture designed for high-speed real-time applications. PIC32 microcontrollers can be used in real-time applications requiring high throughput. Some of the application areas are digital signal processing, digital automatic control, real-time games, and fast communication. The chip employs the industry standard M4K MIPS32 core from MIPS Technologies Inc. PIC32 family offers programming interface similar to other PIC microcontroller families, thus making the programming an easy task if the programmer is already familiar with the basic PIC microcontroller architecture. PIC32 microcontrollers are pin to pin compatible with most members of the PIC24 family of 16-bit microcontrollers, and thus the migration from 16-bit to 32-bit operation should be relatively easy.

Figure 2.1 shows a simplified architectural overview of the PIC32 microcontrollers. At the heart of the microcontroller is a 32-bit M4K MIPS32 core processor that connects to the rest of the chip via a *Bus Matrix* and a *Peripheral Bus*. The Bus Matrix runs at the same speed as the core processor and connects various high-speed modules such as the USB, DMA, memory, ports, and so on. The Peripheral Bus can be programmed to run at slower speeds and it connects to slower modules such as A/D converter, UART, SPI, real-time clock and calendar (RTCC) and so on.

The core processor has the following features:

- 80 MHz clock speed
- 32-bit address bus and 32-bit data bus
- Five-stage pipelining
- Single-cycle ALU
- Single-cycle multiply and high-speed divide module
- 2 × 32 register files.

Other important features of the chip are

- 2.3–3.6 V operation
- Up to 512 K flash program memory

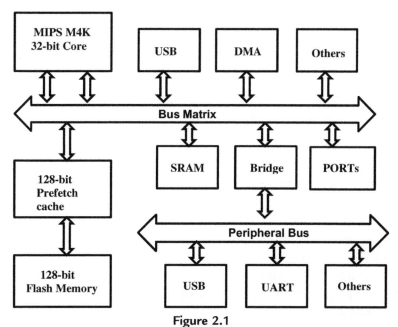

Figure 2.1
Simplified architecture of the PIC32 microcontroller family.

- Up to 32 K SRAM data memory
- Internal oscillators
- Multiple interrupt vectors
- UART, SPI, and I^2C modules
- Configurable watchdog timer and real-time clock
- High-speed I/O with 18 mA current sink/source capability
- Configurable open-drain output on I/O pins
- External interrupt pins
- PWM, capture, and compare modules
- JTAG debug interface
- Fast A/D converter and analog comparator modules
- Timers and counters
- Hardware DMA channels
- USB support
- Large pin count for wide range of peripherals and I/O ports
- Wide operating temperature (−40 to +105 °C).

Different chips in the family may have additional features, such as Ethernet support, CAN bus support and so on.

Perhaps the best way of learning the PIC32 microcontroller family architecture is to look at a typical processor in the family in greater detail. The processor to be considered in this chapter is the PIC32MX360F512L (from the PIC32MX3XX/4XX family). The architectures of the other members of the PIC32 microcontroller family are very similar to the chosen one and should not be too difficult to learn them.

2.1 The PIC32MX360F512L Architecture

The PIC32MX360F512L is a typical PIC32 microcontroller, belonging to the family PIC32MX3XX/4XX. This microcontroller has the following features:

- Hundreds pins (TQFP) package
- Up to 80 MHz clock speed
- 512 K flash program memory (+12 K Boot Flash memory)
- 32 K data memory
- 5 × 16-bit Timers/counters
- 5 × Capture inputs
- 5 × Compare/PWM outputs
- Four programmable DMA channels
- 2 × UARTs (supporting RS232, RS485, LIN bus, and iRDA)
- 2 × SPI bus modules
- 2 × I^2C bus modules

- RTCC module
- 8 MHz and 32 kHz internal clocks
- 16 × A/D channels (10-bits)
- 2 × Comparator modules
- Four-wire JTAG interface.

PIC32MX360F512L is available in a 100-pin TQFP package as shown in Figure 2.2. The pins which are +5 V tolerant are shown in bold.

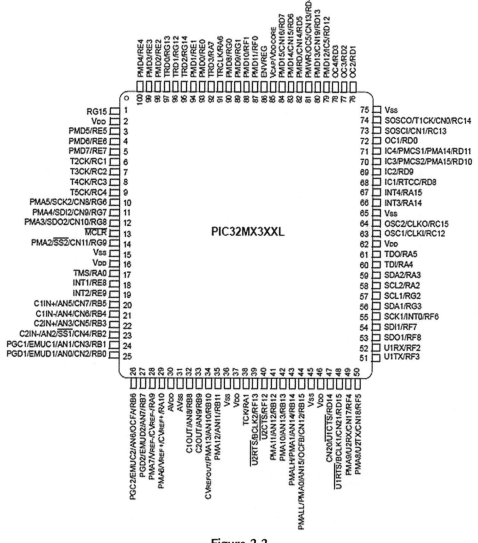

Figure 2.2
Pin configuration of the PIC32MX360F512L.

Figure 2.3 shows the internal block diagram of the PIC32MX360F512L microcontroller. In the middle of the diagram, we can see the MIP32 M4K processor. The 32-bit data memory is directly connected to the processor via the Bus Matrix, offering up to 4 GB addressing space. The 1298-bit flash program memory is also connected to the Bus Matrix via a 32-bit prefetch module. The 32-bit Peripheral Bridge connects the Bus Matrix and the processor to the peripheral modules. The peripheral modules consist of

- 7 × Ports (PORT A to PORT G)
- 5 × Timer modules

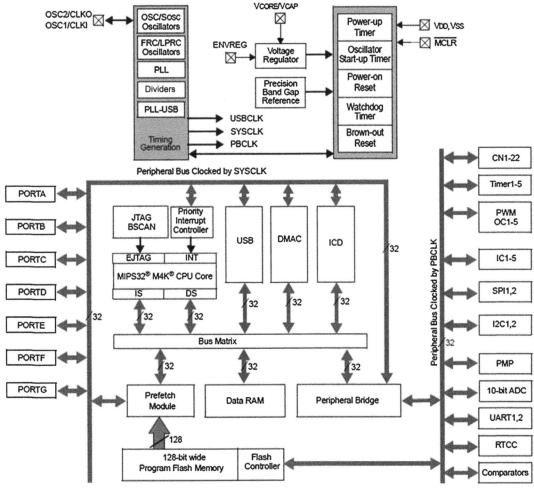

Note   1:   Some features are not available on all device variants.
2:   BOR functionality is provided when the on-board voltage regulator is enabled.

Figure 2.3
Block diagram of the PIC32MX360F512L.

- 5 × PWM modules
- 2 × SPI bus modules
- 2 × I^{2}C modules
- 10 bit ADC module
- 2 × UART modules
- RTCC module
- Comparators
- Change notification inputs.

The system clock and peripheral bus clock are provided by the Timing Generation module which consists of

- Oscillators
- PLL module
- Clock dividers.

The Timing Generation module additionally provides clock to the following modules:

- Power-up timer
- Oscillator start-up timer
- Power-on reset
- Watchdog timer
- Brown-out reset.

2.1.1 The Memory

Figure 2.4 shows the memory structure of the PIC32MX3XX/4XX microcontrollers. The memory structure may look complicated initially but the explanation given in this section should make it simple and easy to understand.

As can be seen from the figure, two address spaces are implemented: *virtual* and *physical*. All hardware resources, such as data memory, program memory, and DMA transfers, are handled by the physical addresses. If we wish to access the memory independent of the CPU (such as the case in DMA), then we must use the physical addresses.

Virtual addresses are important as they are used exclusively by the CPU to fetch and execute instructions. In normal programming, we are only interested with the virtual memory addresses. These addresses are translated into physical addresses by a Fixed Mapping Translation unit inside the processor. The translation process is simply a bitwise AND of the

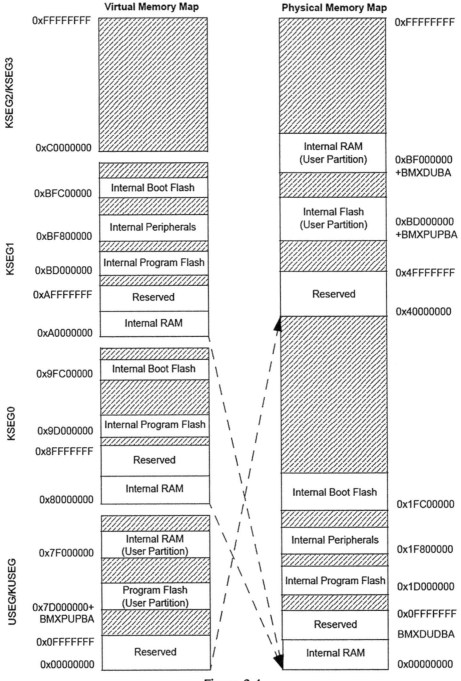

Figure 2.4
The memory structure.

virtual address spaces with fixed number 0x1FFFFFFF. Details of the physical address space are available in the PIC32 Family Reference Manual, or in the individual microcontroller data sheets.

The entire virtual memory address space is 4 GB and is divided into two primary regions: User mode segment and Kernel mode segment. The lower 2 GB of address space forms the User mode segment (called Useg/Kuseg). The upper 2 GB of virtual address space forms the Kernel mode segment. The kernel address space has been designed to be used by the operating system, while the user address space holds a user program that runs under the operating system. This is for safety as a program in the user address space cannot access the memory space of the kernel address space. But, programs in the kernel can access the user address space (this is what the "K" in name Kuser indicates). As most embedded microcontroller applications do not use an operating system, we can place all our programs and data in the kernel address space and do not use the user address space at all. As a result of this, we will have 2 GB of address space for our programs and data.

The kernel address space is divided into four segments of 512 MB each: Kseg 0, Kseg 1, Kseg 2, and Kseg 3. The kernel address space contains the following memory areas:

• Program memory (flash)
• Data memory (RAM)
• Special Function Registers (SFR)
• Boot memory area.

The kernel virtual address space contains two identical subsections, both translated to the same physical memory addresses. One of these subsections is *cacheable*, while the other one is *not cacheable*. Here, cacheable means that the instructions and data stored in the cache can be prefetched and this speeds up the execution time considerably in sequential operations by eliminating the need to fetch data or instructions from the memory (the cache memory is a small very fast memory). Kseg 0 corresponds to the cacheable kernel address space, while Kseg 1 corresponds to the noncacheable kernel address space. Each of Kseg 0 and Kseg 1 contain the program memory (flash), data memory (RAM), SFR, and Boot memory area. Notice that a PIC32 microcontroller can run a program that is stored in the RAM memory (as opposed to the usual case of a program stored in the flash memory).

The cacheable kernel segment is used during normal program executions where instructions and data are fetched with the cache enabled. The noncacheable kernel segment is used during the processor initialization routines where we wish to execute the instructions sequentially and with no cache present. The prefetch cache module can be enabled under software control.

When the PIC32 microcontroller is reset, it goes to the reset address of 0xBFC00000, which is the starting address of the Boot program *in the noncacheable* kernel segment. The code in the Boot location takes care of initialization tasks, sets configuration bits, and then calls the user written program in the cacheable kernel segment of the memory (the user program usually, but not always, starts from address 0x9D001000 in Kseg 0 space). Notice that a new PIC32 microcontroller contains no program or data when it arrives from the factory. The Boot program is normally placed there by the microcontroller programmer device.

The virtual or physical memory addresses are important to programmers writing assembly code, or developing DMA routines. In normal C programs, we are not concerned with the memory addresses as the compiler and linker take care of all the program and data placements.

mikroC Pro for PIC32 Boot program configures

- CP0 (coprocessor registers)
- SFR registers associated with the interrupt
- Stack pointer (R29) and global pointer (R1)
- Executable code allocated in the KSEG0
- Data allocated in the KSEG1.

2.1.2 The Microcontroller Clock

The block diagram of the clock circuit of the PIC32MX microcontroller is shown in Figure 2.5.

There are five clock sources: two of them use internal oscillators, and three require external crystals or oscillator circuits. Three clock outputs are available: CPU system clock, USB clock, and peripheral clock.

Clock Sources

- FRC is an internal oscillator requiring no external components. The clock frequency is 8 MHz and is designed for medium-speed low-power operations. The clock can be tuned via the TUN bits of register OSCTUN. The clock is accurate to about ±2% after calibration.
- LPRC is a low-frequency, low-power internal oscillator with a frequency 32 kHz, requiring no external components.
- POSC is the primary high-speed oscillator, requiring an external crystal up to 20 MHz. The crystal is connected directly to the OSCI and OSCO inputs with two capacitors.

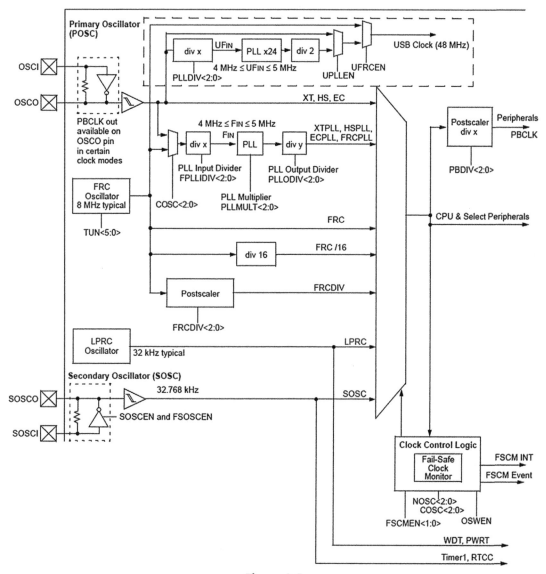

Figure 2.5
The PIC32MX clock block diagram.

- SOSC is the secondary low-speed, high-accuracy oscillator, designed for operation with a crystal of 32,768 Hz. This oscillator can be used for timer and real-time clock modules, or as the main low-speed CPU oscillator.
- EC is the external clock source, requiring no external crystal. A square wave signal is applied to this input at the required frequency.

Clock Outputs

The three clock outputs can be selected in various configurations as described below (Figure 2.5):

USB Clock

The USB operation requires an exact 48 MHz clock. This clock can be derived in many different ways. Some of the methods are given below:

*   External 8 MHz crystal-driven primary clock (POSC), divided by 2 (using PLLDIV), then multiplied by 24 to give 96 MHz. This frequency can then be divided by 2 to give the required 48 MHz clock. UPLLEN and UFRCEN select this clock at the output. It is important to note that the input to PLLx24 must be between 4 and 5 MHz. We can use different external crystals with different PLL divisions. For example, we can use a 4 MHz crystal with POSC, divide by 1 (using PLLDIV), and then multiply by 24 to give 96 MHz. This frequency can then be divided by 2 as above to give the required 48 MHz clock.
*   External 48 MHz primary oscillator. UPLLEN and UFRCEN select this clock at the output.

CPU Clock

The main system clock can be chosen from a variety of sources as shown below (Figure 2.5):

*   Primary clock (POSC) selected at the output directly.
*   Primary clock selected through the PLL input and output dividers, and the PLL multiplier. For example, if we use 8 MHz crystal, divide by 2 (using FPLLIDIV), multiply by 16 (using PLLMULT), and then divide by 2 (using PLLODIV), and we get a 32 MHz clock rate.
*   Internal 8 MHz clock (FRC) can be selected directly at the output to give 8 MHz. Alternatively, the clock can be divided by 16, or it can pass through a postscaler to select the required frequency.
*   Internal 8 MHz clock (FRC) can be selected with the PLL. For example, this clock can be divided by 2 to give 4 MHz at the input of the PLL (remember the input clock to PLL must be between 4 MHz and 5 MHz), it can be multiplied by 16, and then divided by 2 to give 32 MHz at the output.
*   The 32 kHz clock (LPRC) can be selected directly at the output.
*   The 32,768 Hz clock (SOSC) can be selected directly at the output.

Peripheral Clock

The peripheral bus clock (PBCLK) is derived from the CPU system clock by passing it through a postscaler. The postscaler rate can be selected as 1, 2, 4, or 8. Thus, for example, if the CPU system clock is chosen as 80 MHz, then the peripheral clock can be 80, 40, 20, or 10 MHz.

Configuring the Operating Clocks

The clock configuration bits shown in Figure 2.5 can be selected by programming the SFR registers (e.g. OSCCON, OSCTUN, OSCCONCLR and so on) or the device configuration registers (e.g. DEVCFG1 and DEVCFG2) during run time.

Alternatively, the operating clocks can be selected during the programming of the microcontroller chip. Most programming devices give options to users to select the operating clocks by modifying the device configuration registers just before the chip is programmed. The mikroC Pro for PIC32 compiler allows the configuration registers to be modified by selecting *Project → Edit Project* from the drop-down menu of the IDE.

Figure 2.6 shows the clocks that will be used in all the projects in this book (these are the default settings provided by mikroC Pro for PIC32 compiler). The selection is summarized below:

- External 8 MHz crystal connected to OSCI and OSCO pins
- PLL Input Divider: 2
- PLL Multiplier: 20
- USB PLL Input Divider: 12
- USB PLL Enable: Enabled

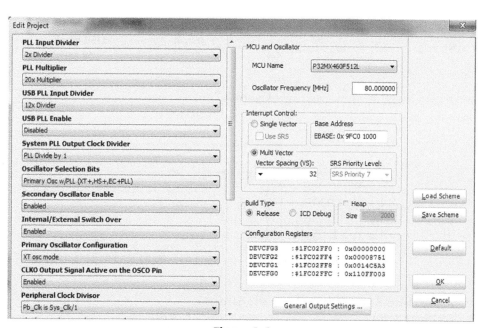

Figure 2.6
Edit project window. (For color version of this figure, the reader is referred to the online version of this book.)

- System PLL Output Clock Divider: 1
- Peripheral Clock Divisor: 1.

The above selection gives (Figure 2.7) the following:

CPU clock	80 MHz
USB clock	48 MHz
Peripheral clock	80 MHz

The crystal connections to OSCI and OSCO pins are shown in Figure 2.8 with two small capacitors. The connection of an external clock source in EC mode is shown in Figure 2.9. In this mode, the OSCO pin can be configured either as a clock output or as an I/O port.

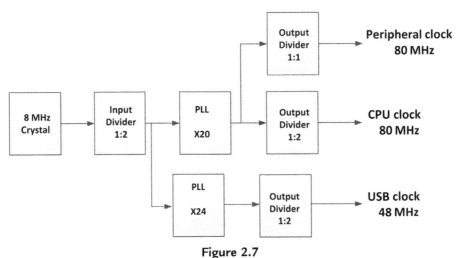

Figure 2.7
Clock selection for the projects.

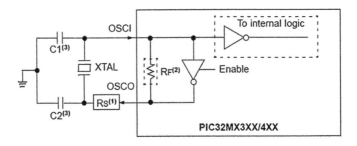

Note   1:   A series resistor, Rs, may be required for AT strip cut crystals.
       2:   The internal feedback resistor, RF, is typically in the range of 2 to 10 MΩ.
       3:   Refer to the *"PIC32MX Family Reference Manual"* (DS61132) for help determining the best oscillator components.

Figure 2.8
Crystal connection.

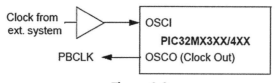

Figure 2.9
External clock connection.

Performance and Power Consumption Considerations

In microcontroller-based applications, the power consumption and performance are in direct conflict with each other. To lower the power consumption in an application, we also have to lower the performance, i.e. the clock rate. Conversely, an increase in the performance also increases the power consumption. Thus, the higher the clock speed, the higher is the power consumption of the device. The designers of the PIC32 microcontroller have spent considerable amount of time to provide a wide range of clock selection mechanisms, so that the user can choose the best clock rate for the required power consumption.

For example, if an application can run at 8 MHz to do a job, there is no point in running the application at 80 MHz. In some applications, the device may be in standby mode and may be waiting for some user action (e.g. pressing a button). In such applications, a low clock rate can be selected while the device is in standby mode, and then full high-speed clock rate can be selected when the device wakes up to run the actual application.

The Flash Wait States

The number of wait states is normally set by default to the highest value as this provides the safest operation. The SFR register CHECON, bits PFMWS, controls the number of wait states and we can reduce its value for higher performance. It is important however that setting wrong number of wait states could cause errors while accessing the flash memory. This register can take values between 0 and 7. For optimum performance, the wait states would be programmed to the minimum possible value. The following example shows how the maximum number of wait states can be set to 4:

```
CHECONbits.PFMWS = 4;
```

By default, the mikroC Pro for PIC32 compiler configures the following:

- Cache is enabled.
- Prefetch is enabled (for executable code and constants).
- Flash Wait states is set for specified oscillator frequency.
- Executable code is allocated in Kseg0.
- Data is allocated in Kseg1.

2.1.3 Resets

There are several sources that can cause the microcontroller to reset. The following is a list of these sources:

- Power-on reset (POR)
- External reset (MCLR)
- Software reset (SWR)
- Watchdog timer reset (WDTR)
- Brown-out reset (BOR)
- Configuration mismatch reset (CMR).

The SFR register RCON is the reset controller register and the source of a reset can easily be found by reading the bits of this register. Any set bit in this register indicates that a reset has occurred, and depending upon the position of this bit, one can tell the actual source of the reset. For example, if bit 7 is set, then the source of reset is the external MCLR input (further details can be obtained from the individual microcontroller data sheets).

In this section, we are interested in external resets which are caused for example by the user pressing a button. The MCLR pin is used for external device reset, device programming, and device debugging. External reset occurs when the MCLR pin is lowered. Figure 2.10 shows a typical external reset circuit where a push-button switch is connected between the points JP. Normally the MCLR pin is at logic 1. Pressing the button forces the MCLR pin to logic 0 and causes the processor to reset. It is recommended by the manufacturers that $R < 10\,K$, $R1 < 470\,\Omega$, and $C = 100\,nF$.

Notice that the microcontroller can be reset in software by executing a specific sequence of operations (see the individual device data sheets). The following mikroC Pro for PIC32 function can initiate a soft reset action:

```
Reset();            // Reset the microcontroller
```

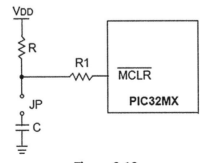

Figure 2.10
External reset.

2.1.4 The Input/Output Ports

The general-purpose I/O ports are very important in many applications as they allow the microcontroller to monitor and control devices attached to it. Although the I/O ports of the PIC32 microcontrollers have some similarities with the 8- and 16-bit devices, PIC32 microcontrollers offer greater functionality and new added features.

Some of the key features of PIC32 I/O ports are the following:

- Open-drain capability of each pin
- Pull-up resistors at each input pin
- Fast I/O bit manipulation
- Operation during CPU SLEEP and IDLE modes
- Input monitoring and interrupt generation on mismatch conditions.

Figure 2.11 shows the basic block diagram of an I/O port.

An I/O port is controlled with the following SFR registers:

- TRISx: Data direction control register for port x
- PORTx: Port register for port x
- LATx: Latch register for port x
- ODCx: Open-drain control register for port x
- CNCON: Interrupt-on-change control register.

TRISx

There is a TRISx register for every port x. The TRISx register configures the port signal directions. A TRISx bit set to 1 configures the corresponding port pin 'x' as an input. Similarly, a TRISx bit cleared to 0 configures the corresponding port pin 'x' as an output. In the example shown in Figure 2.12, odd bits (1, 3, 5, 7, 9, 11, 13, and 15) of PORT B are configured as inputs, and the remaining pins as outputs.

PORTx

PORTx is the actual port register. A write to PORTx sends data to the port latch register LATx and this data appear at the output of the port. A read from a PORTx register reads the actual data at the output pin of the port (the state of the output pin may be affected by a device connected to the pin).

LATx

LATx are the port latch registers. A write to a latch register is the same as sending data to the port register PORTx. A read from the latch register, however, reads the data present at the

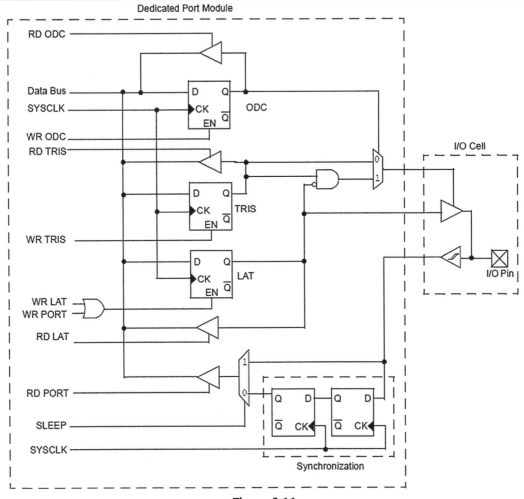

Figure 2.11
Block diagram of an I/O Port.

output latch of the port pin, and this may not be same as the actual state of the port pin. For example, the port pin may be pulled low by an external device. Under such circumstances, the latch register will read the data held at the output latch and not the actual state of the pin.

ODCx

Each I/O output pin can be configured individually as either normal output or open-drain output. SFR register ODCx controls the output state of a pin as follows: an OCDx bit set to 1 configures the corresponding port pin 'x' as open drain. Similarly, an OCDx bit cleared to 0 configures the corresponding port pin 'x' as normal digital output. The open-drain feature

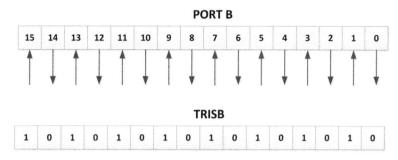

PORT B

15	14	13	12	11	10	9	8	7	6	5	4	3	2	1	0

TRISB

1	0	1	0	1	0	1	0	1	0	1	0	1	0	1	0

TRISB = 0xAAAA
Figure 2.12
Example of TRIS register setting.

allows the generation of outputs higher than the supply voltage (VDD) on any desired digital output pin. Notice that external pull-up resistors are normally used on open-drain output pins. In the example in Figure 2.13, all odd-numbered (1, 3, 5, 7, 9, 11, 13, and 15) PORT B output pins are configured to be open drain.

CNCON

Some of the I/O pins can be configured to generate an interrupt when a change is detected on the pin. The SFR registers that control this change notice are the following:

- CNCON: used to enable or disable the interrupt-on change (change notice) feature.
- CNENx: contains the control bits, where 'x' is the number of the change notice pin.
- CNPUEx: enables or disables pull-up resistors on port pin 'x'.

Notice that bit CNIE (bit 0) of the IEC1 SFR register must be set to 1 to enable interrupt-on-change feature.

SET, CLR, INV I/O Port Registers

In addition to the I/O port registers described in this section, each port register has SET, CLR, and INV registers associated with it that can be useful in bit manipulation operations, allowing faster operations to be carried out.

The following registers are available for bit manipulation:

- TRISxSET
- TRISxCLR

ODCB = 0x5555

Figure 2.13
Example of ODC register setting.

- TRISxINV
- PORTxSET
- PORTxCLR
- PORTxINV
- LATxSET
- LATxCLR
- LATxINV
- ODCxSET
- ODCxCLR
- ODCxINV
- CNCONSET
- CNCONCLR
- CNCONINV
- CNPUESET
- CNPUECLR
- CNPUEINV.

An example is given below to show how the SET, CLR, and INV registers can be used on TRISB register. Use of other registers is similar and is not repeated here.

Register TRISBCLR is the write-only register and it clears selected bits in TRISB register. Writing a 1 in one or more bit positions clears the corresponding bits in TRISB register. Writing a 0 does not affect the register. For example, to clear bits 15, 5, and 0 of TRISB register, we issue the command: TRISBCLR = 0b1000000000100001 or TRSBCLR = 0x08021.

Register TRISBSET is the write-only register and it sets selected bits in TRISB register. Writing a 1 in one or more bit positions sets the corresponding bits in TRISB register. Writing a 0 does not affect the register. For example, to set bits 15, 5, and 0 of TRISB register, we issue the command: TRISBSET = 0b1000000000100001 or TRSBSET = 0x08021.

Register TRISBINV is the write-only register and it inverts selected bits in TRISB register. Writing a 1 in one or more bit positions inverts the corresponding bits in TRISB register. Writing a 0 does not affect the register. For example, to invert bits 15, 5, and 0 of TRISB register, we issue the command: TRISBINV = 0b1000000000100001 or TRSBINV = 0x08021.

Digital/Analog Inputs

By default, all I/O pins are configured as analog inputs. Setting the corresponding bits in the AD1PCFG register to 0 configures the pin as an analog input, independent of the TRIS register setting for that pin. Similarly, setting a bit to 1 configures the pin as digital I/O. For

example, to set all analog pins as digital, we have to issue the command
AD1PCFG = 0xFFFF.

It is recommended that any unused I/O pins that are not used should be set as outputs (e.g. by clearing the corresponding TRIS bits) and cleared to Low in software. It is important to be careful when attaching devices to an I/O port that the normal I/O voltage of a pin is 3.6 V. An input pin can tolerate an input voltage to up to 5 V, but the output voltage from an output pin cannot exceed 3.6 V.

2.1.5 The Parallel Master Port

The Parallel Master Port (PMP) is an 8/16-bit parallel I/O port that can be used to communicate with a variety of parallel devices such as microcontrollers, external peripheral devices, LCDs, GLCDs and so on. In this section, we will briefly look at the operation of the PMP. Further details about the PMP can be obtained from individual device data sheets.

Most microcontroller-based applications require address and data lines, and chip select control lines. The PMP provides up to 16 address and data lines, and up to two chip select lines. Addresses can be autoincremented and autodecremented for greater flexibility. In addition, parallel slave port support, and individual read and write strobes are provided. Figure 2.14 shows a typical application of the PMP to interface to an external EPROM memory.

There are a number of SFR registers used to control the PMP. These are summarized below:

- PMCON: PMP control register
- PMMODE: PMP-mode control register
- PMADDR: PMP address register
- PMDOUT: PMP data-output register

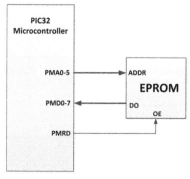

Figure 2.14
Using PMP with external EPROM memory.

- PMDIN: PMP data-input register
- PMAEN: PMP address-enable register
- PMSTAT: PMP status register.

Each register has bit manipulations options. Thus, for example, in addition to PMCON, there are registers named PMCONCLR, PMCONSET, and PMCONINV.

PMCON

Register PMCON controls the PMP module. Figures 2.15 and 2.16 show the PMCON bit configuration and bit definitions, respectively. The bits in PMCON control address multiplexing, enable port control signals, enable chip select signals, and select signal polarity.

PMMODE

Register PMMODE controls the operational modes of the PMP. Figures 2.17 and 2.18 show the PMMODE bit configuration and bit definitions, respectively.

PMADDR

Register PMADDR contains address of the external device and the chip select control bits. Figure 2.19 shows the PMADDR bit configuration and bit definitions.

r-x	r-x	r-x	r-x	r-x	r-x	r-x	r-x
—	—	—	—	—	—	—	—
bit 31							bit 24

r-x	r-x	r-x	r-x	r-x	r-x	r-x	r-x
—	—	—	—	—	—	—	—
bit 23							bit 16

R/W-0	R/W-0	R/W-0	R/W-0	R/W-0	R/W-0	R/W-0	R/W-0
ON	FRZ	SIDL	ADRMUX1	ADRMUX0	PMPTTL	PTWREN	PTRDEN
bit 15							bit 8

R/W-0	R/W-0	R/W-0	R/W-0	R/W-0	r-x	R/W-0	R/W-0
CSF1	CSF0	ALP	CS2P	CS1P	—	WRSP	RDSP
bit 7							bit 0

Legend:			
R = Readable bit	W = Writable bit	P = Programmable bit	r = Reserved bit
U = Unimplemented bit	-n = Bit value at POR: ('0', '1', x = Unknown)		

Figure 2.15
PMCON register bit configuration.

PMDOUT

Register PMDOUT controls the buffered data output in Slave mode. Figure 2.20 shows the bit configuration and bit definitions.

PMDIN

This register controls the I/O data ports in 8/16-bit Master mode, and input data port in 8-bit Slave mode. Figure 2.21 shows the bit configuration and bit definitions.

PMAEN

This register controls the operation of address and chip select pins of the PMP module. Figure 2.22 shows the bit configuration and bit definitions.

bit 31-16   **Reserved:** Write '0'; ignore read

bit 15   **ON:** Parallel Master Port Enable bit

1 = PMP enabled
0 = PMP disabled, no off-chip access performed

Note: When using 1:1 PBCLK divisor, the user's software should not read/write the peripheral's SFRs in the SYSCLK cycle immediately following the instruction that clears the module's ON bit.

bit 14   **FRZ:** Freeze in Debug Exception Mode bit

1 = Freeze operation when CPU is in Debug Exception mode
0 = Continue operation even when CPU is in Debug Exception mode

Note: FRZ is writable in Debug Exception mode only, it is forced to '0' in normal mode.

bit 13   **SIDL:** Stop in IDLE Mode bit

1 = Discontinue module operation when device enters IDLE mode
0 = Continue module operation in IDLE mode

bit 12-11   **ADRMUX<1:0>:** Address/Data Multiplexing Selection bits

11 = All 16 bits of address are multiplexed on PMD<15:0> pins
10 = All 16 bits of address are multiplexed on PMD<7:0> pins
01 = Lower 8 bits of address are multiplexed on PMD<7:0> pins, upper 8 bits are on PMA<15:8>
00 = Address and data appear on separate pins

bit 10   **PMPTTL:** PMP Module TTL Input Buffer Select bit

1 = PMP module uses TTL input buffers
0 = PMP module uses Schmidt Trigger input buffer

bit 9   **PTWREN:** Write Enable Strobe Port Enable bit

1 = PMWR/PMENB port enabled
0 = PMWR/PMENB port disabled

bit 8   **PTRDEN:** Read/Write Strobe Port Enable bit

1 = PMRD/PMWR port enabled
0 = PMRD/PMWR port disabled

Note 1: These bits have no effect when their corresponding pins are used as address lines.

Figure 2.16
(Continued on next page)

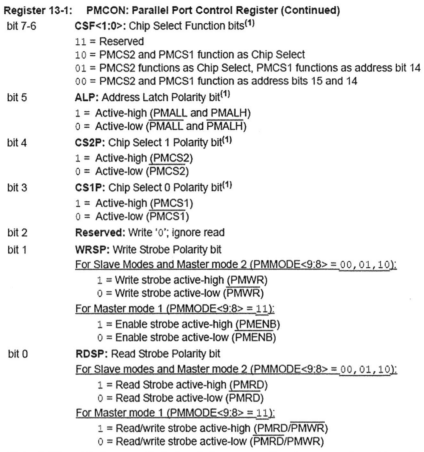

Register 13-1:　PMCON: Parallel Port Control Register (Continued)

bit 7-6　　**CSF<1:0>:** Chip Select Function bits[1]

11 = Reserved
10 = PMCS2 and PMCS1 function as Chip Select
01 = PMCS2 functions as Chip Select, PMCS1 functions as address bit 14
00 = PMCS2 and PMCS1 function as address bits 15 and 14

bit 5　　**ALP:** Address Latch Polarity bit[1]

1 = Active-high (PMALL and PMALH)
0 = Active-low ($\overline{\text{PMALL}}$ and $\overline{\text{PMALH}}$)

bit 4　　**CS2P:** Chip Select 1 Polarity bit[1]

1 = Active-high (PMCS2)
0 = Active-low ($\overline{\text{PMCS2}}$)

bit 3　　**CS1P:** Chip Select 0 Polarity bit[1]

1 = Active-high (PMCS1)
0 = Active-low ($\overline{\text{PMCS1}}$)

bit 2　　**Reserved:** Write '0'; ignore read

bit 1　　**WRSP:** Write Strobe Polarity bit
For Slave Modes and Master mode 2 (PMMODE<9:8> = 00,01,10):

1 = Write strobe active-high (PMWR)
0 = Write strobe active-low ($\overline{\text{PMWR}}$)

For Master mode 1 (PMMODE<9:8> = 11):

1 = Enable strobe active-high (PMENB)
0 = Enable strobe active-low (PMENB)

bit 0　　**RDSP:** Read Strobe Polarity bit
For Slave modes and Master mode 2 (PMMODE<9:8> = 00,01,10):

1 = Read Strobe active-high (PMRD)
0 = Read Strobe active-low ($\overline{\text{PMRD}}$)

For Master mode 1 (PMMODE<9:8> = 11):

1 = Read/write strobe active-high (PMRD/$\overline{\text{PMWR}}$)
0 = Read/write strobe active-low ($\overline{\text{PMRD}}$/PMWR)

Note 1: These bits have no effect when their corresponding pins are used as address lines.

Figure 2.16
PMCON bit definitions.

PMSTAT

This register contains the status bits when operating in buffered mode. Figures 2.23 and 2.24 show the bit configuration and bit definitions, respectively.

2.1.6 Timers

The PIC32 microcontroller supports five timers, namely TIMER1–TIMER5. Timer 1 is 16-bits wide, while the other timers can be combined for 32-bit operation. In this section, we will look at the operation of Timer 1.

r-x	r-x	r-x	r-x	r-x	r-x	r-x	r-x
—	—	—	—	—	—	—	—
bit 31							bit 24

r-x	r-x	r-x	r-x	r-x	r-x	r-x	r-x
—	—	—	—	—	—	—	—
bit 23							bit 16

R-0	R/W-0	R/W-0	R/W-0	R/W-0	R/W-0	R/W-0	R/W-0
BUSY	IRQM<1:0>		INCM<1:0>		MODE16	MODE<1:0>	
bit 15							bit 8

R/W-0	R/W-0	R/W-0	R/W-0	R/W-0	R/W-0	R/W-0	R/W-0
WAITB<1:0>		WAITM<3:0>				WAITE<1:0>	
bit 7							bit 0

Legend:			
R = Readable bit	W = Writable bit	P = Programmable bit	r = Reserved bit
U = Unimplemented bit	-n = Bit value at POR: ('0', '1', x = Unknown)		

Figure 2.17
PMMODE register bit configuration.

Timer 1

Timer 1 is 16-bit wide that can be used in various internal and external timing and counting applications. This timer can be used in synchronous and asynchronous internal and external modes.

Figure 2.25 shows the block diagram of Timer 1. The source of clock for Timer 1 can be from the Low Power secondary Oscillator (SOSC), from the external input pin T1CK, or from the peripheral bus clock (PBCLK). A prescaler is provided with division ratios of 1, 8, 64, and 256 in order to change the timer clock frequency. The operation of the timer is as follows.

Timer register PR1 is loaded with a 16-bit number. Register TMR1 counts up at every clock pulse and when PR1 is equal to TMR1, timer flag T1IF is set. At the same time, register TMR1 is reset to zero so that new count starts from zero again. If Timer 1 interrupts are enabled, then an interrupt will be generated where the program will jump to the interrupt service routine (ISR) whenever the timer flag is set.

TIMER 1 is controlled by three registers: T1CON, PR1, and TMR1. Figures 2.26 and 2.27 show the bit configuration and bit definitions of register T1CON, respectively.

Assuming TMR1 starts counting from zero, the delay before the count reaches to PR1 is given by the following equation:

```
Delay = T × Prescaler × PR1
```

Where T is the clock period and PR1 is the value loaded into register PR1. Rearranging the above equation, we can find the value to be loaded into PR1 as:

```
PR1 = Delay/(T × Prescaler)
```

An example is given here to illustrate the process. Assuming that the clock frequency is 20 MHz, and it is required to generate a timer interrupt every 256 ms. Assuming a prescaler value of 256, the value to be loaded into PR1 is calculates as follows:

```
T = 1/20 MHz = 0.05 × 10⁻³ ms
```

```
PR1 = 256 ms/(0.05 × 10⁻³ × 256) = 20,000
```

Thus, the required settings are as follows:

```
TMR1 = 0
Prescaler = 256
PR1 = 20,000 (0x4E20)
```

bit 31-16	**Reserved:** Write '0'; ignore read
bit 15	**BUSY:** Busy bit (Master mode only)
	1 = Port is busy
	0 = Port is not busy
bit 14-13	**IRQM<1:0>:** Interrupt Request Mode bits
	11 = Reserved, do not use
	10 = Interrupt generated when Read Buffer 3 is read or Write Buffer 3 is written (Buffered PSP mode) or on a read or write operation when PMA<1:0> =11 (Addressable Slave mode only)
	01 = Interrupt generated at the end of the read/write cycle
	00 = No Interrupt generated
bit 12-11	**INCM<1:0>:** Increment Mode bits
	11 = Slave mode read and write buffers auto-increment (PMMODE<1:0> = 00 only)
	10 = Decrement ADDR<15:0> by 1 every read/write cycle[2] [4]
	01 = Increment ADDR<15:0> by 1 every read/write cycle[2] [4]
	00 = No increment or decrement of address
bit 10	**MODE16:** 8/16-bit Mode bit
	1 = 16-bit mode: a read or write to the data register invokes a single 16-bit transfer
	0 = 8-bit mode: a read or write to the data register invokes a single 8-bit transfer
bit 9-8	**MODE<1:0>:** Parallel Port Mode Select bits
	11 = Master mode 1 (PMCSx, PMRD/PMWR, PMENB, PMA<x:0>, PMD<7:0> and PMD<8:15>[3])
	10 = Master mode 2 (PMCSx, PMRD, PMWR, PMA<x:0>, PMD<7:0> and PMD<8:15>[3])
	01 = Enhanced Slave mode, control signals (PMRD, PMWR, PMCS, PMD<7:0>, and PMA<1:0>)
	00 = Legacy Parallel Slave Port, control signals (PMRD, PMWR, PMCS, and PMD<7:0>)

Note 1: Whenever WAITM<3:0> = 0000, WAITB and WAITE bits are ignored and forced to 1 T_{PBCLK} cycle for a write operation; WAITB = 1 T_{PBCLK} cycle, WAITE = 0 T_{PBCLK} cycles for a read operation.

2: Address bit A15 and A14 are not subject to autoincrement/decrement if configured as Chip Select CS2 and CS1.

3: These pins are active when bit MODE16 = 1 (16-bit mode)

4: The PMPADDR register is always incremented/decremented by 1 regardless of the transfer data width.

Figure 2.18
(Continued on next page)

Register 13-5:   PMMODE: Parallel Port Mode Register (Continued)

bit 7-6   **WAITB1:WAITB0:** Data Setup to Read/Write Strobe Wait States bits[1]

11 = Data wait of 4 T$_{PB}$; multiplexed address phase of 4 T$_{PB}$
10 = Data wait of 3 T$_{PB}$; multiplexed address phase of 3 T$_{PB}$
01 = Data wait of 2 T$_{PB}$; multiplexed address phase of 2 T$_{PB}$
00 = Data wait of 1 T$_{PB}$; multiplexed address phase of 1 T$_{PB}$ (DEFAULT)

bit 5-2   **WAITM3:WAITM0:** Data Read/Write Strobe Wait States bits

1111 = Wait of 16 T$_{PB}$
...
0001 = Wait of 2 T$_{PB}$
0000 = Wait of 1 T$_{PB}$ (DEFAULT)

bit 1-0   **WAITE1:WAITE0:** Data Hold After Read/Write Strobe Wait States bits[1]

11 = Wait of 4 T$_{PB}$
10 = Wait of 3 T$_{PB}$
01 = Wait of 2 T$_{PB}$
00 = Wait of 1 T$_{PB}$ (DEFAULT)

for Read operations:
11 = Wait of 3T$_{PB}$
10 = Wait of 2T$_{PB}$
01 = Wait of 1T$_{PB}$
00 = Wait of 0T$_{PB}$ (DEFAULT)

Note 1: Whenever WAITM<3:0> = 0000, WAITB and WAITE bits are ignored and forced to 1 T$_{PBCLK}$ cycle for a write operation; WAITB = 1 T$_{PBCLK}$ cycle, WAITE = 0 T$_{PBCLK}$ cycles for a read operation.

   2: Address bit A15 and A14 are not subject to autoincrement/decrement if configured as Chip Select CS2 and CS1.

   3: These pins are active when bit MODE16 = 1 (16-bit mode)

   4: The PMPADDR register is always incremented/decremented by 1 regardless of the transfer data width.

Figure 2.18
PMMODE bit definitions.

The steps to configure TIMER 1 are summarized below (assuming no timer interrupts are required):

- Disable TIMER 1 (T1CON, bit 15 = 0).
- Select required prescaler value (T1CON, bits 4 through 5).
- Select timer clock (T1CON, bit 1).
- Load PR1 register with required value.
- Clear TMR1 register to 0.
- Enable the timer (T1CON, bit 15 = 1).

An example project is given in the Projects section of this book to show how to configure TIMER 1 with interrupts.

Timers 2, 3, 4, and 5

Although timers 2, 3, 4 and 5 can operate as 16-bit timers, timers 2 and 3, and 4 and 5 can be combined to provide two 32-bit wide internal and external timers.

r-x	r-x	r-x	r-x	r-x	r-x	r-x	r-x
—	—	—	—	—	—	—	—
bit 31							bit 24

r-x	r-x	r-x	r-x	r-x	r-x	r-x	r-x
—	—	—	—	—	—	—	—
bit 23							bit 16

R/W-0	R/W-0	R/W-0	R/W-0	R/W-0	R/W-0	R/W-0	R/W-0
CS2	CS1	ADDR<13:8>					
bit 15							bit 8

R/W-0	R/W-0	R/W-0	R/W-0	R/W-0	R/W-0	R/W-0	R/W-0
ADDR<7:0>							
bit 7							bit 0

Legend:			
R = Readable bit	W = Writable bit	P = Programmable bit	r = Reserved bit
U = Unimplemented bit	-n = Bit value at POR: ('0', '1', x = Unknown)		

bit 31-16   **Reserved:** Write '0'; ignore read

bit 15   **CS2:** Chip Select 2 bit
1 = Chip Select 2 is active
0 = Chip Select 2 is inactive (pin functions as PMA<15>)

bit 14   **CS1:** Chip Select 1 bit
1 = Chip Select 1 is active
0 = Chip Select 1 is inactive (pin functions as PMA<14>)

bit 13-0   **ADDR<13:0>:** Destination Address bits

Figure 2.19
PMADDR register bit configuration and bit definitions.

Figure 2.28 shows the block diagram of these timers in 16-bit mode (note that 'x' represents the timer number and is between 2 and 5). Notice that, compared to TIMER 1, the prescaler has been extended and the low-power secondary oscillator has been removed. The operation of these timers in 16-bit mode is the same as Timer 1.

Figure 2.29 shows the timers in 32-bit mode where timers 2/3 and 4/5 are combined (in this figure, 'x' represents timers 2 through 5 in 16-bit mode, while in 32-bit mode, 'x' represents timers 2 or 4, while 'y' represents timers 3 or 5).

Considering, for example, timers 2 and 3 in 32-bit operation, registers PR3 and PR2 are combined to form a 32-bit register. Similarly, registers TMR3 and TMR2 are combined to form a 32-bit register. A 32-bit comparator is used to compare the two pairs and generate the T3IF flag when they are equal. In 32-bit mode, TIMER 2/3 pair is controlled with register T2CON. Similarly, TIMER 4/5 pair is controlled with register T4CON.

R/W-0	R/W-0	R/W-0	R/W-0	R/W-0	R/W-0	R/W-0	R/W-0
DATAOUT<31:24>							
bit 31							bit 24

R/W-0	R/W-0	R/W-0	R/W-0	R/W-0	R/W-0	R/W-0	R/W-0
DATAOUT<23:16>							
bit 23							bit 16

R/W-0	R/W-0	R/W-0	R/W-0	R/W-0	R/W-0	R/W-0	R/W-0
DATAOUT<15:8>							
bit 15							bit 8

R/W-0	R/W-0	R/W-0	R/W-0	R/W-0	R/W-0	R/W-0	R/W-0
DATAOUT<7:0>							
bit 7							bit 0

Legend:

R = Readable bit	W = Writable bit	P = Programmable bit	r = Reserved bit
U = Unimplemented bit	-n = Bit value at POR: ('0', '1', x = Unknown)		

bit 31-0   **DATAOUT<31:0>:** Output Data Port bits for 8-bit write operations in Slave mode

Figure 2.20
PMDOUT register bit configuration and bit definitions.

The steps to configure TIMER 2/3 are summarized below (assuming no timer interrupts are required):

- Disable TIMER 1 (T2CON, bit 15 = 0).
- Set 32-bit mode (T2CON, bit 3 = 1).
- Select required prescaler value (T2CON, bits 4 through 6).
- Select internal timer clock (T2CON, bit 1 = 0).
- Clear timer registers TMR2 and TMR3.
- Load PR2 and PR3 registers with required value.
- Clear TMR1 register to 0.
- Enable the timer (T2CON, bit 15 = 1).

2.1.7 Real-Time Clock and Calendar

The RTCC module is intended for accurate real-time date and time applications. Some of the features of this module are as follows:

- Provides real-time hours, minutes, seconds, weekday, date, month and year.
- Alarm intervals for half a second, 1 s, 10 s, 1 min, 10 min, 1 h, 1 day, 1 week, 1 month, and 1 year.

R/W-0	R/W-0	R/W-0	R/W-0	R/W-0	R/W-0	R/W-0	R/W-0
DATAIN<31:24>							
bit 31							bit 24

R/W-0	R/W-0	R/W-0	R/W-0	R/W-0	R/W-0	R/W-0	R/W-0
DATAIN<23:16>							
bit 23							bit 16

R/W-0	R/W-0	R/W-0	R/W-0	R/W-0	R/W-0	R/W-0	R/W-0
DATAIN<15:8>							
bit 15							bit 8

R/W-0	R/W-0	R/W-0	R/W-0	R/W-0	R/W-0	R/W-0	R/W-0
DATAIN<7:0>							
bit 7							bit 0

Legend:

R = Readable bit	W = Writable bit	P = Programmable bit	r = Reserved bit
U = Unimplemented bit	-n = Bit value at POR: ('0', '1', x = Unknown)		

bit 31-0   **DATAIN<31:0>:** Input and Output Data Port bits for 8-bit or 16-bit read/write operations in Master mode
Input Data Port for 8-bit read operations in Slave mode.

Figure 2.21
PMDIN register bit configuration and bit definitions.

- Leap-year correction.
- Long-term battery operation.
- Alarm pulse on output pin.
- Requires an external 32,768 Hz crystal.

Figure 2.30 shows a block diagram of the RTCC module. The module is controlled by the following six SFR registers:

- RTCCON
- RTCALRM
- RTCTIME
- RTCDATE
- ALRMTIME
- ALRMDATE.

Interested readers can find the programming details of the RTCC module in the individual microcontroller data sheets.

r-x	r-x	r-x	r-x	r-x	r-x	r-x	r-x
—	—	—	—	—	—	—	—
bit 31							bit 24

r-x	r-x	r-x	r-x	r-x	r-x	r-x	r-x
—	—	—	—	—	—	—	—
bit 23							bit 16

R/W-0	R/W-0	R/W-0	R/W-0	R/W-0	R/W-0	R/W-0	R/W-0
			PTEN<15:8>				
bit 15							bit 8

R/W-0	R/W-0	R/W-0	R/W-0	R/W-0	R/W-0	R/W-0	R/W-0
			PTEN<7:0>				
bit 7							bit 0

Legend:			
R = Readable bit	W = Writable bit	P = Programmable bit	r = Reserved bit
U = Unimplemented bit	-n = Bit value at POR: ('0', '1', x = Unknown)		

bit 31-16   **Reserved:** Write '0'; ignore read

bit 15-14   **PTEN<15:14>:** PMCSx Strobe Enable bits
1 = PMA15 and PMA14 function as either PMA<15:14> or PMCS2 and PMCS1[1]
0 = PMA15 and PMA14 function as port I/O

bit 13-2   **PTEN<13:2>:** PMP Address Port Enable bits
1 = PMA<13:2> function as PMP address lines
0 = PMA<13:2> function as port I/O

bit 1-0   **PTEN<1:0>:** PMALH/PMALL Strobe Enable bits
1 = PMA1 and PMA0 function as either PMA<1:0> or PMALH and PMALL[2]
0 = PMA1 and PMA0 pads functions as port I/O

Figure 2.22
PMAEN register bit configuration and bit definitions.

2.1.8 Analog to Digital Converter

The PIC32MX460F512L microcontroller contains multiplexed 16-channel, 10-bit A/D converters. These converters have the following features:

- Conversion speed of 500 ksps (kilo samples per second)
- Multiplexed 16 channels (two switchable multiplexers to select different analog channels and different reference sources)
- Sample-and-hold amplifier (SHA)
- Automatic input channel scanning
- A 16-word result buffer of 32-bit width

r-x	r-x	r-x	r-x	r-x	r-x	r-x	r-x
—	—	—	—	—	—	—	—
bit 31							bit 24

r-x	r-x	r-x	r-x	r-x	r-x	r-x	r-x
—	—	—	—	—	—	—	—
bit 23							bit 16

R-0	R/W-0	r-x	r-x	R-0	R-0	R-0	R-0
IBF	IBOV	—	—	IB3F	IB2F	IB1F	IB0F
bit 15							bit 8

R-1	R/W-0	r-x	r-x	R-1	R-1	R-1	R-1
OBE	OBUF	—	—	OB3E	OB2E	OB1E	OB0E
bit 7							bit 0

Legend:			
R = Readable bit	W = Writable bit	P = Programmable bit	r = Reserved bit
U = Unimplemented bit	-n = Bit value at POR: ('0', '1', x = Unknown)		

Figure 2.23
PMSTAT register bit configuration.

bit 31-16	Reserved: Write '0'; ignore read
bit 15	IBF: Input Buffer Full Status bit
	1 = All writable input buffer registers are full
	0 = Some or all of the writable input buffer registers are empty
bit 14	IBOV: Input Buffer Overflow Status bit
	1 = A write attempt to a full input byte buffer occurred (must be cleared in software)
	0 = No overflow occurred
	This bit is set (= 1) in hardware; can only be cleared (= 0) in software.
bit 13-12	Reserved: Write '0'; ignore read
bit 11-8	IBnF: Input Buffer n Status Full bits
	1 = Input Buffer contains data that has not been read (reading buffer will clear this bit)
	0 = Input Buffer does not contain any unread data
bit 7	OBE: Output Buffer Empty Status bit
	1 = All readable output buffer registers are empty
	0 = Some or all of the readable output buffer registers are full
bit 6	OBUF: Output Buffer Underflow Status bit
	1 = A read occurred from an empty output byte buffer (must be cleared in software)
	0 = No underflow occurred
	This bit is set (= 1) in hardware; can only be cleared (= 0) in software.
bit 5-4	Reserved: Write '0'; ignore read
bit 3-0	OBnE: Output Buffer n Status Empty bits
	1 = Output buffer is empty (writing data to the buffer will clear this bit)
	0 = Output buffer contains data that has not been transmitted

Figure 2.24
PMSTAT register bit definitions.

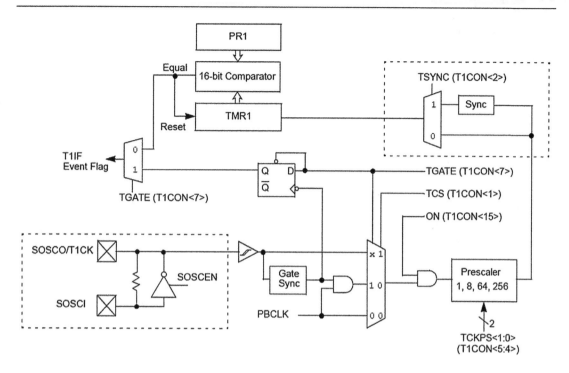

Note 1: The default state of the SOSCEN (OSCCON<1>) during a device Reset is controlled by the FSOSCEN bit in Configuration Word DEVCFG1.

Figure 2.25
Timer 1 block diagram.

r-x	r-x	r-x	r-x	r-x	r-x	r-x	r-x
—	—	—	—	—	—	—	—
bit 31							bit 24

r-x	r-x	r-x	r-x	r-x	r-x	r-x	r-x
—	—	—	—	—	—	—	—
bit 23							bit 16

R/W-0	R/W-0	R/W-0	R/W-0	R-0	r-x	r-x	r-x
ON	FRZ	SIDL	TWDIS	TWIP	—	—	—
bit 15							bit 8

R/W-0	r-x	R/W-0	R/W-0	r-x	R/W-0	R/W-0	r-x
TGATE	—	TCKPS<1:0>		—	TSYNC	TCS	—
bit 7							bit 0

Figure 2.26
T1CON bit configuration.

bit 31-16   **Reserved:** Write '0'; ignore read

bit 15   **ON:** Timer On bit
1 = Timer is enabled
0 = Timer is disabled

bit 14   **FRZ:** Freeze in Debug Exception Mode bit
1 = Freeze operation when CPU is in Debug Exception mode
0 = Continue operation when CPU is in Debug Exception mode

Note: FRZ is writable in Debug Exception mode only, it is forced to '0' in normal mode.

bit 13   **SIDL:** Stop in Idle Mode bit
1 = Discontinue operation when device enters Idle mode
0 = Continue operation in Idle mode

bit 12   **TWDIS:** Asynchronous Timer Write Disable bit
<u>In Asynchronous Timer mode:</u>
1 = Writes to asynchronous TMR1 are ignored until pending write operation completes
0 = Back-to-back writes are enabled (legacy asynchronous timer functionality)
<u>In Synchronous Timer mode:</u>
This bit has no effect.

bit 11   **TWIP:** Asynchronous Timer Write in Progress bit
<u>In Asynchronous Timer mode:</u>
1 = Asynchronous write to TMR1 register in progress
0 = Asynchronous write to TMR1 register complete
<u>In Synchronous Timer mode:</u>
This bit is read as '0'.

bit 10-8   **Reserved:** Write '0'; ignore read

bit 7   **TGATE:** Gated Time Accumulation Enable bit
<u>When TCS = 1:</u>
This bit is ignored and read '0'.
<u>When TCS = 0:</u>
1 = Gated time accumulation is enabled
0 = Gated time accumulation is disabled

bit 6   **Reserved:** Write '0'; ignore read

bit 5-4   **TCKPS<1:0>:** Timer Input Clock prescaler Select bits
11 = 1:256 prescale value
10 = 1:64 prescale value
01 = 1:8 prescale value
00 = 1:1 prescale value

bit 3   **Reserved:** Write '0'; ignore read

bit 2   **TSYNC:** Timer External Clock Input Synchronization Selection bit
<u>When TCS = 1:</u>
1 = External clock input is synchronized
0 = External clock input is not synchronized
<u>When TCS = 0:</u>
This bit is ignored and read '0'.

bit 1   **TCS:** Timer Clock Source Select bit
1 = External clock from T1CKI pin
0 = Internal peripheral clock

bit 0   **Reserved:** Write '0'; ignore read

Figure 2.27
T1CON bit definitions.

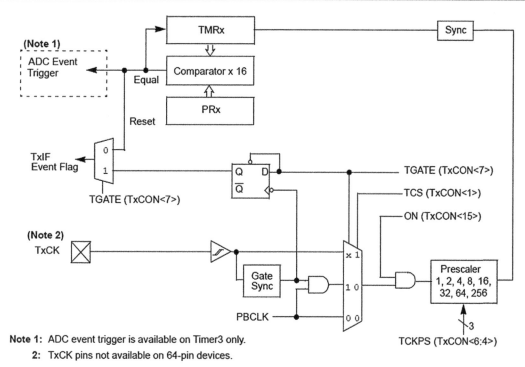

Note 1: ADC event trigger is available on Timer3 only.
    2: TxCK pins not available on 64-pin devices.

Figure 2.28
Timers 2, 3, 4, and 5 in 16-bit mode.

- Various conversion result formats (integer, signed, unsigned, 16-bit or 32-bit output)
- External voltage references
- Operation during CPU SLEEP and IDLE modes.

Figure 2.31 shows a block diagram of the A/D converter module.

The A/D converter module is controlled by the following SFR registers:

- AD1CON1
- AD1CON2
- AD1CON3
- AD1CHS
- AD1PCFG
- AD1CSSL.

In addition, all the above registers have additional bit manipulation registers, e.g. AD1CON1CLR, AD1CON1SET, and AD1CON1INV.

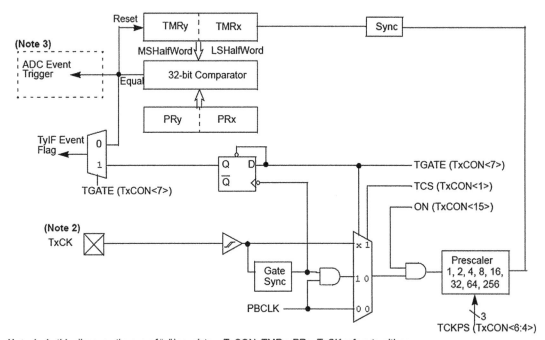

Note 1: In this diagram, the use of "x" in registers TxCON, TMRx, PRx, TxCK refers to either
Timer2 or Timer4; the use of "y" in registers TyCON, TMRy, PRy, TyIF refers to either Timer3 or Timer5.

2: TxCK pins not available on 64-pin devices.

3: ADC event trigger is available only on Timer2/3 pair.

Figure 2.29
Timers 2/3 and 4/5 combined for 32-bit operation.

AD1CON1

This register controls the A/D operating mode, data output format, conversion trigger source
select, sample and hold control, and A/D conversion status. Figures 2.32 and 2.33 show the
AD1CON1 bit configuration and bit definitions, respectively.

AD1CON2

This register controls the voltage reference selection, scan input selection, A/D interrupt
selection bits, A/D result buffer configuration, and alternate input sample mode selection.
Figures 2.34 and 2.35 show the AD1CON2 bit configuration and bit definitions, respectively.

AD1CON3

This register controls the A/D clock selection. Figure 2.36 shows the bit configuration and bit
definitions.

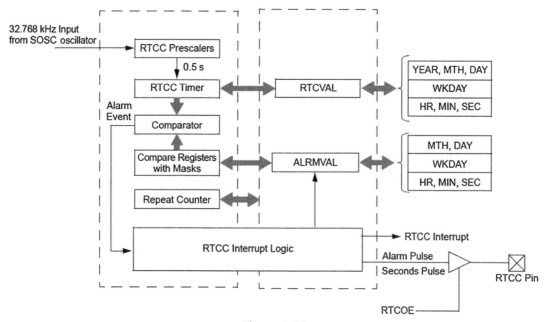

Figure 2.30
Block diagram of the RTCC module.

AD1CHS

This register selects the input channels for multiplexers A and B. Figures 2.37 and 2.38 show the bit configuration and bit definitions, respectively.

AD1PCFG

This register selects the input ports as analog or digital. Setting a bit to 1 makes the corresponding ANx port pin a digital port. Similarly, clearing a bit to 0 makes the corresponding ANx port pin an analog port. To configure all analog ports as digital, issue the command AD1PCFG = 0xFFFF. Similarly, to configure ports as analog, issue the command AD1PCFG = 0.

AD1CSSL

This register controls the A/D input scanning. Setting a bit of the register to1 selects the corresponding ANx port pin for input scan. Similarly, clearing a bit of the register to 0 does not select the corresponding ANx port pin for input scan.

Operation of the A/D Module

The total A/D conversion time consists of the *acquisition time* and the *A/D conversion time*. The A/D has a single SHA amplifier. During the acquisition time, the analog signal is sampled

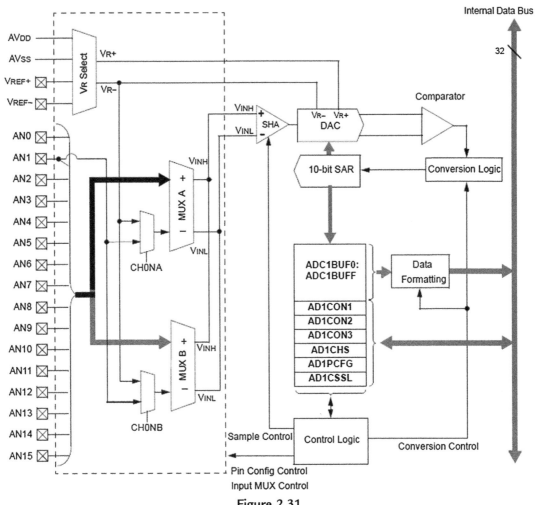

Figure 2.31
A/D converter module.

and held by the SHA. After the input is stable, it is disconnected from the SHA and the A/D conversion starts, which converts the analog signal into digital. The conversion time is the actual time that it takes for the A/D converter to convert the signal held by the SHA. Although there is only one SHA, there are two multiplexers called MUXA and MUXB, controlled by register AD1CHS. The A/D converter can switch between MUXA and MUXB inputs.

The converted data in the result register can be read in eight different formats, controlled by AD1CON1.

Sampling can be started manually or automatically. In manual mode, AD1CON1, bit 2 is cleared. Data acquisition is then started when bit 1 of AD1CON1 is set. This bit must be set to

r-x	r-x	r-x	r-x	r-x	r-x	r-x	r-x
—	—	—	—	—	—	—	—
bit 31							bit 24

r-x	r-x	r-x	r-x	r-x	r-x	r-x	r-x
—	—	—	—	—	—	—	—
bit 23							bit 16

R/W-0	R/W-0	R/W-0	r-x	r-x	R/W-0	R/W-0	R/W-0
ON	FRZ	SIDL	—	—	FORM<2:0>		
bit 15							bit 8

R/W-0	R/W-0	R/W-0	R/W-0	r-x	R/W-0	R/W-0	R/C-0
SSRC<2:0>			CLRASAM	—	ASAM	SAMP	DONE
bit 7							bit 0

Figure 2.32
AD1CON1 Bit Configuration.

restart acquisition. In automatic mode, bit 2 of AD1CON is set, and acquisition starts automatically after a previous sample has completed.

The scan mode enables a number of input channels to be scanned and converted into digital. This mode is enabled by setting bit 10 of AD1CON2. Each bit in the AD1CSSL register corresponds to an analog input channel, and if a bit is set to 1 in AD1CSSL, then the corresponding input channel is in the scan sequence.

The following steps summarize how to configure the A/D module, assuming no interrupts are to be generated (see the individual microcontroller data sheets for more information):

- Configure port pin as analog by clearing the appropriate bits of AD1PCFG.
- Select the required analog input channels using AD1CHS.
- Select the format of the result data using bits 8–10 of AD1CON1.
- Select the sample clock source using bits 5–7 of AD1CON1.
- Select the voltage reference source using bits 13–15 of AD1CON2.
- Select the scan mode (if required), alternating sample mode (if required), and autoconvert sample time (if required).
- Select the result buffer mode using bit 1 of AD1CON2.
- Select the A/D clock source using bit 15 of AD1CON3.
- Select A/D clock prescaler using bits 0–7 of AD1CON3.
- Turn ON the A/D using bit 15 of AD1CON1.

The mikroC Pro for PIC32 language provides a built-in library for simple one-channel A/D conversions, where all the required registers are set automatically.

bit 31-16   **Reserved:** Write '0'; ignore read

bit 15   **ON:** ADC Operating Mode bit

1 = A/D converter module is operating
0 = A/D converter is off

bit 14   **FRZ:** Freeze in Debug Exception Mode bit

1 = Freeze operation when CPU enters Debug Exception mode
0 = Continue operation when CPU enters Debug Exception mode
Note: FRZ is writable in Debug Exception mode only. It reads '0' in Normal mode.

bit 13   **SIDL:** Stop in Idle Mode bit

1 = Discontinue module operation when device enters Idle mode
0 = Continue module operation in Idle mode

bit 12-11   **Reserved:** Write '0'; ignore read

bit 10-8   **FORM<2:0>:** Data Output Format bits

011 = Signed Fractional 16-bit (DOUT = 0000 0000 0000 0000 sddd dddd dd00 0000)
010 = Fractional 16-bit (DOUT = 0000 0000 0000 0000 dddd dddd dd00 0000)
001 = Signed Integer 16-bit (DOUT = 0000 0000 0000 0000 ssss sssd dddd dddd)
000 = Integer 16-bit (DOUT = 0000 0000 0000 0000 0000 00dd dddd dddd)
111 = Signed Fractional 32-bit (DOUT = sddd dddd dd00 0000 0000 0000 0000 0000)
110 = Fractional 32-bit (DOUT = dddd dddd dd00 0000 0000 0000 0000 0000)
101 = Signed Integer 32-bit (DOUT = ssss ssss ssss ssss ssss sssd dddd dddd)
100 = Integer 32-bit (DOUT = 0000 0000 0000 0000 00dd dddd dddd)

bit 7-5   **SSRC<2:0>:** Conversion Trigger Source Select bits

111 = Internal counter ends sampling and starts conversion (autoconvert)
110 = Reserved
101 = Reserved
100 = Reserved
011 = Reserved
010 = Timer 3 period match ends sampling and starts conversion
001 = Active transition on INT0 pin ends sampling and starts conversion
000 = Clearing SAMP bit ends sampling and starts conversion

bit 4   **CLRASAM:** Stop Conversion Sequence bit (when the first A/D converter interrupt is generated)
1 = Stop conversions when the first ADC interrupt is generated. Hardware clears the ASAM bit when the ADC interrupt is generated.
0 = Normal operation, buffer contents will be overwritten by the next conversion sequence

bit 3   **Reserved:** Write '0'; ignore read

bit 2   **ASAM:** ADC Sample Auto-Start bit

1 = Sampling begins immediately after last conversion completes; SAMP bit is automatically set.
0 = Sampling begins when SAMP bit is set

bit 1   **SAMP:** ADC Sample Enable bit

1 = The ADC SHA is sampling
0 = The ADC sample/hold amplifier is holding
When ASAM = 0, writing '1' to this bit starts sampling.
When SSRC = 000, writing '0' to this bit will end sampling and start conversion.

bit 0   **DONE:** A/D Conversion Status bit

1 = A/D conversion is done
0 = A/D conversion is not done or has not started
Clearing this bit will not affect any operation in progress.
Note: The DONE bit is not persistent in automatic modes. It is cleared by hardware at the beginning of the next sample.

Figure 2.33
AD1CON1 bit definitions.

r-x	r-x	r-x	r-x	r-x	r-x	r-x	r-x
—	—	—	—	—	—	—	—
bit 31							bit 24

r-x	r-x	r-x	r-x	r-x	r-x	r-x	r-x
—	—	—	—	—	—	—	—
bit 23							bit 16

R/W-0	R/W-0	R/W-0	R/W-0	r-x	R/W-0	r-x	r-x
VCFG<2:0>			OFFCAL	—	CSCNA	—	—
bit 15							bit 8

R-0	r-x	R/W-0	R/W-0	R/W-0	R/W-0	R/W-0	R/W-0
BUFS	—	SMPI<3:0>				BUFM	ALTS
bit 7							bit 0

Figure 2.34
AD1CON2 bit configuration.

2.1.9 Interrupts

Interrupt control is one of the most complex parts of the PIC32 microcontrollers. There are 96 interrupt sources with up to 64 interrupt vectors, and a large number of interrupt control registers. The full description of the interrupt control is beyond the scope of this book and, if interested, reader should consult the individual microcontroller data sheets for much more information.

In this section, we shall be looking at the basic operation of the interrupt module and see how an interrupt-based program can be written using the mikroC Pro for PIC32 language.

The basic features of the PIC32 interrupt controller module are as follows:

- Up to 96 interrupt sources
- Up to 64 interrupt vectors
- Single and multiple vector interrupt modes
- Seven user selectable priority levels for each interrupt
- Four user selectable subpriority levels within each priority
- User configurable interrupt vector table location and spacing.

The interrupt control module has the following SFR registers:

- INTCON
- INSTAT
- IPTMR
- IFS0, IFS1

- IEC0, IEC1
- IPC0–IPC11.

In addition, all the above registers have additional bit manipulation registers, e.g. INTCONCLR, INTCONSET, and INTCONINV. Register INTCON controls the interrupt vector mode and external interrupt edge mode. Other registers control the individual interrupt sources, such as enabling and disabling them.

bit 31-16   **Reserved:** Write '0'; ignore read

bit 15-13   **VCFG<2:0>:** Voltage Reference Configuration bits

	ADC V$_{R+}$	ADC V$_{R-}$
000	AV$_{DD}$	AV$_{SS}$
001	External V$_{REF+}$ pin	AV$_{SS}$
010	AV$_{DD}$	External V$_{REF-}$ pin
011	External V$_{REF+}$ pin	External V$_{REF-}$ pin
1xx	AV$_{DD}$	AV$_{SS}$

bit 12   **OFFCAL:** Input Offset Calibration Mode Select bit

1 = Enable Offset Calibration mode
    V$_{INH}$ and V$_{INL}$ of the SHA are connected to V$_{R-}$
0 = Disable Offset Calibration mode
    The inputs to the SHA are controlled by AD1CHS or AD1CSSL

bit 11   **Reserved:** Write '0'; ignore read

bit 10   **CSCNA:** Scan Input Selections for CH0+ SHA Input for MUX A Input Multiplexer Setting bit
1 = Scan inputs
0 = Do not scan inputs

bit 9-8   **Reserved:** Write '0'; ignore read

bit 7   **BUFS:** Buffer Fill Status bit
Only valid when BUFM = 1 (ADRES split into 2 x 8-word buffers).
1 = ADC is currently filling buffer 0x8-0xF, user should access data in 0x0-0x7
0 = ADC is currently filling buffer 0x0-0x7, user should access data in 0x8-0xF

bit 6   **Reserved:** Write '0'; ignore read

bit 5-2   **SMPI<3:0>:** Sample/Convert Sequences Per Interrupt Selection bits
1111 = Interrupts at the completion of conversion for each 16th sample/convert sequence
1110 = Interrupts at the completion of conversion for each 15th sample/convert sequence
. . . . .
0001 = Interrupts at the completion of conversion for each 2nd sample/convert sequence
0000 = Interrupts at the completion of conversion for each sample/convert sequence

bit 1   **BUFM:** ADC Result Buffer Mode Select bit
1 = Buffer configured as two 8-word buffers, ADC1BUF(7...0), ADC1BUF(15...8)
0 = Buffer configured as one 16-word buffer ADC1BUF(15...0.)

bit 0   **ALTS:** Alternate Input Sample Mode Select bit
1 = Uses MUX A input multiplexer settings for first sample, then alternates between MUX B and
    MUX A input multiplexer settings for all subsequent samples
0 = Always use MUX A input multiplexer settings

Figure 2.35
AD1CON2 bit definitions.

R/W-0	r-x	r-x	R/W-0	R/W-0	R/W-0	R/W-0	R/W-0
ADRC	—	—	SAMC<4:0>				
bit 15							bit 8

R/W-0	R/W-0	R/W-0	R/W-0	R/W-0	R/W-0	R/W-0	R/W-0
ADCS<7:0>							
bit 7							bit 0

Legend:			
R = Readable bit	W = Writable bit	P = Programmable bit	r = Reserved bit
U = Unimplemented bit	-n = Bit value at POR: ('0', '1', x = Unknown)		

bit 31-16   **Reserved:** Write '0'; ignore read

bit 15   **ADRC:** ADC Conversion Clock Source bit

         1 = ADC internal RC clock
         0 = Clock derived from Peripheral Bus Clock (PBClock)

bit 14-13   **Reserved:** Write '0'; ignore read

bit 12-8   **SAMC<4:0>:** Auto-Sample Time bits

         $11111 = 31\ T_{AD}$
         . . . . .
         $00001 = 1\ T_{AD}$
         $00000 = 0\ T_{AD}$ (Not allowed)

bit 7-0   **ADCS<7:0>:** ADC Conversion Clock Select bits

         $11111111 = T_{PB} \cdot (ADCS<7:0> + 1) \cdot 2 = 512 \cdot T_{PB} = T_{AD}$
         . . . . . .
         $00000001 = T_{PB} \cdot (ADCS<7:0> + 1) \cdot 2 = 4 \cdot T_{PB} = T_{AD}$
         $00000000 = T_{PB} \cdot (ADCS<7:0> + 1) \cdot 2 = 2 \cdot T_{PB} = T_{AD}$

Figure 2.36
AD1CON3 bit configuration and bit definitions.

R/W-0	r-x	r-x	r-x	R/W-0	R/W-0	R/W-0	R/W-0
CH0NB	—	—	—	CH0SB<3:0>			
bit 31							bit 24

R/W-0	r-x	r-x	r-x	R/W-0	R/W-0	R/W-0	R/W-0
CH0NA	—	—	—	CH0SA<3:0>			
bit 23							bit 16

r-x	r-x	r-x	r-x	r-x	r-x	r-x	r-x
—	—	—	—	—	—	—	—
bit 15							bit 8

r-x	r-x	r-x	r-x	r-x	r-x	r-x	r-x
—	—	—	—	—	—	—	—
bit 7							bit 0

Figure 2.37
AD1CHS bit configuration.

PIC32 microcontrollers support both single and multiple vectored interrupts. In single vectored interrupts, all interrupting devices have the same common ISR addresses. The source of the interrupt is then determined by examining the interrupt flags of each interrupting source. This is actually the commonly used method in 8-bit microcontrollers. In multiple vectored interrupt operations, each interrupting device has its unique ISR address. (In fact, PIC32 microcontrollers have 96 interrupt sources and only 64 vectors. As such, some of the interrupts share the same vector.)

PIC32 microcontrollers support seven levels of interrupt priority (ipl1−ipl7). If more than one interrupt occurs at the same time, then the one with higher priority is serviced first. While servicing an interrupt, if a lower priority interrupt occurs, then it will be ignored by the processor. In addition to the standard interrupt priority levels, the PIC32 microcontrollers support four levels of subpriorities. Thus, should two interrupts at the same priority level interrupt at the same time, the one with the higher subpriority level will be serviced first. At reset or power-up, all interrupts are disabled and are all set to priority level ipl0.

In addition to general interrupt configuration bits, each interrupt source has associated control bits in the SFR registers and these bits must be configured correctly for an interrupt to be accepted by the CPU. Some of the important interrupt source control bits are as follows:

- Each interrupting source has an *Interrupt Enable Bit* (denoted by suffix IE) in the device data sheet. This bit must be set to 1 for an interrupt to be accepted from this source (at reset

bit 31	**CH0NB:** Negative Input Select for MUX B bit
	1 = Channel 0 negative input is AN1
	0 = Channel 0 negative input is V$_R$−
bit 30-28	**Reserved:** Write '0'; ignore read
bit 27-24	**CH0SB<3:0>:** Positive Input Select for MUX B bits
	1111 = Channel 0 positive input is AN15
	1110 = Channel 0 positive input is AN14
	1101 = Channel 0 positive input is AN13
	. . . . . .
	0001 = Channel 0 positive input is AN1
	0000 = Channel 0 positive input is AN0
bit 23	**CH0NA:** Negative Input Select for MUX A Multiplexer Setting bit[2]
	1 = Channel 0 negative input is AN1
	0 = Channel 0 negative input is V$_R$−
bit 22-20	**Reserved:** Write '0'; ignore read
bit 19-16	**CH0SA<3:0>:** Positive Input Select for MUX A Multiplexer Setting bits
	1111 = Channel 0 positive input is AN15
	1110 = Channel 0 positive input is AN14
	1101 = Channel 0 positive input is AN13
	. . . . . .
	0001 = Channel 0 positive input is AN1
	0000 = Channel 0 positive input is AN0
bit 15-0	**Reserved:** Write '0'; ignore read

Figure 2.38
AD1CHS bit definitions.

or power-up all interrupt enable bits are cleared). Some commonly used interrupt enable bits are given in the following table:

Interrupt Source	Interrupt Enable Bit	Bit Position	Register
Timer 1	T1IE	4	IEC0
External Int 0	INT0IE	3	IEC0
External Int 1	INT1IE	7	IEC0
External Int 2	INT2IE	11	IEC0
External Int 3	INT3IE	15	IEC0

- Each interrupting source has an *Interrupt Flag* (denoted by IF) in the device data sheet. This bit is set automatically when an interrupt occurs, and must be cleared in software before any more interrupts can be accepted from the same source. The bit is usually cleared inside the ISR of the interrupting source. The interrupt flags of some commonly used interrupt sources are given below:

Interrupt Source	Interrupt Flag Bit	Bit Position	Register
Timer 1	T1IF	4	IFS0
External Int 0	INT0IF	3	IFS0
External Int 1	INT1IF	7	IFS0
External Int 2	INT2IF	11	IFS0
External Int 3	INT3IF	15	IFS0

- Each interrupting source has a priority level from 1 to 7 (priority level 0 disabled the interrupt), and subpriority levels from 0 to 3. The priority levels are denoted by suffix IP, and the subpriority levels by suffix IS. The priority levels of some commonly used interrupt sources are given below:

Interrupt Source	Interrupt Priority Level	Bit Positions	Register
Timer 1	T1IP	2–4	IPC1
External Int 0	INT0IP	26–28	IPC0
External Int 1	INT1IP	26–28	IPC1
External Int 2	INT2IP	26–28	IPC2
External Int 3	INT3IP	26–28	IPC3

Interrupt Source	Interrupt Subpriority Level	Bit Positions	Register
Timer 1	T1IS	0–1	IPC1
External Int 0	INT0IS	24–25	IPC0
External Int 1	INT1IS	24–25	IPC1
External Int 2	INT2IS	24–25	IPC2
External Int 3	INT3IS	24–25	IPC3

Some examples are given below to show how the timer interrupts and external interrupts can be configured. Notice that the statement *EnableInterrupts()* must be called after configuring the interrupts, so that any interrupt can be accepted by the microcontroller. The Projects section of this book gives real examples on configuring and using both timer and external interrupts.

Configuring Timer 1 Interrupts

Timer 1 counts up until the value in TMR1 matches the one in period register PR1 and then the interrupt flag T1IF is set automatically. If the Timer 1 interrupt enabled flag T1IE is set, then an interrupt will be generated to the processor. In multi-interrupt operations, it is recommended to set the priority and subpriority levels of the timer interrupt.

The steps in configuring the Timer 1 interrupts are given below:

- Configure for single- or multivector interrupt mode (by default, after reset or power-up, the single vector mode is selected).
- Disable Timer 1, ON (bit 15 in T1CON).
- Clear timer register TMR1 to 0.
- Select timer prescaler (bits 4—5 in T1CON).
- Select timer clock source (bit 1 in T1CON).
- Load period register PR1 as required.
- Clear interrupt flag, T1IF (bit 4 in IFS0).
- Set Timer 1 priority level, T1IP (bit 2-4 in IPC1).
- Set Timer 1 subpriority level, T1IS (if required).
- Enable Timer 1 interrupts, T1IE (bit 4 in IEC0).
- Enable Timer 1, ON (bit 15 in T1CON).
- Write the ISR.
- Call EnableInterrupts().

Configuring External Interrupt 0

The PIC32 microcontroller supports four external interrupts inputs, INT0 through INT3. Interrupts can be recognized either on the low-to-high or on the high-to-low transition of the interrupt pin, selected by INTCON. External interrupt flag INT0IF must be cleared before an interrupt can be accepted. In addition, INT0IE bit must be set to enable interrupts from external interrupt INT0 pin. In multi-interrupt operations, it is recommended to set the priority and subpriority levels of the interrupt.

The steps in configuring the external interrupt 0 are given below:

- Configure for single- or multivector interrupt mode (bit 12, INTCON). By default, after reset or power-up, the single-vector mode is selected.
- Clear external interrupt 0 flag, INT0IF (bit 3 in IFS0).

- Set required interrupt edge, INT0EP (bit 0 in INTCON).
- Set external interrupt 0 priority level, INT0IP (bits 26–28 in IPC0).
- Set external interrupt subpriority level, INT0IS (if required).
- Enable external interrupt 0, INT0IE (bit 3 IEC0).
- Write the ISR.
- Call EnableInterrupts().

Writing mikroC Pro for PIC32 ISRs

The PIC32 microcontroller interrupt controller can be configured to operate in one of the two modes:

- **Single-Vector mode**—all interrupt requests will be serviced at one vector address (mode out of reset).
- **MultiVector mode**—interrupt requests will be serviced at the calculated vector address.

In single-vector mode, the CPU always vectors to the same address. This means that only one ISR can be defined. The single-vector mode address is calculated using the Exception Base (EBase) address (its address default is 0x9FC01000E). The exact formula for single-vector mode is as follows (see the PIC32 microcontroller individual data sheets for more information):

```
Single-Vector Address = EBase + 0x200.
```

In multivector mode, the CPU vectors to the unique address for each vector number. Each vector is located at a specific offset, with respect to a base address specified by the EBase register in the CPU. The individual vector address offset is determined by the following equation:

```
EBase + (Vector_Number × Vector_Space) + 0x200.
```

By default, the mikroC Pro for PIC32 compiler configures interrupts in the multivector mode, with the EBase address set to 0x9FC01000 and vector spacing of 32.

Configuring the interrupt operating can be selected in the compiler edit window. As we shall see in later chapters, this is selected by clicking *Project → Edit Project* in the drop-down menu of the compiler IDE. An example is shown in Figure 2.39 (see middle right-hand side of the window).

The PIC32 family of devices employs two register sets: a *primary register set* for normal program execution and a *shadow register set* for highest priority interrupt processing.

- Register Set Selection in Single-Vector Mode
  In single-vector mode, you can select which register set will be used. By default, the interrupt controller will instruct the CPU to use the first register set. This can be changed later in the code.

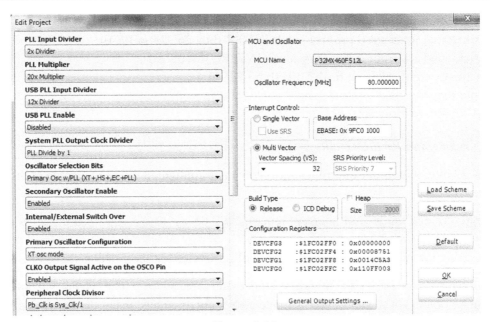

Figure 2.39
Project edit window showing the interrupt vector mode. (For color version of this figure, the reader is referred to the online version of this book.)

- Register Set Selection in Multivector Mode
  When a priority level interrupt matches a shadow set priority, the interrupt controller instructs the CPU to use the shadow set. For all other interrupt priorities, the interrupt controller instructs the CPU to use the primary register set.

In order to correctly utilize interrupts and correctly write the ISR code, the user will need to take care of the following:

1. Write the ISR. The mikroC Pro for PIC32 Interrupt Assistant window can be used to simplify this process.
2. Initialize the module which will generate an interrupt.
3. Set the correct priority and subpriority for the used module according to the priorities set in the ISR.
4. Enable Interrupts.

In mikroC Pro for PIC32 compiler, an ISR is defined in this way:

```
void interrupt() iv IVT_ADC ilevel 7 ics ICS_SOFT
{
        // Interrupt service routine code
}
```

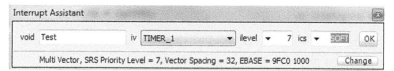

Figure 2.40
Using the interrupt assistant to generate the ISR. (For color version of this figure,
the reader is referred to the online version of this book.)

where:

- iv—reserved word that inform the compiler that it is an ISR.
- IVT_ADC—appropriate Interrupt Vector.
- ilevel 7—Interrupt priority level 7.
- ics—Interrupt Context Saving; Interrupt Context Saving can be performed in several ways:
  - ICS_SOFT—Context saving is carried out by the software.
  - ICS_SRS—Shadow Register set is use for context saving.
  - ICS_OFF—No context saving
  - ICS_AUTO—Compiler chooses whether the ICS_SOFT or ICS_SRS will be used.

The Interrupt Assistant is accessed by clicking *Tools → Interrupt Assistant* in the drop-down menu of the compiler IDE. Figure 2.40 shows how the ISR can be generated for Timer 1. In this example, the ISR is given the name *Test*, priority level 7 is chosen, and context switching is required to be carried out by software (Figure 2.41).

The actual code generated by the Interrupt Assistant for the above example is given below. The user ISR code should be written inside the body of the function:

```
void Test() iv IVT_TIMER_1 ilevel 7 ics ICS_SOFT
{
}
```

Figure 2.41
Code generated by the interrupt assistant.

2.2 Summary

This chapter has described the architecture of the PIC32 family of microcontrollers. PIC32MX360F512L was taken as a typical example microcontroller in the family.

Various important parts and peripheral circuits of the PIC32 series of microcontrollers have been described, including the data memory, program memory, clock circuits, reset circuits, general purpose timers, A/D converter, and the very important topic of interrupt structure. Steps are given to show how the timer and external interrupts can be configured. Finally, the mikroC Pro for PIC32 Interrupt Assistant has been described for creating an ISR for Timer 1.

2.3 Exercises

1. Describe the memory structure of the PIC32 series of microcontrollers. What is the difference between physical and virtual addresses?

2. Explain the functions of the Boot memory in a PIC32 series of microcontroller.

3. Explain the differences between a general-purpose register and an SFR.

4. Explain the various ways that the PIC32 series of microcontrollers can be reset. Draw a circuit diagram to show how an external push-button switch can be used to reset the microcontroller.

5. Describe the various clock sources that can be used to provide clock to a PIC32 series of microcontroller. Draw a circuit diagram to show how a 10 MHz crystal can be connected to the microcontroller.

6. Explain how an external clock can be used to provide clock pulses to a PIC32 series of microcontroller.

7. What are the registers of a typical I/O port in a PIC32 series of microcontroller? Explain the operation of the port by drawing the port block diagram.

8. Explain the differences between the PORT and LAT registers.

9. Explain the structure of the A/D converter of a PIC32 series of microcontroller.

10. An LM35DZ type analog temperature sensor is connected to analog port AN0 of a PIC32MX460F512L microcontroller. The sensor provides an analog output voltage proportional to the temperature, i.e. Vo = 10 mV/°C. Show the steps required to read the temperature.

11. Explain the differences between a priority interrupt and a nonpriority interrupt.

12. Explain the differences between a single-vector interrupt and a multivector interrupt.

13. Show the steps required to configure Timer 1 to generate an interrupt. Assuming a 10 MHz clock frequency, what will be loaded into register PR1 to generate an interrupt every second?

14. Show the steps required to set up an external interrupt input INT0 to generate interrupts on its falling edge.

15. Show the steps required to set up Timer 1 to generate interrupts every millisecond having a priority of 7, and subpriority 3.

16. In an application, the CPU registers have been configured to accept interrupts from external sources INT0, INT1, and INT2. An interrupt has been detected. Explain how you can find the source of the interrupt.

C Programming for 32-Bit PIC Microcontrollers

The number of C compilers available in the market place for the PIC32 series of microcontrollers is very limited. Most PIC microcontroller C compilers are available only for the PIC16, PIC18, PIC24, or dsPIC series of microcontrollers.

Some of the popular C compilers used in the development of commercial, industrial, and educational PIC32 microcontroller applications are listed below. Most of the features of these compilers are similar and they can all be used to develop complex and powerful PIC32 projects:

- mikroC Pro for PIC32 compiler
- Hi-Tech C compiler for PIC32
- MPLAB C compiler for PIC32.

mikroC Pro for PIC32 C compiler has been developed by MikroElektronika (website: www.microe.com) and is one of the easy methods to learn compilers with rich resources, such as a large number of library functions and an integrated development environment with built-in simulator, and an in-circuit debugger (e.g. mikroICD). A demo version of the compiler with a 2 K program limit is available from MikroElektronika. This demo version should be sufficient for many small projects. The full version of the compiler costs $299. Users can either purchase a dongle to use the compiler on any computer they like, or the compiler can be registered to a single computer. In this book, we shall be mainly concentrating on the use of this compiler and all the projects will be based on this compiler.

Hi-Tech C compiler for PIC32 is another popular C compiler, developed by Hi-Tech Software (website: www.htsoft.com). At the time of writing, this compiler has three versions: the PRO compiler, the Standard compiler, and the Lite compiler. The limited functionality Lite version is free, the single user Standard version costs $895, and the single-user PRO version costs $1495. The PRO compiler is available for 45 days free evaluation. All three versions of the compiler support all members of the PIC32 microcontrollers. The Standard and the PRO versions create highly optimized code with dynamic register allocation and improved speed. The full C library source code is provided with all three versions, and they are all compatible with the MPLAB IDE. In addition, all three compilers run on all platforms, including Windows, Linux, and Mac OS X.

MPLAB C compiler for PIC32 is a product of the Microchip Inc. (website: www.microchip. com). The evaluation version of the compiler is available with full functionality for 60 days. Although the compiler is usable after 60 days, certain optimization levels are disabled. This compiler is fully compatible with the MPLAB IDE. All the members of the PIC32 microcontroller are supported. The compiler is distributed with extensive libraries and support for PIC32 microcontrollers with integrated Ethernet, CAN bus, and USB. High-performance maths and Digital Signal Processing routines are also supported. A special Lite version of the compiler for academic use is available. Although this compiler is not used in the projects in this book, its main features and the development of simple projects using this compiler are given in detail in Appendix A.

In this book, we shall be looking at the features of the popular and powerful mikroC Pro for PIC32 C programming language and also see how to use the compiler, as this language will be used in all the projects in this book.

3.1 Structure of a Simple mikroC Pro for PIC32 Program

Figure 3.1 shows the simplest structure of a mikroC Pro for PIC32 program. This program flashes an LED connected to port RB0 (bit 0 of PORTB) of a PIC microcontroller with 1 s intervals. Do not worry if you do not understand the operation of the program at this stage as all will be clear as we progress through this chapter. Some of the programming concepts used in Figure 3.1 are described below in the following sections.

```
/*========================================================================
                           LED FLASHING PROGRAM
                           ====================

This program flashes the LED connected to port pin RB0 of a PIC32 microcontroller every
second.

A P32MX460F512L type 32-bit PIC microcontroller is used in this example. The clock
frequency is 8MHz, provided using a crystal.


Author:      Dogan Ibrahim
Date:        February, 2012
File:        LED.C
Processor:   P32MX460F512L

========================================================================*/


void main()
{
  TRISB = 0;                          // Initialize PORTB as output

  for(;;)                             // DO FOREVER
  {
    RB0_bit = 0;                      // Turn OFF LED at port  pin RB0
    Delay_Ms(1000);                   // Wait 1 second
    RB0_bit = 1;                      // Turn ON LED at port pin RB0
    Delay_Ms(1000);                   // Wait 1 second
  }
}
```

Figure 3.1
Structure of a simple mikroC Pro for PIC32 program.

All C programs start with the reserved word "void main". This shows the beginning of the main program code that should be executed when the program is run. In a large program, there could be global variables, symbols, definitions, functions, or subprograms before the main program, but still the first code executed after a power-on, or after the microcontroller is Reset, is the code just after the "void main" statement. A pair of brackets is used after the "void main" statement. The code inside the "void main" is the body of the main program, and it must be enclosed within a pair of curly brackets:

```
void main()
{
  Body of the main program
}
```

3.1.1 Comments

Comments are used by programmers to clarify the operation of the program or a program statement. Comment lines are ignored and are not compiled by the compiler and their use is optional. Two types of comments can be used in mikroC Pro for PIC32 programs: long comments and short comments.

Long comments start with characters "/*" and end with characters "*/". These comments are usually used at the beginning of a program to describe what the program does, the type of processor used, brief description of the algorithm, and the interface to any external hardware used. In addition, the name of the author, the date on which the program was written, and the program filename are usually given in long comment lines to help make any future development of the program easier. As shown in Figure 3.1, long comment lines can extend over many lines.

Short comment lines start with characters "//" and are usually used at the end of statements to describe the operation performed by the statements. These comments can be inserted at any other places of programs as well, for example:

```
//
// Increment the counter
//
Cnt = Cnt + 1;
```

Short comments can only occupy one line and there is no need to terminate them.

3.1.2 White Spaces

White spaces in programs are blanks, spaces, tabs, and new-line characters. All white spaces are ignored by the C compiler. Thus, the following statements are all identical:

```
        char i,j;
or
        char i,  j;
or
        char i,
        j;
or
        char i,j
        ;
```

3.1.3 Terminating Program Statements

In C programs, all statements must be terminated with the semicolon ";" character, otherwise a compilation error occurs:

```
char i,j,k        // error
char i,j,k;       // correct
j = i + 2         // error
j = i + 2;        // correct
```

3.1.4 Case Sensitivity

In C programs, all names are case sensitive with variables with lowercase names being different to those with uppercase names. Thus, the following variables are all different and they represent different locations in memory:

```
Count    count    CounT    count    COUNT    count
```

3.1.5 Variable Names

In C programs, variable names can begin with an underscore character, or with an alphabetical character. Thus, valid variable names can start with characters a–z, A–Z, and the "_" character. Digits 0–9 can be used after valid characters. Variable names must be unique and can extend to up to 31 characters. Examples of valid variable names are the following:

```
Total    Total99    Sum1  _Name    SumOfNumbers    UserName    x123
```

Examples of invalid variable names are the following:

```
9Sum    ?Name    +Count    £Pound    45Total    /Address
```

3.1.6 Reserved Names

Some names are reserved for the compiler itself and these names must not be used in our programs as variable names. A list of the reserved names is given in Figure 3.2. For example, the following variable names are not valid:

```
for   while   if   switch   char   struct   do   float   main   static
```

3.1.7 Variable Types

The mikroC Pro for PIC32 language supports a large number of variable types. Table 3.1 gives a list of the supported variable types. Examples are given below to show how these variables can be used in programs.

bit: This is a 1-bit variable which can take values 0 or 1. In the following example, variable flag is declared as a bit and is assigned 1:

```
bit flag;
flag = 1;
```

Note that variables can be assigned values during their declarations. Thus, the above statement can also be written as follows:

```
bit flag = 1;
```

absolute asm at auto
bit bool break
case catch char class code const continue
data default delete dma do double
else enum explicit extern
false far float for friend
goto
if inline int iv
long
mutable
namespace near
operator org
pascal private protected public
register return rx
sfr short signed sizeof static struct switch
template this throw true try typedef typeid typename
union unsigned using
virtual void volatile
while
xdata
ydata

Figure 3.2
List of reserved names.

char or **unsigned char**: These are 8-bit unsigned character or integer variables ranging from 0 to 255 and occupying 1-byte in memory. In the following example, variable **Sum** is assigned value 125:

```
    char Sum = 125;
or
    unsigned char Sum = 125;
or
  char Sum;
  Sum = 125;
```

These variables can also be declared using the keywords **unsigned short** or **unsigned short int**.

signed char: These are 8-bit signed character or integer variables ranging from −128 to +127 and occupying 1-byte in memory. In the following example, variable **Cnt** is assigned value −25:

```
signed char Cnt;
Cnt = −25;
```

These variables can also be declared using keywords **short**, **signed short**, or **short int**, or **signed short int**.

int or **signed int**: These are 16-bit signed variables ranging from −32768 to +32767 and occupying 2 bytes in memory. In the following example, variable **Total** is assigned value −23000:

```
signed int Total = −23000;
```

Table 3.1: mikroC for PIC32 Variable Types

Type	Size (bits)	Range
Bit	1	0 or 1
Unsigned char	8	0 to 255
Signed char	8	−128 to 127
Unsigned int	16	0 to 65535
Signed int	16	−32768 to +32767
Unsigned long	32	0 to 4294967295
Signed long	32	−2147483648 to 2147483647
Unsigned long long	64	0 to 18446744073709551615
Signed long long	64	−9223372036854775808 to 9223372036854775807
Float	16	-1.5×10^{45} to 3.4×10^{38}

unsigned or **unsigned int**: These are 16-bit unsigned variables ranging from 0 to 65535 and occupying 2 bytes in memory. In the following example, variable **Sum** is assigned value 33000:

```
unsigned int Sum;
Sum = 33000;
```

long or **signed long**: These are 32-bit signed variables ranging from −2147483648 to +2147483647 and occupy 4 bytes in memory. In the following example, variable **Total** is assigned value −200:

```
signed long Total = -200;
```

These variables can also be declared using keywords **signed long int** or **long int**.

unsigned long or **unsigned long int**: These are 32-bit unsigned variables ranging from 0 to 4294967295 and occupy 4 bytes in memory. In the following example, variable **x1** is assigned value 1250:

```
unsigned long x1 = 1250;
```

long long or **signed long long**: These are 64-bit signed variables ranging from −9223372036854775808 to +9223372036854775807 and occupying 8 bytes in memory. In the following example, variable **Sum** is assigned value −237867510:

```
signed long long Sum = -237867510;
```

unsigned long long: These are 64-bit unsigned variables ranging from 0 to 18446744073709551615 and occupying 8 bytes in memory. In the following example, variable **Sum** is assigned value 12000000:

```
unsigned long long Sum = 12000000;
```

float or double or long double: These are floating point numbers in the range -1.5×10^{45} to $+3.4 \times 10^{38}$ occupying 4 bytes in memory. In the following example, variable **Max** is assigned value 12.52:

```
float Max = 12.52;
```

Variables with the following names will be used in this book to specify the type and sign of the variable in order to avoid any confusion:

```
bit
unsigned char
signed char
unsigned int
signed int
unsigned long
signed long
unsigned long long
signed long long
float
```

Bits of a Variable

In many applications, we may wish to access individual bits of a variable. This can easily be achieved by using the direct member selector operator ".", followed by one of the identifiers B0, B1, ..., B31, or F0, F1, ..., F31. For example, bit 2 of variable Test can be cleared as follows:

```
Test.B2 = 0;    or Test.F2 = 0;
```

The bits of the processor Special Function Registers (SFR) can be accessed in several ways. For example, the following statement sets bit 0 of PORTB to 1:

```
PORTB.RB0 = 1;
```

or

```
PORTBbits.RB0 = 1;
```

Alternatively, we can assign a symbol to an SFR register bit and then use that symbol to manipulate the bits. In the following example, symbol LED is assigned to bit 0 of PORTB, and is then set to 1:

```
sbit LED at PORTB.B0;
LED = 1;
```

3.1.8 Number Bases

mikroC Pro for PIC32 supports the decimal, binary, hexadecimal, and octal number bases. Variables can be declared using any of these bases.

Decimal numbers are written as they are without any special symbols. Binary numbers are written by preceding the number with characters "0b" or "0B". Hexadecimal numbers are written by preceding the number with characters "0x" or "0X". Finally, octal numbers are written by preceding the number with number "0". The example below shows how the decimal number 55 can be represented in different bases:

```
55              decimal
0b00110111      binary
0x37            hexadecimal
067             octal
```

3.1.9 Constants

Constants are variables that have fixed values, and whose values cannot be changed in a program. Normally, variables are stored in the RAM memory of the microcontroller. Constants are stored in the flash program memory of the microcontroller. Variables with fixed values should be declared as constants to save the limited and valuable RAM space. A constant is declared by preceding its name with the keyword **const** or **const code**.

mikroC Pro for PIC32 language supports:

- Character constants
- Integer constants
- Floating point constants
- String constants
- Enumeration constants.

Constants must be assigned fixed values during their declarations. The following example shows how an unsigned character constant named **Min** can be declared and assigned to decimal value 25. In this example, one byte will be allocated in the program memory to constant **Min** and the byte will be initialized to decimal 25 during the compilation time:

```
const unsigned char Min = 25;
```

or

```
const code unsigned char Min = 25;
```

Similarly, a floating point constant is declared as

```
const code float Pi = 3.14159;
```

A constant character array is declared as

```
const code char P[ ] = {'B', 'O', 'O', 'K'};
```

or

```
const code char X = "This is a constant string";
```

Enumeration constants are of integer type, and they are used to make a program easier to follow. In the following example, the Weekdays constant stores the names of the weekdays. The first element is given the value 0 by default but it can be changed if desired:

```
const code enum Weekdays {Monday, Tuesday, Wednesday, Thursday, Friday, Saturday};
```

where Monday is assigned 0, Tuesday is assigned 1, Wednesday is assigned 2, and so on.

or

```
   const code enum Weekdays (Monday = 1, Tuesday = 2, Wednesday, Thursday, Friday,
Saturday);
```

where, Monday is assigned 1, Tuesday is assigned 2, Wednesday is assigned 3, and so on.

3.1.10 Escape Sequences

Escape sequences are commonly used in C languages to represent nonprintable ASCII characters. For example, the character combination "\n" represents the new-line character. An escape sequence can also be represented by specifying its hexadecimal code after a backslash character. For example, the new-line character can also be represented as "\x0A" (Table 3.2).

3.1.11 Static Variables

Static variables are local variables in functions whose values are preserved on entry to functions. Normally, local variables declared at the beginning of a function are reinitialized every time the function is called. By using the keyword **static**, we make sure that the value of the variable is preserved every time the function is called, i.e. throughout the lifetime of the program. As shown below, these variables are declared using the keyword static:

```
   static unsigned char Max;
```

3.1.12 Volatile Variables

Volatile variables are declared using the keyword **volatile**. These variables are important in input—output routines and in interrupt service routines (ISR). The value of a volatile variable may change during the lifetime of a program, independent from the normal flow of the

Table 3.2: Some Commonly Used Escape Sequences

Escape Sequence	Hex Value	Characters
\a	0x07	BEL (bell)
\b	0x08	BS (backspace)
\f	0x0C	FF (formfeed)
\n	0x0A	LF (line feed)
\r	0x0D	CR (carriage return)
\t	0x09	HT (horizontal tab)
\v	0x0B	VT (vertical tab)
\nH		String of hex digits
\\	0x5C	\ (backslash)
\'	0x27	' (single quote)
\"	0x22	" (double quote)
\?	0x3F	? (question mark)

program. For example, the value may change in an ISR. Variables declared as volatile are not optimized by the compiler since their values can change at any unexpected time. In the following example, variable **Tst** is declared as volatile:

```
volatile unsigned char Tst;
```

In the following program code, we have a loop which repeats as long as variable **Cnt** is not equal to 100:

```
int Cnt;
Cnt = 0;
while(Cnt != 100)
{
    Body of loop
};
```

since the value of **Cnt** is not changed inside the while loop, the compiler recognizes this and changes the program code to the following equivalent code, where the loop is repeated forever:

```
int Cnt;
Cnt = 0;
while(true)
{
    Body of loop;
};
```

But in actual fact, the value of variable **Cnt** can change somewhere else in the program, for example, in an ISR, and we may want the first code, which loops until **Cnt** becomes 100. By using the keyword **volatile**, we tell the compiler not to optimize the code:

```
volatile int Cnt;
Cnt = 0;
while(Cnt != 100)
{
    Body of loop;
};
```

3.1.13 External Variables

A variable is declared as external by using the keyword **extern** before the name of the variable. This keyword tells the compiler that the variable is declared outside this program, in a separate program file. In the following example, variables Cnt1 and Cnt2 are declared as external:

```
extern char Cnt1, Cnt2;
```

3.1.14 Memory Type Specifiers

mikroC Pro for PIC32 supports the following memory type specifiers:

- code
- data
- rx
- sfr

The **code** specifier, as we have seen before, forces a variable to be stored in the flash program memory of the microcontroller. The **data** specifier, on the other hand, forces a variable to be stored in the data RAM memory (the default storage area). The **rx** specifier allows a variable to be stored in the working registers of the microcontroller. The **rx** specifier is reserved for compiler usage only. The **sfr** memory specifier allows a variable to be stored in an SFR of the microcontroller.

Some examples are given below:

```
const code unsigned char Max = 250;    // Store Max in program memory
data unsigned char Temp;               // Store Temp in RAM memory
rx char Tmp;                           // Store Tmp in working register
sfr char Tmp;                          // Store Tmp in SFR register
```

3.1.15 Arrays

Arrays are objects of the same type, collected under the same name. An array is declared by specifying its type, name and the number of objects it contains. A pair of square brackets is used to specify the number of objects, starting from 0. For example,

```
unsigned char  MyVector[5];
```

creates an array called MyVector of type unsigned char, having five objects (or elements). The first element of the array is indexed with 0. Thus, MyVector[0] refers to the first element of the array, and MyVector[4] refers to the last element of the array. The array MyVector occupies five consecutive bytes in memory as follows:

| MyVector[0] |
| MyVector[1] |
| MyVector[2] |
| MyVector[3] |
| MyVector[4] |

Data can be stored in an array by specifying the array name and the index. For example, to store 50 in the second element of the above array, we have to write as follows:

```
MyVector[1] = 50;            // Store 50 in the second element
```

Similarly, any element of an array can be copied to a variable. In the following example, the last element of the array is copied to variable **Temp**:

```
Temp = MyVector[4];          // Copy last element to Temp
```

The elements of an array can be initialized either during the declaration of the array or inside the program.

The array initialization during the declaration is done before the beginning of the program where the values of array elements are specified inside a pair of curly brackets, and are separated with commas. An example is given below where the array **Months** has 12 elements and each element is initialized to the length of a month:

```
unsigned char Months[12] = {31, 28, 31, 30, 31, 30, 31, 31, 30, 31, 30, 31};
```

Thus, for example, Months[0] = 31, Months[1] = 28 and so on.

The above array could also be initialized without specifying its size. The size is filled in automatically by the compiler:

```
unsigned char Months[ ] = {31, 28, 31, 30, 31, 30, 31, 31, 30, 31, 30, 31};
```

Character arrays are declared similarly with or without specifying the size of the array. In the following example, array **Temp** stores characters of letter "COMPUTER", and the array size is set to 8 by the compiler:

```
unsigned char Temp[ ] = {'C', 'O', 'M', 'P', 'U', 'T', 'E', 'R'};
```

Strings are special cases of characters arrays, which are terminated by the NULL character. The NULL character is represented by characters "\0", or by the hexadecimal number "0x0". An example string named **MyString** is shown below which is initialized to string "COMPUTER":

```
unsigned char MyString[ ] = "COMPUTER";
```

or

```
unsigned char MyString[9] = "COMPUTER";
```

notice that the above string declarations are automatically terminated by the compiler with a NULL character, and they are equivalent to the following character array declaration:

```
unsigned char MyString[ ] = {'C', 'O', 'M', 'P', 'U', 'T', 'E', 'R', '\0'};
```

In the case of string declarations, the size of the array is always one more than the number of characters declared in the string.

Arrays are normally one dimensional and such arrays are also known as *vectors*. In C language, we can also declare arrays with *multiple dimensions*. Two-dimensional arrays are also known as *matrices*. The declaration of arrays with multiple dimensions is similar to the declaration of arrays with single dimensions. For example, a two-dimensional array is declared by specifying the array type, array name, and the size of each dimension. In the following example, **M** is declared as an integer array having the dimension 2×2 (i.e. two rows, two columns):

```
int M[2][2];          // A 2-dimensional array
```

The structure of this array is as follows, where the first element is M[0][0], and the last element is M[1][1]:

M[0][0]	M[0][1]
M[1][0]	M[1][1]

Similarly, a three-dimensional array of size $2 \times 2 \times 4$ can be declared as follows:

```
int P[2][2][4];       // A 3-dimensional array
```

The elements of a multidimensional array can be copied to a variable by using an assignment operator. In the example below, the data at row 1, column 2 of a two-dimensional array P is copied to variable **Temp**:

```
Temp = P[1][2];
```

Multidimensional arrays are initialized as the single dimensional arrays where the elements are specified inside curly brackets and are separated by commas. In the following example, a 3×2 integer array named **P** is declared and its elements are initialized:

```
int P[3][2] = { {3, 5}, {2, 1}, {1, 1} };
```

The structure of this array is as follows:

3	5
2	1
1	1

The above array could also be declared by omitting the first dimension (number of rows):

```
int P[ ][2] = { {3, 5}, {2, 1}, {1, 1} };
```

or

```
int P[ ][2] = {
               {3, 5},
               {2, 1},
               {1, 1}
              };
```

An interesting and useful application of multidimensional arrays is in creating two-dimensional text strings. For example, as shown below, we can create a two-dimensional text string called **wdays** to store the names of the week:

```
unsigned char wdays[ ] [10] = {
```

```
                              "Monday",
                              "Tuesday",
                              "Wednesday",
                              "Thursday",
                              "Friday",
                              "Saturday",
                              "Sunday"
                        };
```

Notice here that the first dimension (number of rows) of the array is not declared and we expect the compiler to make this dimension seven as there are seven rows. The second dimension is set 10, because **Wednesday** is the longest string having nine characters. If we allow one character for the string terminating NULL character, then a total of 10 characters will be required (each day name is terminated with a NULL character automatically).

3.1.16 Pointers

Pointers in C language are very important concepts and they are used almost in all large and complex programs. Most students have difficulty in understanding the concept of pointers. In this section, we shall be looking at the basic principles of pointers and see how they can be used in C programs.

A pointer in C language is a variable that holds the memory address of another variable. Another simple definition of a pointer is the following: a pointer points to the memory address of a variable. Thus, by knowing where a variable is actually located in memory, we can carry out various types of operations on this variable.

Pointers in C language are declared the same as any other variables, but the character "*" is inserted before the name of the variable. In general, pointers can be declared to hold the addresses of all types of variables, such as character, integer, long, floating point, and even they can hold the addresses of functions.

The following example shows how a character pointer named **Pnt** can be declared:

```
char *Pnt;
```

When a new pointer is declared, its content is not defined and in general we can say that it does not point to any variable. We can assign the address of a variable to a pointer using the "&" character. In the following example, character pointer **Pnt** holds the address of (or points to the address of) character variable **Temp**:

```
Pnt = &Temp;                    // Pnt holds the address of Temp
```

Assuming variable Temp is at memory location 1000, pointer Pnt contains the number 1000, which is the address where the variable Temp is located at. We can now assign value to our variable Temp using its pointer (see Figure 3.3):

```
*Pnt = 25;                      // Assign 25 to variable Temp
```

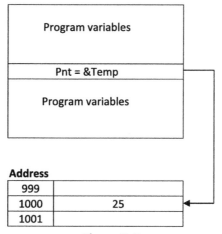

Figure 3.3

This is same as the following assignment:

```
Temp = 25;                          // Assign 25 to Temp
```

Similarly, the value of **Temp** can be copied to variable **Total** as follows:

```
Total = *Pnt;                       // Copy Temp to Total
```

This is same as the following:

```
Total = Temp;                       // Copy Temp to Total
```

In C language, we can perform pointer arithmetic that may involve the following operations:

- Adding or subtracting a pointer and an integer value.
- Adding two pointers.
- Subtracting two pointers.
- Comparing two pointers.
- Comparing a pointer to a Null.

For example, assuming that the pointer **Pnt** holds address 1000, the statement

```
Pnt = Pnt + 1;
```

Will increment pointer **Pnt** so that it now points to address 1001.

Pointers can be very useful when arrays are used in a program. In C language, the name of an array is also a pointer to the array. Thus, for example, consider the following array:

```
char Sum[5];
```

The name **Sum** is also a pointer to this array and it holds the address of the first element of the array. Thus, the following two statements are equal:

```
Sum[2] = 100;
```

and

```
*(Sum + 2) = 100;
```

Since **Sum** is also a pointer to array Sum[ ], the following statement is also true:

```
&Sum[p] = Sum + p;
```

The following example shows how pointer arithmetic can be used in array-based operations. Assuming that **P** is a pointer, it can be set to hold the address of the second element of an array **Sum**:

```
P = Sum[2];
```

Suppose now that we want to clear second and third elements of this array. This can be done as follows:

```
*P = 0;                    // Sum[2] = 0
P = P + 1;                 // point to next location
*P = 0;                    // Sum[3] = 0;
```

or

```
*P = 0;                    // Sum[2] = 0
*(P + 1) = 0;              // Sum[3] = 0
```

or

```
Sum[2] = 0;                // Sum[2] = 0
Sum[3] = 0;                // Sum[3] = 0
```

3.1.17 Structures

Structures are similar to arrays where they store related items, but they can contain data with different data types. A structure is created using the keyword struct. The following example creates a structure called Family:

```
struct Family
{
    unsigned char name[20];
    unsigned char surname[30];
    unsigned char age;
    float weight;
    float height;
}
```

It is important to realize that the above is just a template for a structure and it does not occupy any space in memory. It is only when variables of the same type as the structure are created that the structure occupies space in memory. For example, variables **John** and **Peter** of type **Family** can be created by the statement:

```
struct Family John, Peter;
```

Variables of type structure can also be created during the declaration of the structure:

```
struct Family
{
        unsigned char name[20];
        unsigned char surname[30];
        unsigned char age;
        float weight;
        float height;
} John, Peter;
```

We can assign values to members of a structure by specifying the name of the structure, followed by a ".", and the value to be assigned. In the following example, the age of John is set to 25 and his weight is set to 82.5:

```
John.age = 25;
John.weight = 82.5;
```

The members of a structure can also be initialized during the declaration of the structure, by specifying their values enclosed in a pair of curly brackets at the end of the structure definition, and are separated with commas. In the following example, a structure called **Cube** is created with variable **MyCube** having sides A, B, and C are initialized to 2.5, 4.0, and 5.3, respectively:

```
struct Cube
{
        float sideA;
        float sideB;
        float sideC;
} MyCube = {2.5, 4.0, 5.3};
```

It is also permissible to use pointers to assign values to members of a structure. An example is given below:

```
struct Cube
{
        float sideA;
        float sideB;
        float sideC;
} *MyCube;
MyCube → sideA = 2.5;
MyCube → sideB = 4.0;
MyCube → sideC = 5.3;
```

The size of a structure depends on the data types of its members. Although we can calculate the size by adding the known size of each data member, it is easier to use the operator **sizeof** to calculate the size of a structure. For example, the following statement stores the size of structure **MyCube** in variable **z**:

```
z = sizeof(MyCube);
```

Another use of structures, especially in microcontroller-based applications, is the separation of bits of a variable so that each bit or group of bits can be manipulated separately. This is also known as bit fields. With bit fields, we can assign identifiers to each bit or to groups of bits, and then manipulate them easily. An example is given below where a byte variable called **Flag** is separated into three bit groups: the two least significant bits (bit 0 and bit 1) are called **LSB2**, the two most significant bits (bit 6 and bit 7) are called **MSB2**, and the remaining bits (bits 2–5) are called **Middle**:

```
struct Bits
{
        unsigned char LSB2: 2;
        unsigned char Middle: 4
        unsigned char MSB2: 2;
} Flag;
```

We can then assign values to bits of the following variable:

```
Flag.LSB2 = 3;
Flag.Middle = 6;
```

3.1.18 Creating New Data Types

In C programming, we can create new data types by using the existing data types. This can be useful in many applications where we may want to give our own names to variable types. The keyword **typedef** is used to create new variable types. In the following example, a new data type called **INTEGER** is created from the data type **unsigned int**:

```
typedef unsigned int INTEGER;
```

Which simply means that the new data type **INTEGER** is a synonym for the data type **unsigned int** and can be used whenever we wish to declare an **unsigned int** type variable. In the following example, **x**, **y**, and **z** are declared as **INTEGER** which also means that they are of type **unsigned int**:

```
INTEGER x, y, z;
```

As another example, consider the following definition:

```
typedef char *string;
```

Here, string is a new data type, which is really a character pointer. We can now use this data type in our programs to create string arrays:

```
string Surname = "Jones";
```

The **typedef** keyword can also be used in structures. For example, we can create a new structure called **Person** using **typedef**

```
typedef struct
{
      unsigned char name[20];
      unsigned char surname[30];
      unsigned char age;
} Person;
```

New variables of type Person can now be created:

```
Person John, Peter, Mary;
```

The contents of one structure can be copied to another one if both structures have been derived from the same template. An example is given below where a structure is created with two variables **Circle1** and **Circle2**. Circle1 is initialized with data and is then copied to **Circle2**:

```
struct Circle
{
      float radius;
} Circle1, Circle2;
Circle1.radius = 2.5;
Circle2 = Circle1;
```

3.1.19 Unions

Unions are similar to structures and they are even declared and used in a similar way. The difference of a union is that all the member variables in a union share the same memory space. For example, the following union declaration creates two member variables **x** and **y**. **y** is an **unsigned integer** and occupies 2 bytes in memory. **x** is an **unsigned char** and occupies only 1 byte in memory. Because the member variables share the same memory space, the low byte of **y** shares the same memory byte as **x**:

```
union Test
{
      unsigned char x;
      unsigned int y;
}
```

If, for example, **y** is loaded with the hexadecimal data 0x2EFF, then **x** will automatically be loaded with data 0xFF.

3.1.20 Operators in mikroC Pro for PIC32

Operators are used in mathematical and logical operations to produce results. Some operators are *unary* where only one operand is required, some are *binary* where two operands are required, and an operator is *tertiary* where it requires three operands.

mikroC Pro for PIC32 language supports the following operators:

- Arithmetic operators
- Relational operators
- Bitwise operators
- Logical operators
- Conditional operators
- Assignment operators
- Preprocessor operators.

Arithmetic Operators

Arithmetic operators are used in mathematical expressions. Table 3.3 gives a list of the arithmetic operators. All these operators, except the autoincrement and autodecrement, require at least two operands. Autoincrement and autodecrement operators are unary as they require only one operand.

The arithmetic operators "+, −, *, and /" are obvious and require no explanation. Operator "%" gives the remainder after an integer division is performed. Some examples are given below:

```
12 % 3      gives 0 (no remainder)
12 % 5      gives 2 (remainder is 2)
−10 % 3     gives −1 (remainder is −1)
```

The autoincrement operator is used to increment the value of a variable by one. Depending on how this operator is used, the value is incremented before (preincrement) or after (postincrement) an assignment. Some examples are given below:

```
i = 5;              // i is equal to 5
i++;                // i is equal to 6
i = 5;              // i = 5
j = i++;            // i = 6 and j = 5
i = 5;              // i = 5
```

Table 3.3: Arithmetic Operators

Operator	Operation
+	Addition
−	Subtraction
*	Multiplication
/	Division
%	Remainder (integer division only)
++	Autoincrement
−−	Autodecrement

```
j = ++i;                        // i = 6 and j = 6
```

Similarly, the autodecrement operator is used to decrement the value of a variable by one. Depending on how this operator is used, the value is decremented before (predecrement) or after (postdecrement) an assignment. Some examples are given below:

```
i = 5;                          // i is equal to 5
i--;                            // i is equal to 4

i = 5;                          // i = 5
j = i--;                        // i = 4 and j = 5

i = 5;                          // i = 5
j = --i;                        // i = 4 and j = 4
```

Relational Operators

The relational operators are used in comparisons. All relational operators are evaluated from left to right and they require at least two operands. Table 3.4 gives a list of the valid mikroC Pro for PIC32 relational operators.

Notice that the equal operator is written with two equal signs, as "==", and not as "=". If an expression evaluates to TRUE, 1 is returned, otherwise 0 is returned. Some examples are given below:

```
i = 5
i > 0           // returns 1
i == 5          // returns 1
i < 0           // returns 0
i != 10         // returns 1
```

Bitwise Operators

Bitwise operators are used to modify bits of a variable. These operators (except the bitwise complement) require at least two operands. Table 3.5 gives a list of the bitwise operators.

Table 3.4: Relational Operators

Operator	Operation
==	Equal to
!=	Not equal to
>	Greater than
<	Less than
>=	Greater than or equal to
<=	Less than or equal to

Table 3.5: Bitwise Operators

Operator	Operation
&	Bitwise AND
|	Bitwise OR
^	Bitwise exclusive-OR
~	Bitwise complement
≪	Shift left
≫	Shift right

Bitwise AND returns 1 if the corresponding two bits of the variables are 1, otherwise it returns 0. An example is given below where the first variable is 0xFE and the second variable is 0x4F. The result of the bitwise AND operation is 0x4E:

```
0xFE: 1111 1110
0x4F: 0100 1111
      =========
      0100 1110 = 0x4E
```

Bitwise OR returns 0 if the corresponding two bits of the variables are 0, otherwise it returns 1. An example is given below where the first variable is 0x1F and the second variable is 0x02. The result of the bitwise OR operation is 0x1F:

```
0x1F: 0001 1111
0x02: 0000 0010
      =========
      0001 1111 = 0x1F
```

Bitwise exclusive-OR returns 0 if the corresponding two bits of the variables are the same, otherwise it returns 1. An example is given below where the first variable is 0x7F and the second variable is 0x1E. The result of the Exclusive-OR operation is 0x61:

```
0x7F: 0111 1111
0x1E: 0001 1110
      =========
      0110 0001 = 0x61
```

The complement operator requires only one operand and it complements all bits of its operand. An example is given below where variable 0x2E is complemented to give 0xD1:

```
0x2E: 0010 1110
      1101 0001 = 0xD1
```

The left-shift operator requires two operands: the variable to be shifted, and the count specifying the number of times the variable is to be shifted. When a variable is shifted left, 0 is filled in to the last significant bit position. Shifting a variable left by one digit is same as multiplying the variable by two. An example is given below where the variable 0x1E is shifted left by two digits to give the number 0x78:

```
0x1E: 0001 1110
      0011 1100              // shift left by one digit
      0111 1000 = 0x78       // shift left by two digits
```

The right-shift operator requires two operands: the variable to be shifted, and the count specifying the number of times the variable is to be shifted. When a variable is shifted right, 0 is filled in to the most significant bit position. Shifting a variable right by one digit is same as dividing the variable by two. An example is given below where the variable 0x1E is shifted right by one digit to give the number 0x0F:

```
0x1E: 0001 1110
      0000 1111 = 0x0F            // shift right by one digit
```

Logical Operators

The logical operators are used in comparisons, and they return TRUE (or 1) if the comparison evaluates to nonzero, or FALSE (or 0) is the comparison evaluates to zero. Table 3.6 gives a list of the logical operators.

Some examples are given below to show how the logical operator can be used:

```
a = 10;
a > 0 && a < 100            // returns 1
a > 0 || a < 20            // returns 1
a == 10 || a > 0            // returns 1
a > 0 && a > 10            // returns 0
```

Conditional Operator

mikroC Pro for PIC32 language supports a conditional operator with the following syntax:

```
result = expression1 ? expression2: expression3
```

Here, **expression1** is evaluated and if its value is TRUE (nonzero), then **expression2** is assigned to **result**, otherwise **expression3** is assigned to **result**. An example of use of the conditional operator is given below which assigns the larger of variables **a** or **b** to **max**. If **a** is greater than **b**, then **a** is assigned to **max**, otherwise **b** is assigned to **max**:

```
max = (a > b) ? a: b;
```

Another common application of the conditional operator is to convert a lowercase character to uppercase as shown below. In this example, if the character **c** is lowercase

Table 3.6: Logical Operators

Operator	Operation
&&	AND
||	OR
!	NOT

(between 'a' and 'z'), then it is converted to uppercase by subtracting hexadecimal 0x20 from its value:

```
c = (c >= 'a' && c <= 'z') ? (c - 0x20): c;
```

Similarly, we can write a conditional operator-based statement to convert an uppercase character to lowercase:

```
c = (c >= 'A' && c <= 'Z') ? (c + 0x20): c;
```

Assignment Operators

The basic assignment operator is the "=" sign where a constant or an expression is assigned to a variable:

```
i = 200;
j = i * 2 + p;
```

The C language also supports compound assignment operators in the following format:

```
result compound operator = expression
```

where the compound operator can be one of the following:

```
+=    -=    *=    /=    %=
&=    |=    ^=    >>=   <<=
```

Some examples are given below to show how the compound operator can be used:

```
i += j;          // equivalent to i = i + j
i += 1;          // equivalent to i = i + 1
j *= p;          // equivalent to j = j * p;
i <<= 1;         // equivalent to i = i << 1;
```

Preprocessor Operators

The preprocessor operators are an important part of the C language as these operators allow a programmer to:

*   compile a program segment conditionally;
*   replace symbols with other symbols;
*   insert source files to a program.

A preprocessor operator is identified by the character "#", and any program line starting with this character is treated as a preprocessor operator. All preprocessor operator operations are handled by the preprocessor, just before the program is compiled. The preprocessor operators are not terminated with a semicolon.

mikroC Pro for PIC32 language supports the following preprocessor operators:

```
#define  #undef  #ifdef  #ifndef
```

```
#if     #elif   #endif
#line
#error
#include
```

#define preprocessor operator is a Macro expansion operator where every occurrence of an identifier in a program is replaced with the specified value of the identifier. In the following example, **MAX** is defined to be number 1000, and wherever **MAX** is encountered in the program, it will be replaced with 1000:

```
#define MAX 1000
```

It is important to realize that an identifier which has been defined cannot be defined again before its definition is removed. The preprocessor operator **#undef** is used to remove the definition of an identifier. Thus, we could redefine **MAX** to be 10 as follows:

```
#undef MAX
#define MAX 10
```

We can also use the preprocessor operator **#ifdef** to check whether or not an identifier has been declared:

```
#ifdef MAX
        #undef MAX
        #define MAX 10
#endif
```

Notice that the **#ifdef** preprocessor operator must be terminated with the **#endif** operator.

The **#define** preprocessor operator can also be used with parameters, as in a Macro expansion with parameters. These parameters are local to the operator and it is not necessary to use the same parameter names when the operator is used (or called) in a program. An example is given below where the Macro **INCREMENT** is defined to increment the value of its parameter by one:

```
#define INCREMENT(a) (a + 1)
```

When this Macro is used in a program, the value of the variable representing the parameter will be incremented by one as in the following example:

```
z = INCREMENT(x);  will change by the preprocessor to z = (x + 1);
```

Similarly, we can define a Macro to return the larger of two numbers as

```
#define MAX(a, b) ((a > b) ? a: b)
```

And then use it in a program as

```
p = MAX(x, y);
```

The above statement will change by the preprocessor to

```
p = ((x > y) ? x: y);
```

It is important to make sure that the macro expansion is written with brackets around it in order to avoid confusion. Consider the example below which is written with no brackets:

```
#define ADD(a) a + 1
```

Now, we can use this Macro in an expression as follows:

```
z = 2 * ADD(x);
```

The above statement will be expanded into $z = 2 * a + 1$, which is not same as $z = 2 * (a + 1)$.

The **#include** preprocessor operator is used to include a sources file (text file) in a program. Usually program header files with extension **.h** are included in C programs. There are two ways of using the **#include** operator:

```
#include <MyFile>
```

and

```
#include "MyFile"
```

In the first option, the specified file is searched in the following order:

- mikroC Pro for PIC32 installation folder
- User search path.

In the second option, the specified file is searched in the following order:

- mikroC Pro for PIC32 project folder
- mikroC Pro for PIC32 installation folder
- user search path.

We can also specify the full path to a file as in:

```
#include "C:\Test\MyFile.h"
```

The **#if**, **#elf** and **#else** preprocessor operators are used in conditional compilation of parts of a program. For example, we may wish to selectively compile part of a program depending upon the clock frequency selected. An example is given below where the code with variables **A** and **B** is only compiled if **CLK** is nonzero, otherwise part of the program with variables **C** and **D** are compiled. Notice that the **#if** operator must always be terminated with **#endif** operator:

```
#if CLK
      A = 1
      B = 1
#else
      C = 1
      D = 1
#endif
```

The mikroC for PIC32 language also supports character merging using the "##" operator. An example is given below where the character "-" in inserted between two identifiers:

```
#define MIX(a, b) (a ## - ## b)
```

Which will expand into (a - b) when used in a program.

3.1.21 Modifying the Flow of Control

Statements in a program are normally executed in sequence one after the other until the end of the program is reached. In almost all programs we wish to compare variables with each other, and then execute different parts of the program based on the outcome of these comparisons. We may also want to execute part of a program several times. C language provides several programming tools for modifying the normal flow of control in a program.

The following flow control statements are available in the mikroC for PIC32 language:

- Selection statements
- Iteration statements
- Unconditional modification of the flow of control statements

Selection Statements

There are two basic selection statements: **if** and **switch**.

if Statement

The **if** statement is used to carry out comparisons in a program and then force one or another part of the program to execute based on the outcome of the comparison. The basic format of the **if** statement is as follows:

```
if(condition)expression;
```

Thus, the **expression** will be executed if the **condition** evaluates to TRUE.

or

```
if(condition)
   expression;
```

or

```
if(condition)
        expression1;
else
        expression2;
```

Thus, **expression1** will execute if the **condition** evaluates to TRUE, otherwise **expression2** will be executed.

An example is given below where if **x** is equal to five, then **z** is incremented by one, otherwise **z** is decremented by one:

```
if(x == 5)
        z++;
else
        z--;
```

In many applications we may wish to include more statements if a comparison evaluates TRUE or FALSE. This can be done by enclosing the statements in a pair of curly brackets as shown in the following example:

```
if(x > 10 && y > 0)
{
        a++;
        b = a;
        p = 2 * b + 1;
}
```

or

```
if(x > 10 && y > 0)
{
        a++;
        b = a;
        p = 2 * b + 1;
}
else
{
        q = 2 * p - 2;
        a++;
}
```

The **if** statements can be nested in complex comparison operations. An example is given below:

```
if(x == 1)
        a++;
else if(x == 2)
        b++;
else if(x == 3)
        c++;
else
        d++;
```

A more practical example is given below to illustrate how the **if** statements can be nested.

Example 3.1

In an experiment, it was observed that the relationship between X and Y is found to be as follows:

X	Y
1	3.5
2	7.1
3	9.4
4	15.8

Write the **if** statements that will return Y value, given the X value.

Solution 3.1

The required program code is as follows:

```
unsigned char x;
float Y;
if(x == 1)
        Y = 3.5;
else if(x == 2)
        Y = 7.1;
else if(x == 3)
        Y = 9.4;
else if(x == 4)
        Y = 15.8;
```

switch Statement

The **switch** statement is the second selection statement. This statement is used when we want to compare a given variable to a different number of conditions and then execute different code based on the outcome of each comparison. The general format of the switch statement is as follows:

```
switch(variable)
{
    case condition1:
            statements;
            break;
    case condition2:
            statements;
            break;
    ..............................
    ..............................
    case condition:
            statements;
            break;
    default:
            statements;
}
```

Here, the **variable** is compared to **condition1** and if the result evaluates to TRUE, the statements in that block are executed until the **break** keyword is encountered. This keyword forces the program to exit the switch statement by moving to the statement just after the closing curly bracket. If the **variable** is not equal to **condition1**, then it is compared to **condition2** and if the result is TRUE, the statements in that block are executed. If none of the conditions are satisfied, then the statements under the **default** keyword are executed and then the switch statement terminates. The use of the **default** keyword is optional and can be omitted if we are sure that one of the comparisons will always evaluate to TRUE.

Perhaps the easiest way to learn how the switch statement operates is to look at some examples.

Example 3.2

Rewrite the program code in Example 3.1 using a switch statement.

Solution 3.2
The required program code is the following:

```
unsigned char x;
float Y;
switch (x)
{
  case 1:
        Y = 3.5;
        break;
  case 2:
        Y = 7.1;
        break;
  case 3:
        Y = 9.4;
        break;
  case 4:
        Y = 15.8;
}
```

Example 3.3

Write the program code to convert hexadecimal numbers A–F to decimal using a switch statement.

Solution 3.3
Assuming that the hexadecimal number (letter) is in variable **c**, the required program code is as follows:

```
switch(c)
{
```

```
        case 'A':
                Y = 10;
                break;
        case 'B':
                Y = 11;
                break;
        case 'C':
                Y = 12;
                break;
        case 'D':
                Y = 13;
                break;
        case 'E':
                Y = 14;
                break;
        case 'F'
                Y = 15;
                break;
}
```

Iteration Statements

Iteration statements are used in most programs as they enable us to perform loops in our programs and repeat part of the code specified number of times. For example, we may want to calculate the sum of numbers between 1 and 10. In such a calculation, we will form a loop and keep adding the numbers from 1 to 10. mikroC Pro for PIC32 language supports the following iteration statements:

- **for** statements
- **while** statements
- **do while** statements
- **goto** statements

for Statements

The **for** statement is perhaps the most widely used statement to create loops in programs. The general format of this statement is as follows:

```
for(initial loop counter; conditional expression; change in loop counter)
{
    statements;
}
```

A loop variable is used in **for** loops. The initial expression specifies the starting value of the loop variable, and this variable is compared to some condition using the conditional expression before the looping starts. The statements enclosed within the pair of curly brackets are executed as long as the condition evaluates to TRUE. The value of the loop counter is changed (usually incremented by one) at each iteration. Looping stops when the condition evaluates to FALSE.

An example is given below where a loop is formed and inside this loop the statements are executed 10 times:

```
for(i = 0; i < 10; i++)
{
    statements;
}
```

or

```
for(i = 1; i <= 10; i++)
{
    statements;
}
```

Example 3.4

Write the program code using the **for** statement to calculate the sum of integer numbers from 1 to 10. Store the result in an integer variable called **Sum**.

Solution 3.4

The required program code is given below:

```
unsigned char i, Sum;
Sum = 0;
for(i = 1; i <= 10; i++)
{
    Sum = Sum + i;
}
```

All the parameters in the **for** loop are optional and can be omitted if required. For example, if the condition expression (middle expression) is omitted, it is assumed to be always TRUE. This creates an endless loop. An example is given below where the loop never terminates and the value of **i** is incremented by one at each iteration:

```
for(i = 0; ; i++)
{
    statements;
}
```

An endless loop can also be created if all the parameters are omitted as is in the following example:

```
for(;;)
{
    statements;
}
```

Endless loops are used frequently in microcontroller programs. An example of creating an endless loop using a preprocessor command and a **for** statement are given in the following example.

Example 3.5

Write the program code to create an endless loop using a **for** statement. Create a Macro called **DO_FOREVER** to implement the loop.

Solution 3.5
The required program code is given below:

```
#define DO_FOREVER for(;;)
................
................
DO_FOREVER
{
  statements;
}
```

If we omit the second and third parameters of a **for** loop, the loop repeats endless as in the earlier example, but the value of the initial loop counter never changes. An example is given below where the value of **i** is always 0:

```
for(i = 0;;)
{
  statements;
}
```

for loops can be nested such that in a two-level loop, the inner loop is executed for each iteration of the outer loop. Two-level nested loops are commonly used in matrix operations, such as adding two matrices and so on. A two-level nested **for** loop is given below. In this example, the statements in the inner loop are executed 100 times:

```
for(i = 0; i < 10; i++)
{
  for(j = 0; j < 10; j++)
  {
    statements in inner loop;
  }
}
```

An example of nested loop to add all the elements of a 2 × 2 matrix M is given below.

Example 3.6

Write the program code using **for** loops to add elements of a 2 × 2 matrix called M.

Solution 3.6

The required program code is given below:

```
Sum = 0;
for(i = 0; i < 2; i++)
{
  for(j = 0; j < 2; j++)
  {
    Sum = Sum + M[i][j];
  }
}
```

while Statement

while statement is another statement that can be used to create loops in programs. The format of this statement is as follows:

```
while (condition)
{
    statements;
}
```

Here, the loop enclosed by the curly brackets executes as long as the specified **condition** evaluates to TRUE. Notice that if the **condition** is FALSE on entry to the loop, then the loop never executes. An example is given below where the statements inside the **while** loop execute 10 times:

```
i = 0;
while (i < 10)
{
    statements;
    i++;
}
```

Here, at the beginning of the loop, **i** is equal to 0, which is less than 10 and therefore the loop starts. Inside the loop **i** is incremented by one. The loop terminates when **i** becomes equal to 10. It is important to realize that the condition specified at the beginning of the loop should be satisfied inside the loop, otherwise an endless loop is formed.

In the following example **i** is initialized to 0, but is never changed inside the loop and is therefore always less than 10. Thus, this loop will never terminate:

```
i = 0;
while (i < 10)
{
    statements;
}
```

It is possible to have a **while** loop with no statements. Such a loop is commonly used in microcontroller applications to wait for certain action to be completed. For example, to wait until a button is pressed. An example is given below, where the program waits until variable CLK becomes 0:

```
while(CLK == 1);
```

or

```
while(CLK);
```

while loops can be nested inside one another. An example is given below:

```
i = 0;
j = 0;
while (i < 10)
{
    while (j < 10)
    {
        statements;
        j++;
    }
    statements;
    i++;
}
```

do while Statement

do while statement is similar to the while statement, but here the condition for the existence of the loop is at the end of the loop. The format of the **do while** statement is as follows:

```
do
{
    statements;
} while (condition);
```

The loop executes as long as the **condition** is TRUE. That is, until the **condition** becomes FALSE. Another important difference of the **do while** loop is that because the condition is at the end of the loop, the loop is always executed at least once. An example for **do while** loop is given below where the statements inside the loop are executed 10 times:

```
i = 0;
do
{
    statements;
    i++;
} while (i < 10);
```

The loop starts with i = 0 and the value of **i** is incremented inside the loop. The loop repeats as long as **i** is less than 10, and it terminates when **i** becomes equal to 10.

As with the **while** statement, an endless loop is formed if the condition is not satisfied inside the loop. For example, the following **do while** loop never terminates:

```
i = 0;
do
{
  statements;
} while (i < 10);
```

goto Statement

The **goto** statement can be used to alter the flow of control in a program. Although the **goto** statement can be used to create loops with finite repetition times, use of other loop structures such as **for**, **while**, and **do while** is recommended. The use of the **goto** statement requires a label to be defined in the program. The format of the **goto** statement is as follows:

....

Label:

....
```
goto Label;
```

The label can be any valid variable name and it must be terminated with a colon character. The following example shows how the **goto** statement can be used in a program to create a loop to iterate 10 times:

```
i = 0;
```
Loop:
```
statements;
i++;
if(i < 10) goto Loop;
......................
......................
```

Here, label **Loop** identifies the beginning of the loop. Initially i = 0 and is incremented by one inside the loop. The loop continues as long as i is less than 10. When i becomes equal to 10, program terminates and continues with the statement just after the **goto** statement.

Unconditional Modification of Flow of Control

As described in above, the **goto** statement is used to unconditionally modify the flow of control in a program. Although use of the **goto** statement is not recommended as it may give rise to unreadable and unmaintainable code, its general format is as follows:

....

Label:

....

Table 3.7: Methods of Creating an Infinite Loop

for loop	while loop	do while loop	goto loop

for(;;)
{
    statements;
}
	while (1)
	{
	    statements;
	}
		do
		{
		    statements;
		} while(1);
			Loop:
			    statements;
			goto Loop;

```
....................
goto Label;
```

Creating Infinite Loops

As discussed earlier in this chapter, infinite loops are commonly used in microcontroller-based applications. Such loops can be created using all the flow control statements we have seen. Table 3.7 summarizes the ways that an infinite loop can be created.

Premature Termination of a Loop

There are applications where we may want to terminate a loop prematurely, i.e. before the loop-terminating condition is satisfied. The **break** statement can be used for this purpose. An example is given below where the **for** loop is terminated when the loop counter **i** becomes equal to10:

```
for(i = 0; i < 100; i++)
{
        statements;
        if(i == 10)break;
}
```

Another example is given below where the **while** statement is used to create the loop:

```
while (i < 10)
{
        statements;
        if(i == 3)break;
}
```

Skipping an Iteration

The **continue** statement is used to skip an iteration inside a loop. This statement, rather than terminating a loop, forces a new iteration of the loop to begin, by skipping any statements left in that iteration. An example is given below using the **for** loop where iteration 3 is skipped:

```
for(i = 0; i < 10; i++)
{
        if(i == 3)continue;    // skip iteration 3
        statements;
}
```

3.2 Functions

It is widely accepted that a large program should, wherever possible, be expressed in terms of a number of smaller, separate functions. Each function should be a self-contained section of the code written to perform a specific task. Such an approach makes the programming, and more importantly the testing and maintenance of the overall program, much easier. For example, during program development, each function can be developed and tested independently. When the individual functions are tested and working, they can then be included in the overall program. The main program can be a few lines long, and can consist of function calls to carry out the required operations.

In general, most functions perform some tasks and then return data to the calling program. It is not however a requirement that a function must return data. Some functions only do certain tasks and do not return any data. For example, a function to turn on an LED is not expected to return any data.

3.2.1 mikroC Pro for PIC32 Functions

In mikroC for PIC32 language, the format of a function is as follows:

```
type name (parameter1, parameter2,......)
{
     function body
}
```

At the beginning of a function declaration, we first declare the data type to be returned by the function. If the function is not expected to return any data, the keyword void should be inserted before the function name. Then, we specify the unique name of the function. A function name can be any valid C variable name. This is followed by a set of brackets where function parameters are specified, separated by commas. Some functions may not have any parameters and empty brackets should be used while declaring such functions. The body of the function contains the code to be executed by the function and this code must be enclosed within a pair of curly brackets.

An example of function declaration with the name **INC** is shown below. This function receives a parameter of type **int**, multiplies the value of this parameter by 2 and then returns the result to the calling program.

```
int INC(int a)
{
   return (2 * a);
}
```

Notice that the keyword **return** must be used if it is required to return value to the calling program. The data type of the returned number must match to the data type specified before the function name.

The above function can be called from a program as follows:

```
z = INC(x);
```

Which will increment the value of variable **x** by one. Notice that the number of parameters supplied by the calling statement must match with the number of parameters the function expects. Also, the variables used in a function are local to the function and do not have any connection with variables with the same name in the main program. As a result of this, variable names used in the parameters do not have to be the same as the names used while calling the function.

If a function is not expected to return any value, then the **return** statement can be omitted. In such an application, the function will automatically return to the calling program when it terminates. An example function with no return data is given below. This function simply sets LED1 to 1:

```
void LED()
{
      LED1 = 1;
}
```

void functions are called with no parameters:

```
LED();
```

Although functions are usually declared before the main program, this is not always the case as we shall see later.

Some example programs are given below to illustrate how functions may be used in main programs.

Example 3.7

Write a program to calculate the area of a circle whose radius is 3.5 cm. The program should call a function named **Area_Of_Circle** to calculate and return the area. The radius of the circle should be passed as a parameter to the function.

Solution 3.7
The area of a circle with radius **r** is given by the following:

```
Area = πr2
```

Since the radius and the area are floating point numbers, the function type and the parameter must be declared as **float**.

The required function is shown in Figure 3.4. This function is used in a main program as shown in Figure 3.5. Notice that comments are used throughout the program to clarify the operation of the program. The function is called with the parameter set to 3.5 to calculate the area of the circle.

```
float Area_Of_Circle(float radius)
{
        float area;

        area = 3.14 * radius * radius;
        return (area);
}
```

Figure 3.4
Function to calculate the area of a circle.

```
/***********************************************************

                    AREA OF A CIRCLE
                    ===============

This program calculates the area of a circle with a radius of 3.5 cm.


Programmer:     Dogan Ibrahim
Date:           February, 2012
File:           Circle.C
***********************************************************/

//
// This function calculates and returns the area of a circle, given its radius
//
float Area_Of_Circle(float radius)
{
        float area;

        area = 3.14 * radius * radius;
        return (area);
}


//
// Start of MAIN program
//
void main()
{
        float r, a;

        r = 3.5;
        a = Area_Of_Circle(radius);
}
```

Figure 3.5
Program to calculate the area of a circle.

Example 3.8

Write a program to calculate the area and circumference of a rectangle whose sides are 2.4 and 6.5 cm. The program should call functions named **Area_Of_Rectangle** and **Circ_Of_Rectangle** to calculate and return the area and the circumference of the rectangle, respectively.

Solution 3.8
The circumference and the area of a rectangle with sides a and b are given by the following:

```
Circumference = 2 * (a + b)
Area = a * b
```

Since the inputs and outputs are floating point numbers, we shall be declaring the variable as float.

The required functions are shown in Figure 3.6. These functions are used in a main program as shown in Figure 3.7. Notice that comments are used throughout the program to clarify the operation of the program. The function is called with the sides of the rectangle set to 2.4 and 6.5 cm.

Example 3.9

Write a function called **HEX_CONV** to convert a single-digit hexadecimal number "A"–"F" to decimal integer. Use this function in a program to convert hexadecimal "E" to decimal.

Solution 3.9
The required function code is shown in Figure 3.8. The hexadecimal number to be converted is passed as a parameter to the function. A switch statement is used in the function to convert and return the given single-digit hexadecimal number to decimal. The main program listing is given in Figure 3.9. Notice that the hexadecimal number to be converted is passed as a parameter when calling the function **HEX_CONV**.

```
float Circ_Of_Rectangle(float a, float b)
{
        float circumference;

        circumference = 2 * (a + b);
        return (circumference);
}


float Area_Of_Rectangle(float a, float b)
{
        float area;

        area = a * b;
        return (area);
}
```

Figure 3.6
Function to calculate the circumference and area of a rectangle.

```
/********************************************************************

                    CIRCUMFERENCE AND AREA OF A RECTANGLE
                    =======================================

This program calculates the circumference and area of a rectangle with sides
2.4 cm and 6.5 cm

Programmer:    Dogan Ibrahim
Date:          February, 2012
File:          Rectangle.C
********************************************************************/
//
// This function calculates and returns the circumference of a rectangle, given its sides
//
float Circ_Of_Rectangle(float a, float b)
{
        float circumference;

        circumference = 2 * (a + b);
        return (circumference);
}


//
// This function calculates and returns the area of a rectangle, given its sides
//
float Area_Of_Rectangle(float a, float b)
{
        float area;

        area = a * b;
        return (area);
}
// This function calculates and returns the area of a circle, given its radius



//
// Start of MAIN program
//
void main()
{
        float a = 2.4, b = 6.5;
        float circ, area;

        circ = Circ_Of_Rectangle(a, b);
        area = Area_Of_Rectangle(a, b);
}
```

Figure 3.7
Program to calculate the circumference and area of a rectangle.

```
unsigned char HEX_CONV(unsigned char c)
{
        unsigned char y;

        switch(c)
        {
                case 'A':
                        y = 10;
                        break;
                case 'B':
                        y = 11;
                        break;
                case 'C':
                        y = 12;
                        break;
                case 'D':
                        y = 13;
                        break;
                case 'E':
                        y = 14;
                        break;
                case 'F'
                        y = 15;
                        break;
        }
        return (y);
}
```

Figure 3.8
Function to convert hexadecimal to decimal.

3.2.2 Passing Parameters to Functions

Functions are the building blocks of programs where we divide a large program into manageable smaller independent functions. As we have seen earlier, functions generally have parameters supplied by the calling program. There are many applications where we may want to pass arrays to functions and let the function carry out the required array manipulation operations.

Passing single elements of an array to a function is simple and all we have to do is specifying the required elements when calling the function. In the following statement, the second and third elements of array **A** are passed to a function called **Sum**, and the result returned by the function is assigned to variable **b**:

```
b = Sum(A[1], A[2]);
```

Passing an Array by Reference

In most array-based operations, we may want to pass an entire array to a function. This is normally done by specifying the address of the first element of the array in the calling program. Since the address of the first element of an array is the same as the array name itself,

```
/*****************************************************************

                    HEXADECIMAL TO DECIMAL CONVERSION
                    ===================================

This program converts the single digit hexadecimal number "E" to decimal.
Function HEX_CONV is used to convert a given single digit hexadecimal number
to decimal.

Programmer:    Dogan Ibrahim
Date:          February, 2012
File:          Hex.C
*****************************************************************/
//
// This function converts a given single digit hexadecimal number A–F to decimal
//
unsigned char HEX_CONV(unsigned char c)
{
        unsigned char y;

        switch(c)
        {
                case 'A':
                        y = 10;
                        break;
                case 'B':
                        y = 11;
                        break;
                case 'C':
                        y = 12;
                        break;
                case 'D':
                        y = 13;
                        break;
                case 'E':
                        y = 14;
                        break;
                case 'F'
                        y = 15;
                        break;
        }
        return (y);
}


//
// Start of MAIN program
//

void main()
{
        unsigned char hex_number = 'E';
        unsigned char r;

        r = HEX_CONV(hex_number);
}
```

Figure 3.9
Program to convert hexadecimal number "E" to decimal.

we can simply specify the name of the array. In the function declaration, we have to specify the array name with a set of brackets where the size of the array is not specified. This way, all the elements of the array will be available to the function. Notice that here we are passing the entire array by **reference**, which means that the original array elements can be modified inside the function.

The following statements show how the entire array called **A** is passed to a function called **Sum**;

```
In the calling program:   Sum(A);
In the function header:    type Sum(A[ ])
                           {
                                 body of the function
                           }
```

Some complete examples are given below.

Example 3.10

Write a program to store numbers 1—10 in an integer array called **A**. Then, call a function to calculate and return the sum of all the array elements.

Solution 3.10

The required program listing is shown in Figure 3.10. In the main program, a **for** loop is used to store integer numbers 1—10 in an array called **A**. Then, function **Sum** is called to calculate and return the sum of all the array elements. The returned value is stored in variable **Sum_Of_Numbers**. The array is passed to the function by specifying its starting address, i.e. its name.

Example 3.11

Repeat Example 3.10, but this time, define the array size at the beginning of the program using the **#define** preprocessor operator and then pass both the array and its size to the function.

Solution 3.11

The required program listing is shown in Figure 3.11. Here, the array size is defined to be 50. Notice also that the data types are changed to **unsigned integer** as the sum could be larger than 255.

Passing an Array Using Pointers

It is also possible to pass an entire array to a function using pointers. In this method, the address of the array (i.e. its name) is specified in the calling program. In the function header, a pointer declaration is used and the elements of the array are manipulated using pointer operations as we have seen earlier. An example is given below to illustrate the concept.

```
/***************************************************************

                PASSING AN ARRAY TO A FUNCTION BY REFERENCE
                =============================================
```

This program shows how an entire array can be passed to a function by reference. In this example, the name of the array (i.e. the address of the first array element) is passed as a parameter to the array.

The array calculates and returns the sum of all the array elements.


```
Programmer:    Dogan Ibrahim
Date:          February, 2012
File:          Array1.C
***************************************************************/
//
// This function calculates and returns the sum of all the array elements
//
unsigned char Sum(unsigned char Nums[ ])
{
        unsigned char i, Sum;

        Sum = 0;
        for(i = 0; i < 10; i++)Sum = Sum + Nums[i];
        return (Sum);
}


//
// Start of MAIN program
//
void main()
{
        unsigned char  A[10];
        unsigned char i, Sum_Of_Numbers;

        for(i = 0; i < 10; i++)A[i] = i;
        Sum_Of_Numbers = Sum(A);
}
```

Figure 3.10
Program passing an array by reference.

Example 3.12

Repeat Example 3.10, but this time, pass the array to the function using pointers.

Solution 3.12

The required program listing is given in Figure 3.12. Initially, array **A** is loaded with integer numbers 0–10. Then, function **Sum** is called to calculate and return the sum of all array elements. The address of array **A** is passed in the calling program by specifying its name. In the function header, a pointer declaration is made so that the elements of the array can be accessed in the function body using pointer operations.

```
/****************************************************************

                 PASSING AN ARRAY TO A FUNCTION BY REFERENCE
                 ===========================================

This program shows how an entire array can be passed to a function by reference.
In this example, the name of the array (i.e. the address of the first array element) is
passed as a parameter to the array.

In addition to passing the entire array, the array size is also passed to the function.
The array calculates and returns the sum of all the array elements.


Programmer:    Dogan Ibrahim
Date:          February, 2012
File:          Array2.C
****************************************************************/
#define Array_Size 50

//
// This function calculates and returns the sum of all the array elements
//
unsigned int Sum(unsigned int Nums[ ], unsigned int N)
{
        unsigned int i, Sum;

        Sum = 0;
        for(i = 0; i < N; i++)Sum = Sum + Nums[i];
        return (Sum);
}



//
// Start of MAIN program
//
void main()
{
        unsigned int  A[Array_Size];
        unsigned int i, Sum_Of_Numbers;

        for(i = 0; i < Array_Size; i++)A[i] = i;
        Sum_Of_Numbers = Sum(A, Array_Size);
}
```

Figure 3.11
Another program passing an array by reference.

```
/*******************************************************************

                 PASSING AN ARRAY TO A FUNCTION USING POINTERS
                 ================================================
```

This program shows how an entire array can be passed to a function by using pointers.
In this example, the name of the array (i.e. the address of the first array element) is
passed as a parameter to the array.

The array calculates and returns the sum of all 10 array elements.

```
Programmer:   Dogan Ibrahim
Date:         February, 2012
File:         Array3.C
*******************************************************************/
//
// This function calculates and returns the sum of all the array elements
//
unsigned char Sum(unsigned char *Nums)
{
        unsigned char i, Sum;

        Sum = 0;
        for(i = 0; i < 10; i++)Sum = Sum + *(Nums + i);
        return (Sum);
}


//
// Start of MAIN program
//
void main()
{
        unsigned char  A[10];
        unsigned char i, Sum_Of_Numbers;

        for(i = 0; i < 10; i++)A[i] = i;
        Sum_Of_Numbers = Sum(A);
}
```

Figure 3.12
Program passing an array using pointers.

Passing a String Using Pointers

As we have seen before, a string is a character array, terminated with a NULL character.
An array can be passed to a function using pointers as in the earlier example. An example
is given below to illustrate the concept.

Example 3.13

Write a program to declare a string with the text "**Learning to program in C**" in the main program. Then, call a function to calculate the number of times character **a** is used in the string and return this number to the main program.

Solution 3.13

The required program listing is given in Figure 3.13. The text is stored in a string called **Txt**. This string and the character to be matched are passed to a function called **COUNT**. Pointer operations are used inside the function to find out how many times the required character is used in the string. This count is returned to the main program and stored in variable **Sum**. Notice that in the main program, the string is declared using a character array where the size of the array is automatically initialized by the compiler:

```
unsigned char Txt[ ] = "Learning to program in C";
```

we could also have declared the same string using a pointer as:

```
unsigned char *Txt = "Learning to program in C";
```

3.2.3 Passing Variables by Reference to Functions

In C language when we call a function, the variables in function parameters are passed **by value** by default. This means that only the value of the variables are passed to the function and these values cannot be modified inside the function. There are applications however where we may want to modify the values of the variables passed as parameters. In such applications, we have to pass the parameters **by reference**. Passing a parameter **by reference** requires the address of the parameter to be passed to the function so that the value of the parameter can be modified inside the function. It is important to realize that such modifications are permanent and affect the actual variable in the main program that is used while calling the function. In order to pass parameters **by reference** we have to use pointers. An example is given below to illustrate the concept of how a parameter can be passed to a function **by reference**.

Example 3.13

Write a function to receive an integer number and then increment its value by one. Use this function in a program and pass the value to be incremented to the function **by reference**.

Solution 3.13

The required function listing is given in Figure 3.14. In the main program, variable **Total** is initialized to 10. Function **INCR** is then called and the address of **Total** is passed to the function. The function increments **Total** by one. Notice that the function type is declared as **void** since it does not return any value to the calling program.

```
/********************************************************************

                   PASSING A STRING USING POINTERS
                   ===============================

This program shows how a string can be passed to a function by using pointers. In this
example, a string is declared in the main program and then a function is called to find
out how many times the character "a" is used in the string. This count is returned to the
main program.


Programmer:    Dogan Ibrahim
Date:          February, 2012
File:          Array4.C
********************************************************************/

//
// This function calculates how many times the character c is used in the string and
// returns this count to the calling program
//
unsigned char Count(unsigned char *str, unsigned char c)
{
        unsigned char Cnt = 0;

        while(*str)                             // do while not a NULL character
        {
                if(*str == c)Cnt++;             //if  a matching character
                str++;                          // point to next character in string
        }
        return (Cnt);                           // return the count
}




//
// Start of MAIN program
//
void main()
{
        unsigned char  Txt[ ] = "Learning to program in C";
        unsigned  char Sum;

        Sum = Count(Txt, 'a');
}
```

Figure 3.13

Passing a string to a function.

```
/********************************************************************

                 PASSING A VARIABLE TO A FUNCTION BY REFERENCE
                 ================================================

This program shows how a variable can be passed to a function by reference. The
address of the variable is passed to the function. In this example the function
increments the value of the variable by one. It is important to realize that since the
address of the variable is passed the change to the variable is permanent. i.e. the
value of the actual variable used in the calling program is modified by the function.


Programmer:    Dogan Ibrahim
Date:          February, 2012
File:          Byref.C
********************************************************************/

//
// This function increment the value of its parameter by one. The parameter is passed
// by reference. i.e. the actual address of the variable is passed to the function. Notice
// the function is declared as void as it does not return any value to the calling program
//
void INCR(int *a)
{
        *a = *a + 1;
}



//
// Start of MAIN program
//
void main()
{
        int Total;

        Total = 10;                             // initialize p to 10
        INCR(&Total);                           // Increment value of Total by one
}
```

Figure 3.14
Passing a variable by reference.

3.2.4 Passing Variable Number of Arguments to Functions

In the mikroC for PIC32 language, we can pass variable number of arguments to a function, or arguments with varying types. The characters "…" (three ellipsis) with no spaces are used in function parameter list to indicate that we wish to pass variable number of arguments. For example, consider the function is declared as follows:

```
int Test(char a, char b,...)
```

It means that the function has at least two parameters, but it can have more than two. An example is given below to show how variable number of arguments can be passed to a function.

Example 3.14

Write a program to call a function and to calculate the sum of two and three numbers. The function should be declared to expect at least one parameter which will be the number of parameters to follow.

Solution 3.14

The required program listing is shown in Figure 3.15. Notice that a header file called **stdarg.h** must be declared at the beginning of the program. This header file defines a new data type called **va_list**, which is a character pointer. In addition, Macro **va_start** initializes an object of type **va_list** to point to the address of the first additional parameter presented to the function. The second parameter of **va_start** is the number of parameters the function will have (this must be known). To extract the parameters, we have to call to Macro **va_arg** successively with the character pointer and the type of the parameter as the arguments of **va_arg**. It is important that we know the data types of the parameters.

In the main program, variable **Total** is declared as an integer and function **Sum** is called twice to calculate the sum of its arguments: in the first case, there are two arguments with values 5 and 8; and in the second case, there are three parameters with values 3, 6, and 8. Function **Sum** is declared with variable number of parameters where the first parameter (**num_of_parameters**) is the number of parameters the function is expected to have at any time. Then, Macros **va_list** and **va_start** are defined. The actual parameters are extracted inside a **for** loop. The **for** loop is executed as many times as there are parameters. At each execution, Macro **va_arg** is called with the expected data type of the parameter to be extracted. The value of the parameter is extracted and added to temporary variable **tot**. After all the parameters are extracted, Macro **va_end** is called to clean up the stack and return from the Macro. The function then terminates by returning the sum of parameter values to the calling program.

3.2.5 Pointers to Functions

mikroC Pro for PIC32 language supports pointers to functions. In C language, a function's name is simply same as the address of that function. Thus, in the following function call,

```
a = Test();
```

name **Test** is same as the address of the function, while the function invocation operator () is an identifier that the function at the given address will be invoked. Since we know the address of the function, we can assign this address to a pointer and then invoke the function via the pointer. Consider the following function which adds two integer numbers, passed as parameters:

```
int Add(int a, int b)
{
        return (a + b);
}
```

```
/********************************************************************

        PASSING VARIABLE NUMBER OF PARAMETERS TO A FUNCTION
        =====================================================
```

This program shows how a variable number of parameters can be passed to a function.
In this program the function expects at least one parameter, which specifies the actual
number of parameters to follow. The function in this example calculate the value of the
sum of its parameters and returns it to the calling program.


```
Programmer:   Dogan Ibrahim
Date:         February, 2012
File:         Varparams.C
********************************************************************/
#include <stdarg.h>

//
// This function accepts variable number of parameters. At least one parameter must be
// supplied to the function. This parameter indicates how many additional parameters to
// expect. The function calculates and returns the sum of the values of its parameters
//
int Sum(int num_of_parameters,...)
{
        va_list ap;
        unsigned char i;
        int tot = 0;

        va_start(ap, num_of_parameters);
        for(i = 0; i < num_of_parameters; i++)
        {
                tot = tot + va_arg(ap, int);
        }
        va_end(ap);
        return (tot);
}




//
// Start of MAIN program
//
void main()
{
        int Total;

        Total = Sum(2, 5, 8);                   // 2 arguments, p = 5 + 8
        Total = Sum(3, 3, 6, 8);                // 3 arguments, p = 3 + 6 + 8

}
```

Figure 3.15
Passing variable number of parameters.

We can define a pointer called **fpointer** to a function having two integer parameters as follows:

```
int (*fpointer)(int, int);
```

and then, point to our function as follows:

```
fpointer = &Add; or simply as fpointer = Add;
```

Now that we have a pointer to the function, and we can use it as follows:

```
Sum = (*fpointer)(3, 4);      // Sum becomes 7
```

An example is given below to illustrate how we can call a function using a pointer.

Example 3.15

Write a function to multiply two integer numbers. Show how you can call this function from a main program using a pointer to the function.

Solution 3.15

The required program listing is shown in Figure 3.16. Function **Mult** simply multiplies its two parameters and returns the result to the calling program. Inside the main program, **fpointer** is declared as a function pointer with two integer parameters. Then, **fpointer** is set to point to the address of function **Mult**. The function is then called and the result of the multiplication is stored in variable **Total**.

3.2.6 Static Variables in Functions

Normally, the variables used in functions are local to the function and they are reinitialized (usually to zero) when the function is called. There are some applications however where we may want to preserve the values of some or all of the variables inside a function declaration. This is done using the static statement. An example function code is given below. Here, variable **Total** is initialized to 0 using the static keyword. Notice that the variable is incremented by one before returning from the function. What happens now is that every time the function is called, variable **Total** will retain its previous value. Thus, on the second call, **Total** will be 2, and on the third call, it will be 3 and so on:

```
void Sum(void)
{
      static Total = 0;
      ..................
      ..................
      Total++;
}
```

3.2.7 Function Prototypes

If a function is defined after a main program, then it is not visible to the compiler at the time the function is called first time in the main program. The same will happen if there are several

```
/*********************************************************************

                    POINTER TO A FUNCTION
                    =====================

This program shows how we can create a pointer to a function and then use this pointer
To invoke the function. In this example the function simply multiplies two integer numbers.


Programmer:    Dogan Ibrahim
Date:          February, 2012
File:          Funcpointer.C
*********************************************************************/

//
// This function multiplies two integer numbers and returns the result. The function is
// invoked using a function pointer.
//
int Mult(int a, int b)
{
        return (a * b);                              // return the multiplication
}



//
// Start of MAIN program
//
void main()
{
        int x, y, Total;
        int (*fpointer) (int, int);                  // fpointer is a function pointer

        x = 5;
        y = 8;
        fpointer = Mult;                             // fpointer points to function Mult
        Total = (*fpointer)(x, y);                   // invoke the function and get the result
}
```

Figure 3.16
Invoking a function with a pointer.

functions in a program and a function makes call to another function which is defined at a later point in the program. In order to avoid compilation errors arising these compiler visibility issues, we construct function prototypes and declare them at the beginning of the program. A function prototype simply includes a function name and a list of the data types of its parameters. Function prototypes must be terminated with a semicolon. A typical function prototype statement is given below where the function is named **Test** and it has two integer and one floating point parameters:

```
int Test(int, int, float);
```

An example is given below to illustrate how function prototypes can be used in programs.

Example 3.16

Write a function to add two integer numbers and return the result to a calling program. Declare the function after the main program and use function prototype declaration at the beginning of the program.

Solution 3.16

The required program listing is shown in Figure 3.17. Notice that the function prototype for function **Add** is declared at the beginning of the program and the function declaration is after the main program is declared.

```
/*********************************************************************

                        USING FUNCTION PROTOTYPES
                        =========================

This program shows how we can use a function prototype in a program. In this program
a function is created which adds two integer numbers.


Programmer:   Dogan Ibrahim
Date:         February, 2012
File:         Funcproto.C
*********************************************************************/
int Add(int, int);                                    // Declare function prototype

//
// Start of MAIN program
//
void main()
{
        int x, y, Total;

        x = 5;
        y = 8;
        Total = Add(x, y);
}



//
// This function adds two integer numbers and returns the result.
//
int Add(int a, int b)
{
        return (a + b);                               // return the sum
}
```

Figure 3.17
Declaring a function prototype.

3.3 PIC32 Microcontroller Specific Features

mikroC Pro for PIC32 language differs from the ANSI C standard in certain ways. The modifications have been implemented to make the language more specific to the PIC32 series of microcontrollers. In this section, we shall be looking at some of the important features of the mikroC Pro for PIC32 language which are specific to the architecture of the PIC32 microcontrollers.

mikroC Pro for PIC32 language includes the following keywords which have specifically included to support the PIC32 microcontrollers:

code: This memory type is used for allocating **constants** in program memory. An example is given below:

```
const code x, y, z;
```

data: This memory type is used for storing variables in data RAM. An example is given below:

```
data int x, y, z;
```

rx: This memory type is used to store variables in the working registers. An example is given below:

```
rx char x;
```

sfr: This memory type allows the programmer to access the SFRs. An example is given below:

```
char Temp at PORTB;
```

at: This is used as an alias to a variable. An example is given below:

```
char x at PORTA;
```

sbit: This data type is used to provide access to processors registers and SFR. An example is given below:

```
sbit LCD_RS at LATB2_bit;
```

bit: This data type is used to declare single bit variables. An example is given below:

```
bit flag;
```

iv: Used to inform the compiler that this is an ISR.

3.3.1 SFR Registers

All PIC32 microcontroller SFR registers are defined as global variables of type **volatile int**. Access to all these registers and their bits are available in our programs. Some examples are given below:

```
PORTB
PORTB.RB0
INTCONbits.TPC
```

3.3.2 Linker Directives

mikroC for PIC32 linker supports a number of useful directives as explained below.

Directive Absolute

Specifies the starting address of a variable in RAM, or a constant in ROM. Some examples are given below:

```
char x absolute 0xA0000000;    // x occupies a byte in 0xA0000000
int x absolute 0xA0000000;     // x occupies 2 bytes staring from 0xA0000000
```

Directive org

This directive specifies the starting address of a routine in ROM. **Org** is appended to a function definition. An example is shown below:

```
void Func(void) org 0xBA000000    // Function starts at 0xBA000000
{
}
```

Directive orgall

This directive is used to place all routines, constants, etc. above a specified ROM address.

Directive funcorg

With this directive, we can specify the starting address of a routine in ROM using the routine name. An example is given below:

```
#pragma funcorg <function name> <starting address>
```

3.3.3 Built-in Utilities

mikroC Pro for PIC32 language provides a number of utilities that could be useful in programs. Some of the important utilities are summarized in this section. Further details can be obtained from the mikroC Pro for PIC32 User Guide.

Lo: Returns (or sets) the low byte of a number. For example,

```
If x = 0x12345678 then z = Lo(x) returns 0x78
```

or

```
Lo(x) = 0x11 sets x to x = 0x12345611
```

Hi: Returns (or sets) the high byte of a number. For example,

```
If x = 0x12345678 then z = Hi(x) returns 0x56
```

Higher: Returns (or sets) the higher byte of a number. For example,

```
If x = 0x12345678 then z = Higher(x) returns 0x34
```

Highest: Returns (or sets) the highest byte of a number. For example,

```
If x = 0x12345678 then z = Highest(x) returns 0x12
```

LoWord: Returns (or sets) the low word of a number. For example,

```
If x = 0x12345678 then z = LoWord(x) returns 0x5678
```

HiWord: Returns (or sets) the high word of a number. For example,

```
If x = 0x12345678 then z = HiWord(x) returns 0x1234
```

Delay_us: Creates delay in microseconds.

Delay_ms: Creates delay in milliseconds.

Delay_Cyc: Creates delay based on the clock frequency.

Clock_MHz: Returns the clock frequency in MHz.

EnableInterrupts: Enables interrupts.

DisableInterrupts: Disables interrupts.

3.3.4 Start-up Initialization

Upon reset the MCU jumps to address 0xBFC00000 where the **BootStartUp** function is located. Before the processor jumps to the main program, the following actions take place:

The **BootStartUp** function configures:

* CP0 coprocessor register
* SFR registers associated with the interrupt
* Stack pointer and global pointer.

The processor is configured by default as

* cache and prefetch are enabled;
* flash wait states are set;
* executable code is allocated in the KSEG0;
* data are allocated in the KSEG1.

3.3.5 Read—Modify—Write Cycles

There are many applications where we may want to read the value of a port pin, modify it, and then write it back. This is commonly known as the read—modify—write cycle. There are some issues that we should be careful about when performing read—modify—write cycles.

When data are read from a port pin, we read the actual physical state of that port pin. Now, the problem arises if the port pin is loaded by the external hardware such that the state of the port pin may be affected. For example, if the port pin is connected to a capacitive load, the state of the port pin may be affected if the capacitor is charged or discharged. Thus, what we are reading is not the true state of the port pin, but rather the physical state of the port pin.

The read—modify—write cycle problem can be avoided by using the **LATx** registers when writing to and reading from ports, rather than **PORTx** registers. Writing to a **LATx** register is equivalent to writing to a **PORTx** register, but reading from a **LATx** register returns the data held in the port latch register, regardless of the physical state of the port pin. Thus, reading will not be affected by the external hardware connected to the port pin. It is recommended to use the **LATx** port register instead of the **PORTx** register, especially while reading from port pins connected to external devices such as LCDs, GLCDs, capacitive and inductive loads.

3.3.6 Interrupts

Interrupts are an important and complex topic in microcontroller programming. The interrupt handling mechanisms of the mikroC Pro for PIC32 language are described in detail in the projects sections of this book.

3.4 Summary

This chapter presented an introduction to the highly popular mikroC Pro for PIC32 language. The basic structure of a C program is given with an explanation of the basic elements that make a C program.

The data types of the mikroC Pro for PIC32 language have been described with examples. Various data structures, such as arrays, structures, and unions, have been described with examples.

Functions are important building blocks of all complex programs. The creation and use of functions in programs have been described with many examples.

Various flow control statements such as *if*, *switch*, *while*, *do*, *break*, and so on have been explained in this chapter with examples on their use in C programs.

Pointers are important elements of all professional C programs. The basic principles of pointers have been described with various examples to show how they can be used in programs and in functions.

Finally, the basic features of the mikroC Pro for PIC32 language, developed specifically for the PIC32 microcontroller series, have been explained briefly.

3.5 Exercises

1. What does program repetition mean? Describe how program repetition can be created using **for** and **while** statements.

2. What is an array? Write statements to define arrays with the following properties:
    a. Array of 10 integers
    b. Array of five floating point numbers
    c. Array of 3 characters.

3. Given that $x = 10$ and $y = 0$, list the outcome of the following expressions as either TRUE or FALSE:
    a. $x > 0$ && $y > 0$
    b. $y <= 0$
    c. $x == 10$ || $y == 5$

4. Assuming that $a = 0x2E$ and $b = 0x6F$, determine the results of the following bitwise operations:
    a. a | b
    b. b & 0xFF
    c. a^b
    d. b^0xFF

5. How many times does each of the following loops iterate?
    a. for(j = 0; j < 5; j++)
    b. for(i = 0; i <= 5; i++)
    c. for(i = 10; i > 0; i--)
    d. for(i = 10; i > 0; i -= 2)

6. Write a program to calculate the sum of all positive integer numbers between 1 and 5.

7. Write a program to calculate the average of all the numbers from 1 to 100.

8. Derive equivalent **if-else** statements for the following:
    a. $(x > y)$? 1: 0
    b. $(x > y)$? $(a - b)$: $(a + b)$

9. What can you say about the following **while** loop?
    ```
    i = 0;
    Count = 0;
    while (i < 10)
    {
       Count++;
    }
    ```

10. What will be the value of **Sum** at the end of the following **do–while** loop?

```
k = 0;
Sum = 0;
do
{
  Sum++;
  k++;
}while(k < 20);
```

11. What can you say about the following program?

```
Sum = 0;
for(;;)
{
  Sum++;
}
```

12. Rewrite the following code using a **while** statement:

```
k = 1;
for(i = 1; i <= 10; i++)
{
  k = k + i;
}
```

13. Write a function to receive two integer parameters and to return the larger one to the calling program.

14. Write a program to store the even numbers between 0 and 20 in an integer array. Then call a function to calculate and store the square of each array element in the same array.

15. Write a function to convert inches to centimeters. The function should receive inches and return the equivalent centimeters. Show how you can call this function in a main program to convert 5.2 inches to centimeters.

16. Write a program to add two matrices A and B, having dimensions 2 × 2 and store the result in a third matrix called C.

17. Write a function to receive three matrices A, B, and C. Perform the matrix operation C = A × B. Show how the function can be called from a main program.

18. Explain what a function prototype is, where can it be used?

19. Write a function to convert a two-digit hexadecimal number 0x00 to 0xFF into decimal. Show how you can call this function to convert 0x1E into decimal.

20. In an experiment, the relationship between the x and y values is calculated to be as in the following table:

x	y
1.2	4.0
3.0	5.2
2.4	7.0
4.5	9.1

Write a function using the **switch** statement to find y, given x. Show how the function can be called from a main program to return the value of y when x is 2.4.

21. Explain how can you pass variable number of arguments to a function. Write a function to find the average of the numbers supplied as parameters. Assume that the first parameter is the number of parameters to be supplied. Show how this function can be called from a main program to calculate the average of numbers 2, 3, 5, and 7.

22. Repeat Exercise 19, using **if-else** statements.

mikroC Pro for PIC32 Built-in Library Functions

mikroC Pro for PIC32 language provides a large number of built-in library functions that can be called from anywhere in a program, and there is no need to include header files at the beginning of the program. Most of the library functions are known as hardware functions and they have been developed for peripheral devices. In addition, standard ANSI C and miscellaneous software libraries are provided. mikroC Pro for PIC32 user manual gives a detailed description of each built-in function. In this chapter, we shall be looking at the details of some commonly used important functions.

Table 4.1 gives a list of the mikroC Pro for PIC32 built-in functions. Some of the frequently used library functions are as follows:

- Analog to digital converter (ADC) module library
- LCD library

Table 4.1: mikroC Pro for PIC32 Library Functions

Library	Description
ADC	Analog to digital conversion functions
CAN	CAN Bus functions
CANSPI	SPI-based CAN Bus functions
Compact Flash	Compact Flash memory functions
Epson Graphic LCD	Epson Graphic LCD functions
Flash Memory	Flash Memory functions
Graphics LCD	Standard Graphics LCD functions
T6963C Graphics LCD	T6963-based Graphics LCD functions
I^2C	I^2C bus functions
Keypad	Keypad functions
LCD	Standard LCD functions
Manchester Code	Manchester Code functions
Memory Manager	Memory Management functions
Multi Media	Multi Media functions
One Wire	One Wire functions
PS/2	PS/2 functions
PWM	PWM functions
RS-485	RS-485 communication functions
Sound	Sound functions
SPI	SPI bus functions
USART	USART serial communication functions
Touch Panel	Touch Panel LCD functions
SPI Graphics LCD	SPI-based Graphics LCD functions
Port Expander	Port expander functions
TFT Display	TFT display functions
USB	USB functions
SPI LCD	SPI-based LCD functions
ANSI C Ctype	C Ctype functions
ANSI C Math	C Math functions
ANSI C Stdlib	C Stdlib functions
ANSI C string	C string functions
Button	Button functions
Conversion	Conversion functions
Trigonometry	Trigonometry functions
Time	Time functions

- Software Universal Asynchronous Receiver Transmitter (UART) library
- Hardware UART library
- Sound library
- ANSI C library
- Miscellaneous library.

4.1 ADC Library

The ADC module library offers a number of functions that enable the programmer to use the ADC module easily. Four functions are provided in this library:

- ADCx_Init
- ADCx_Init_Advanced
- ADCx_Get_Sample
- ADCx_Read

"x" refers to the number of the ADC module used in microcontrollers with multiple ADC units.

The *ADCx_Init* function is used to configure an ADC module to work with default settings. By default, when this function is called the ADC module, it is set to work as follows:

- Single-channel conversion
- Ten-bit resolution
- Unsigned integer format
- Power supply as the positive reference, and ground as the negative reference
- 32*Tcy conversion clock

The function requires no parameters and a typical call to initialize ADC module 1 is as follows:

```
ADC1_Init();
```

The *ADCx_Init_Advanced* function is used to configure an ADC to work with user defined settings. This function is called with one parameter, the reference voltage, which can take two possible values:

- _ADC_INTERNAL_REF (internal voltage reference)
- _ADC_EXTERNAL_REF (external voltage reference)

In the following example, ADC module 1 is configured to operate with external reference voltage:

```
ADC1_Init_Advanced(_ADC_EXTERNAL_REF);
```

The *ADCx_Get_Sample* function reads data from the specified channel. The 10-bit unsigned value is returned by the function. The ADC module must be initialized before using this function. In addition, the appropriate TRISx port bit must be configured as input. In the following example, data are read from channel 5 of ADC module 1 and the converted data are stored in variable *Temp*:

```
unsigned Temp;
.................
ADC1_Init();
Temp = ADC1_Get_Sample(5);
```

The *ADCx_Read* function initializes the ADC module and reads data from the specified channel input. There is no need to call function *ADCx_Init* to initialize the ADC module. In the following example, the ADC module 1 is initialized and data are read from channel 10:

```
unsigned Temp;
Temp = ADC1_Read(10);
```

Example 4.1

Write a function to read the voltage from analog port AN0 of a microcontroller. Convert the voltage to millivolts and return to a main program. Repeat the above process forever with 1 s delay between each sample.

Solution 4.1

The required program is given in Figure 4.1. The ADC module is initialized in the main program. Function *Read_Voltage* reads the analog voltage, converts to millivolts and returns to the main program. The process is repeated forever using a *for* loop.

4.2 LCD Library

One thing all microcontrollers lack is some kind of video display. A video display would make a microcontroller much more user-friendly as it will enable text messages, graphics, and numeric values to be output in a more versatile manner than the seven-segment displays, LEDs, or alphanumeric displays. Standard video displays require complex interfaces and their cost is relatively high. LCDs are alphanumeric (or graphical) displays which are frequently used in microcontroller-based applications. These display devices come in different shapes and sizes. Some LCDs have 40 or more character lengths with the capability to display several lines. Some other LCD displays can be programmed to display graphic images. Some modules offer color displays while some others incorporate backlighting so that they can be viewed in dimly lit conditions.

There are basically two types of LCDs as far as the interfacing technique is concerned: parallel LCDs and serial LCDs. Parallel LCDs (e.g. Hitachi HD44780 series) are connected to the microcontroller circuitry such that data are transferred to the LCD using more than one line, and usually four or eight data lines are used. Serial LCD is connected to a microcontroller using one data line only and data are transferred using the RS232 asynchronous data communications protocol. Serial LCDs are generally much easier to use but they are more costly than the parallel ones. In this book, only the parallel LCDs will be considered as they are used cheaper and are used more commonly in microcontroller-based projects.

Low-level programming of a parallel LCD is usually a complex task and requires a good understanding of the internal operation of the LCD, including an understanding of the timing

```
/*************************************************************************

                    READ ANALOG VOLTAGE FROM PORT AN0
                    ==================================

This program reads the analog voltage from port AN0, converts the voltage
to millivolts and returns the value to the main program.

It is assumed that ADC module 1 is used with 10-bit resolution.


Programmer:   Dogan Ibrahim
Date:         March, 2012
File:         ADC.C
*************************************************************************/
//
// This function reads the analog voltage from port AN0, converts to millivolts and returns
// to the main program
//
float Read_Voltage()
{
        unsigned t;
        float mV;

        t = ADC1_Get_Sample(0);                         // Read from AN0
        mV = t*5000.0 / 1024.0;                         // convert to millivolts
        return (mV);                                    // Return
}



//
// Start of MAIN program
//
void main()
{
        float voltage;

        TRISA = 1;
        ADC1_Init();                                    // Initialize ADC 1
        for(;;)                                         // DO FOREVER
        {
                Voltage = Read_Voltage();               // Read voltage
                Delay_Ms(1000);                         // Wait 1 second
        }

}
```

Figure 4.1
Program to read from the ADC module.

diagrams. Fortunately, mikroC for PIC32 language provides functions for both text-based and graphical LCDs, which simplify the use of LCDs in PIC microcontroller-based projects.

HD44780 controller is commonly used in LCD-based microcontroller applications. A brief description of this controller and information on some commercially available LCD modules is given below.

4.2.1 HD44780 LCD Controller

HD44780 is one of the most popular LCD controllers used in many LCD modules in industrial and commercial applications, and also by hobbyists. The module is monochrome and comes in different shapes and sizes. Modules with character lengths of 8, 16, 20, 24, 32, and 40 characters can be selected. Depending on the model chosen, the display provides a 14-pin or a 16-pin connector to interface to the external world. Table 4.2 shows the pin configuration and pin functions of a typical 14-pin LCD.

V_{SS} is the 0 V supply or ground. V_{DD} pin should be connected to the positive supply. Although the manufacturers specify a 5 V DC supply, the modules will usually work with as low as 3 V or as high as 6 V.

Pin 3 is named as V_{EE} and this is the contrast control pin. This pin is used to adjust the contrast of the display and it should be connected to a DC supply. A potentiometer is usually connected to the power supply with its wiper arm connected to this pin and the other leg of the potentiometer connected to the ground. This way the voltage at the V_{EE} pin and hence the contrast of the display can be adjusted as desired.

Pin 4 is the Register Select (RS) and when this pin is LOW, data transferred to the LCD are treated as commands. When RS is HIGH, character data can be transferred to and from the module.

Table 4.2: Pin Configuration of the HD44780 LCD Module

Pin No.	Name	Function
1	V_{SS}	Ground
2	V_{DD}	positive supply
3	V_{EE}	Contrast
4	RS	Register select
5	R/W	Read/write
6	EN	Enable
7	D0	Data bit 0
8	D1	Data bit 1
9	D2	Data bit 2
10	D3	Data bit 3
11	D4	Data bit 4
12	D5	Data bit 5
13	D6	Data bit 6
14	D7	Data bit 7

Pin 5 is the Read/Write (R/W) pin. This pin is pulled LOW in order to write commands or character data to the LCD module. When this pin is HIGH, character data or status information can be read from the module.

Pin 6 is the Enable (EN) pin which is used to initiate the transfer of commands or data between the module and the microcontroller. When writing to the display, data are transferred only on the HIGH to LOW transition of this pin. When reading from the display, data become available after the LOW to HIGH transition of the enable pin and this data remain valid as long as the enable pin is at logic HIGH.

Pins 7–14 are the eight data bus lines (D0–D7). Data can be transferred between the microcontroller and the LCD module using either a single eight-bit byte, or a two four-bit nibbles. In the latter case, only the upper four data lines (D4–D7) are used. The four-bit mode has the advantage that fewer I/O lines are required to communicate with the LCD.

mikroC Pro for PIC32 LCD library provides large number of functions to control text-based LCDs with 4-bit and 8-bit data interface, and for graphics LCDs. Four-bit interface-based text LCDs are the most commonly used ones and this section describes the available mikroC for PIC32 functions for these LCDs. Further information on other text- or graphics-based LCD functions can be obtained from the mikroC manual.

Following is a list of the LCD functions available for four-bit interface text-based LCDs:

- Lcd_Init
- Lcd_Out
- Lcd_Out_Cp
- Lcd_Chr
- Lcd_Chr_Cp
- Lcd_Cmd.

A brief description of these functions is given below.

The *Lcd_Init* function is called to configure the interface between the microcontroller and the LCD. The following global variables must be declared before calling this function:

LCD_RS: LCD register select pin
LCD_EN: LCD signal enable pin
LCD_D4: LCD data pin 4
LCD_D5: LCD data pin 5
LCD_D6: LCD data pin 6
LCD_D7: LCD data pin 7
LCD_RS_Direction: Direction of the register select pin
LCD_EN_Direction: Direction of the enable signal pin
LCD_D4_Direction: Direction of data pin 4

 LCD_D5_Direction: Direction of data pin 5
 LCD_D6_Direction: Direction of data pin 6
 LCD_D7_Direction: Direction of data pin 7

The interface between the LCD and the microcontroller is defined using the *sbit* instructions. The following example shows the interface definitions when the LCD is connected to PORT B of the microcontroller (RS and EN pins are connected to port pins RB2 and RB3, respectively; D4–D7 are connected to port pins RB4–RB7, respectively):

```
sbit LCD_RS at LATB2_bit;
sbit LCD_EN at LATB3_bit;
sbit LCD_D4 at LATB4_bit;
sbit LCD_D5 at LATB5_bit;
sbit LCD_D6 at LATB6_bit;
sbit LCD_D7 at LATB7_bit;

sbit LCD_RS_Direction at TRISB2_bit;
sbit LCD_EN_Direction at TRISB3_bit;
sbit LCD_D4_Direction at TRISB4_bit;
sbit LCD_D5_Direction at TRISB5_bit;
sbit LCD_D6_Direction at TRISB6_bit;
sbit LCD_D7_Direction at TRISB7_bit;
Lcd_Init();
```

Figure 4.2 shows the connection between the LCD and the microcontroller for the above example configuration. Notice that the brightness of the LCD is controlled by using a 10 K potentiometer connected to pin VEE of the LCD. A PIC32MX360F512L microcontroller is used in this example.

The function *Lcd_Out* displays text at the specified *row* and *column* position of the LCD. The function should be called with the parameters in the following order:

```
row, column, text
```

For example, to display text "Computer" at row 1 and column 2 of the LCD, we should call the function as follows:

```
Lcd_Out(1, 2, "Computer");
```

The *Lcd_Out_Cp* function displays text at the *current* cursor position. For example, to display text "Computer" at the current cursor position the function should be called as follows:

```
Lcd_Out_Cp("Computer");
```

The function *Lcd_Chr* displays a character at the specified *row* and *column* position of the cursor. The function should be called with the parameters in the following order:

```
row, column, character
```

For example, to display character "K" at row 2 and column 4 of the LCD, we should call the function as follows:

```
Lcd_Chr(2, 4, 'K');
```

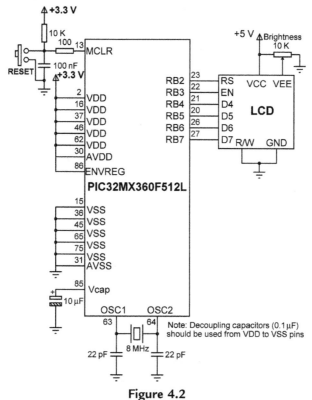

Figure 4.2
LCD connections to the microcontroller.

The function *Lcd-Chr_Cp* displays a character at the current cursor position. For example, to display character "M" at the current cursor position, the function should be called as follows:

```
Lcd_Chr_Cp('M');
```

This *Lcd_Cmd* function is used to send a command to the LCD. With the commands, we can move the cursor to any required row, clear the LCD, blink the cursor, shift display, etc. A list of the most commonly used LCD commands is given in Table 4.3. For example, to clear the LCD, we should call the function as follows:

```
Lcd_Cmd(_LCD_CLEAR);
```

An example is given here which illustrates how the LCD can be initialized and used.

Example 4.2

A text-based LCD is connected to a PIC32MX360F512L microcontroller in the default mode as shown in Figure 4.2. Write a program to send the text "My Computer" to row 1, column 5 of the LCD.

Solution 4.2

The required program listing is given in Figure 4.3 (program LCD.C). At the beginning of the program, PORTB is configured as output with the TRISB = 0 statement. The LCD is then initialized, display is cleared, and the text message "My Computer" is displayed on the LCD.

Table 4.3: LCD Commands

LCD Command	Description
LCD_CLEAR	Clear display
LCD_RETURN_HOME	Return cursor to home position
LCD_FIRST_ROW	Move cursor to first row
LCD_SECOND_ROW	Move cursor to second row
LCD_THIRD_ROW	Move cursor to third row
LCD_FOURTH_ROW	Move cursor to fourth row
LCD_BLINK_CURSOR_ON	Blink cursor
LCD_TURN_ON	Turn display on
LCD_TURN_OFF	Turn display off
LCD_MOVE_CURSOR_LEFT	Move cursor left
LCD_MOVE_CURSOR_RIGHT	Move cursor right
LCD_SHIFT_LEFT	Shift display left
LCD_SHIFT_RIGHT	Shift display right

4.3 Software UART Library

UART is used for RS232-based serial communication between two electronic devices. In serial communication, only two cables (plus a ground cable) are required to transfer data in either direction. Data are sent in serial format over the cable bit by bit. Normally the receiving device is in idle mode with its transmit (TX) pin at logic 1, also known as MARK. Data transmission starts when this pin goes to logic 0, also known as SPACE. The first bit sent is the **start** bit at logic 0. Following this bit, 7 or 8 **data bits** are sent followed by an optional **parity bit**. The last bit sent is the **stop bit** at logic 1. Serial data are usually sent as a 10-bit frame consisting of a start bit, eight data bits, and a stop bit, and no parity bits. Figure 4.4 shows how character "A" can be sent using serial communication. Character "A" has the ASCII bit pattern "01000001". As shown in the figure, first the start bit is sent, this is followed by eight data bits "01000001", and finally the stop bit is sent.

The bit timing is very important in serial communication and both the transmitting (TX) and receiving (RX) devices must have the same bit timings. The bit timing is measured by the **baud rate** which specifies the number of bits transmitted or received each second.
Typical baud rates are 4800, 9600, 19200, 38400 and so on. For example, when operating at 9600 baud rate with a frame size of 10 bits, 960 characters are transmitted or received each second. The timing between each bit is then about 104 ms.

In RS232-based serial communication, the two devices are connected to each other (Figure 4.5) using either a 25-way connector or a 9-way connector. Normally only the TX, RX, and GND pins are required for communication. The required pins for both types of connectors are given in Table 4.4.

```
/****************************************************************************

                        DISPLAY TEXT ON LCD
                        ==================

This program displays text "My Computer" at row 1, column 5 of the LCD.

The LCD is connected to PORT B of the microcontroller.

Programmer:    Dogan Ibrahim
Date:          March, 2012
File:          LCD.C
****************************************************************************/
//
// Define LCD interface
//
sbit LCD_RS at LATB2_bit;
sbit LCD_EN at LATB3_bit;
sbit LCD_D4 at LATB4_bit;
sbit LCD_D5 at LATB5_bit;
sbit LCD_D6 at LATB6_bit;
sbit LCD_D7 at LATB7_bit;
//
// LCD pin directions
//
sbit LCD_RS_Direction at TRISB2_bit;
sbit LCD_EN_Direction at TRISB3_bit;
sbit LCD_D4_Direction at TRISB4_bit;
sbit LCD_D5_Direction at TRISB5_bit;
sbit LCD_D6_Direction at TRISB6_bit;
sbit LCD_D7_Direction at TRISB7_bit;

void main()
{
        TRISB = 0;
        Lcd_Init();
        Lcd_Cmd(_LCD_CLEAR);
        Lcd_Out(1, 5, "My Computer");
}
```

Figure 4.3
LCD program listing.

The voltage levels specified by the RS232 protocol are ± 12 V. A logic HIGH signal is at -12 V and a logic LOW signal is at $+12$ V. PIC microcontrollers on the other hand normally operate at $0-5$ V voltage levels and it is required to convert the RS232 signals to $0-5$ V when input to a microcontroller. Similarly, the output of the microcontroller must be converted to

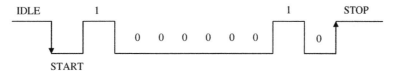

Figure 4.4
Sending character "A" in serial communication.

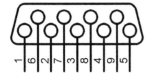 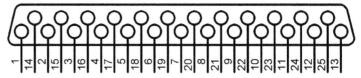

Figure 4.5
25-way and 9-way RS232 connectors.

Table 4.4: Pins Required for Serial Communication

Pin	9-Way Connector	25-Way Connector
TX	2	2
RX	3	3
GND	5	7

± 12 V voltage level before sending to the receiving RS232 device. The voltage conversion is usually carried out using RS232 converter chips such as the MAX232, manufactured by Maxim Inc.

Serial communication is either implemented in hardware using a specific pin of a microcontroller, or the required signals can be generated in software from any required pin of a microcontroller. Hardware implementation requires either an on-chip UART (or USART) circuit or an external UART chip can be connected to the microcontroller. Software-based UART on the other hand is more commonly used and it does not require any special circuits. Serial data are generated by delay loops in the software-based UART applications. In this section, only the software-based UART functions will be described.

mikroC Pro for PIC32 compiler supports the following software UART functions. Software UART is implemented in software and there is no need to have a UART module. The reason for using soft UART functions is when it is required to have more serial communications ports than supported by the target microcontroller. For example, the PIC32MX360F512L microcontroller provides two hardware UARTs. In this case, the soft UART functions can be used when we wish to have three or more serial ports. mikroC for PIC32 soft UART offers the following functions:

- Soft_Uart_Init
- Soft_Uart_Read
- Soft_Uart_Write
- Soft_UART_Break.

The advantage of using the soft UART functions in serial communication is that any microcontroller pin can be configured as either the transmitter or the receiver pin.

The *Soft_UART_Init* function is used to specify the serial communications parameters and the pin to be used to send and receive data. The parameters are specified in the following order:

```
port, rx pin, tx pin, baud rate, mode
```

port is the port used as the software UART (e.g. PORTB), **rx** is the receive pin number, **tx** is the transmit pin number, and **baud rate** is the chosen baud rate where the maximum value depends on the clock rate of the microcontroller. **Mode** specifies whether or not the data should be inverted at the output of the port. A zero indicates that the data should not be inverted, and a one indicates that the data should be inverted at the output of the port. When an RS232 voltage level converter chip is used (e.g. MAX232), the mode must be set to 0. **Soft_Uart_Init** must be the first function to be called before software-based serial communication is to be established. The function returns a byte with the following values:

```
0:   successful initialization
1:   error (requested baud rate is too high)
2:   error (requested baud rate is too low)
```

An example is given below which configures the software UART to use PORT B as the serial port with RB0 as the RX pin, and RB1 as the TX pin. The baud rate is set to 9600 with the mode noninverted:

```
error = Soft_Uart_Init(PORTB, 0, 1, 9600, 0);
```

The *Soft_URT_Read* function is used to receive a byte from the specified serial port pin. Soft UART must be initialized before this function can be used. The function returns an error condition and the data read from the serial port. The function is blocking, i.e. it waits for a start bit to be available. The function can be terminated by using the *Soft_UART_Break* function. The returned **error** byte can have the following values:

```
0:   no error
1:   stop bit error
255: function aborted (by Soft_UART_Break)
```

An example is given below which reads a byte from the serial port configured by calling the function **Soft_Uart_Init**. The code waits until data are received from the serial port and the received byte is stored in variable **Temp**:

```
do
  Temp = Soft_Uart_Read(&Rx_Error);
while (Rx_Error);
```

The *Soft_UART_Write* function transmits a byte to the configured serial port pin. Soft UART must be initialized before this function can be used. The data to be sent must be specified as a parameter in the call to the function.

An example is given below which sends character "A" to the serial port pin:

```
char MyData = 'A';
Soft_Uart_Write(MyData);
```

The *Soft_UART_Break* function is used to unblock the *Soft_UART_Read* function when the function is waiting for a character. The *Soft_UART_Break* function is usually called from an interrupt service routine. The function has no parameters.

An example is given below to illustrate the use of software UART functions.

Example 4.3

The serial port of a PC (e.g. COM1) is connected to a PIC32MX360F512L microcontroller and a terminal emulation software (e.g. **Hyperterminal**) is operated on the PC to send and receive data via the serial port. Pins RB0 and RB1 of the microcontroller are to be used as the RX and TX pins, respectively. The required baud rate is 9600. Write a program to read data from the terminal, then increase these data by one and send it back to the terminal. For example, if the user enters character "A", then character "B" should be displayed on the terminal. Assume that an MAX232-type voltage level converter chip is to be used to convert the microcontroller signals to RS232 levels. Figure 4.6 shows the circuit diagram of this example.

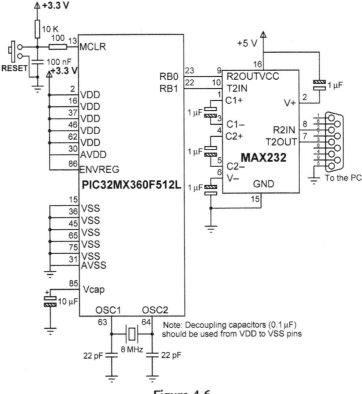

Figure 4.6
Circuit diagram of example 4.3.

```
/*****************************************************************************

                        READING AND WRITING TO SERIAL PORT
                        ==================================

This program uses soft UART to read and write serial data. PORT B pins RB0 and RB1 are
configured as serial RX and TX pins respectively. The baud rate is selected as 9600. A
character is received from a serial terminal, incremented by one and is then sent back to
the terminal. Thus, if for example, character "A" is received, character "B" will be sent to
the terminal.


Programmer:    Dogan Ibrahim
Date:          March, 2012
File:          SERIAL.C
*****************************************************************************/

void main()
{
        unsigned char Temp, Error;

        Soft_UART_Init(PORTB, 0, 1, 9600, 0);       // Configure serial port
        for(;;)
        {
                do
                {
                        Temp = Soft_UART_Read(&error);   // Read a byte
                } while(Error);
                Temp++;                                   // Increment byte
                Soft_UART_Write(Temp);                    // Send the byte
        }
}
```

Figure 4.7
Program listing of example 4.3.

Solution 4.3
MAX232 chip receives the TX signal from pin RB1 of the microcontroller and converts to RS232 levels. Similarly, the serial data received by the MAX232 chip is converted into micro-controller voltage levels and then sent to pin RB0. Notice that correct operation of the MAX232 chip requires four capacitors to be connected to the chip.

The required program listing is shown in Figure 4.7 (program SERIAL.C). At the beginning of the program, function **Soft_Uart_Init** is called to configure the serial port. Then an endless loop is formed using the **for** statement. **Soft_Uart_Read** function is called to read a character from the terminal. After reading a character, the data byte is incremented by one and then sent back to the terminal by calling function **Soft_Uart_Write**.

4.4 Hardware UART Library

Hardware UART library contains a number of functions to transmit and receive serial data using the UART modules built on the PIC microcontroller chips. Most PIC32 microcontrollers provide two UART modules. Hardware UART has the advantage over the software-implemented UART that in general higher baud rates can be obtained with the hardware UART and also the microcontroller can perform other tasks while data are received and sent to the UART.

mikroC Pro for PIC32 hardware UART library provides the following functions:

- UARTx_Init
- UARTx_Init_Advanced
- UARTx_Data_Ready
- UARTx_Tx_Idle
- UARTx_Read
- UARTx_Read_Text
- UARTx_Write
- UARTx_Write_Text
- UART_Set_Active

where "x" refers to the UART module used.

UARTx_Init function is used to initialize the hardware UART with the specified baud rate. The UART is configured with the following default settings:

- Default hardware TX and RX pins
- Loopback mode disabled
- Eight-bit data are used
- Parity bit is not used
- One stop bit is used
- An interrupt is generated on reception end and transmission end (if enabled).

This function should be called before any other UART functions are called. The only parameter required by this function is the baud rate. An example call to this function is given below which sets the baud rate of UART module 1—9600:

```
UART1_Init(9600);
```

The function *UARTx_Init_Advanced* is used to configure and initialize a UART module. The function has the following parameters:

```
Baud rate, bus clock frequency (in kHz), speed selection, parity, stop bits
```

The speed selection can take one of the following values:

```
Low speed:   _UART_LOW_SPEED
High speed:  _UART_HI_SPEED
```

The parity parameter is used to select the parity and the data width. Valid values are as follows:

```
8-bit data, no parity:      _UART_8BIT_NOPARITY
8-bit data, even parity:    _UART_8BIT_EVENPARITY
8-bit data, odd parity:     _UART_8BIT_ODDPARITY
9-bit data, no parity:      _UART_9BIT_NOPARITY
```

The stop bits can be one or two and is selected as follows:

```
One stop bit:  _UART_ONE_STOPBIT
Two stop bits: _UART_TWO_STOPBITS
```

The function *UARTx_Data_Ready* tests if data are available in the receive buffer. This function must be called after the UART module is initialized. The function has no parameters, but returns a byte to show whether or not data are available:

```
0:   there is no data in the receive buffer
1:   there is data in the receive buffer and it can be read
```

This function can be used as shown in the following code where data are read if it has already been received:

```
If(UART1_Data_Ready())
   Rec = UART1_Read();
```

The function *UARTx_Tx_Idle* is used to test whether or not the transmit shift register is empty or not. The UART module must be initialized before this function can be used. The function returns the following values:

```
0:   transmit shift register is not empty
1:   transmit shift register is empty
```

The status of the transmit shift register should be checked before a new byte is sent to UART. The following example shows how the status of the transmit shift register should be checked before sending a new byte to UART:

```
If(UART1_Tx_Idle() == 1)
{
   UART1_Write(new_data);
}
```

The function *UARTx_Read* is called to read a data byte from the UART module. The received byte is returned by the function. Function *UARTx_Data_Ready* should be called to test if data are ready before attempting to read data. This function must be called after the UART module has been initialized.

In the following example, UART is checked and if a data byte has been received, it is copied to variable **MyData**:

```
char MyData;
if(UART_Data_Ready( )) MyData = UART_Read();
```

The function UARTx_Read_Text is used to read a number of characters (text) until a delimiter is detected. The function has three parameters:

```
output, delimiter, attempts
```

The received text is stored in parameter *output*. The *delimiter* byte specifies the character at the end of the text. Parameter *attempts* defines the number of received characters, in which delimiter sequence is expected. If *attempts* is set to 255, the function will continuously receive characters until the delimiter character is detected.

The following example shows how a text delimited by characters "OK" can be received and stored in variable *MyTxt*:

```
UART1_Init(9600);

while(1)
{
    if(UART1_Data_Ready == 1)
    {
      UART1_Read_Text(MyTxt, "OK", 255);
    }
}
```

In the following example, again the text is delimited by characters "OK", but this time only 10 characters are read and the delimiter is expected to be inside these 10 characters:

```
UART1_Init(9600);

while(1)
{
    if(UART1_Data_Ready == 1)
    {
      UART1_Read_Text(MyTxt, "OK", 10);
    }
}
```

The function *UARTx_Write* sends a data byte to the UART and thus serial data are sent out of the UART. The UART module must be initialized before using this function. The data byte to be sent must be supplied as a parameter to the function. In the following example, character "A" is sent to the UART:

```
char Temp = 'A';
UART1_Write(Temp);
```

The function *UARTx_Write_Text* is used to send a number of characters (text) to the UART module. The text must be zero terminated (i.e. it must be a string). This function must be called after the UART module has been initialized. The following example shows how this function can be used:

```
UART1_Write_Text(output);
```

The function *UART_Set_Active* is available for microcontrollers with multiple UART modules and the function sets the active UART module that will be used in subsequent read and write operations. The used UART module must be initialized before this function is called. This function has the following parameters:

```
Read pointer, write pointer, ready pointer, tx_idle pointer
```

An example is given below:

```
UART_Set_Active(&UART1_Read, &UART1_Write, &UART1_Data_Ready,
               &UART1_Tx_Idle);
```

An example is given below to illustrate how the hardware UART functions can be used in a program.

Example 4.4

The serial port of a PC (e.g. COM1) is connected to a PIC32MX360F512L microcontroller and a terminal emulation software (e.g. Hyperterminal) is operated on the PC to use the serial port. Hardware USART pins U1RX/RF2 (USART receive pin, RX) and U1TX/RF3 (USART transmit pin, TX) of the microcontroller are connected to the PC via a MAX232-type RS232 voltage level converter chip. The required baud rate is 9600. Write a program to read data from the terminal, increase these data by one and then send it back to the terminal. For example, if the user enters character "A", then character "B" should be displayed on the terminal. Figure 4.8 shows the circuit diagram of this example.

Solution 4.4

The required program listing is shown in Figure 4.9 (program SERIAL2.C). At the beginning of the program, function **UART1_Init** is called to initialize UART1 and also set the baud rate to 9600. Then an endless loop is formed using the **for** statement. Inside this loop, function

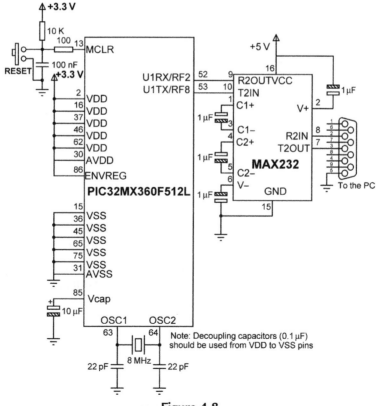

Figure 4.8
Circuit diagram of example 4.4.

```
/**************************************************************************

                    READING AND WRITING TO SERIAL PORT
                    ==================================

This program uses hardware UART to read and write serial data. In this program the first UART
Is used where the receive (RX) pin is U1RX/RF2, and the transmit pin (TX) is U1TX/RF3.

The baud rate is selected as 9600. A character is received from a serial terminal, incremented by
one and is then sent back to the terminal. Thus, if for example, character "A" is received, character
"B" will be sent to the terminal.



Programmer:    Dogan Ibrahim
Date:          March, 2012
File:          SERIAL2.C
**************************************************************************/

void main()
{
        unsigned char Temp, Error;

        UART1_Init(9600);                              // Initialise and set baud rate
        for(;;)                                        // DO FOREVER
        {
                if(UART1_Data_Ready())
                {
                        Temp = UART1_Read();           // Read a byte
                        Temp++;                        // Increment byte
                        UART1_Write(Temp);             // Send the byte
                }
        }
}
```

Figure 4.9
Program listing of example 4.4.

UART1_Data_Ready is called to check if a character is available at the UART buffer and if so, this character is read by calling function **UART1_Read**. The read character is then incremented by one and then sent back to the terminal by calling function **UART1_Write**.

In PIC microcontrollers where there is more than one UART, the second UART is accessed by appending a "2" to the end of keyword UART, e.g. **UART2_Write**, **UART2_Read**, etc.

4.5 Sound Library

The mikroC PRO for PIC32 provides a sound library that includes functions enabling users to generate various sounds in their applications. Sound generation needs additional hardware, such as piezospeaker, which should be connected to the required I/O ports of the microcontroller.

Two functions are offered in the sound library:

- Sount_Init
- Sound_Play

The *Sound_Init* function is used to initialize the sound library and this function must be called before playing a sound. The function requires two parameters to define the interface between the sound generating device and the microcontroller. These are the port address and the output pin number. For example, assuming that the sound generating device is connected to pin 3 of PORT D, the initialization function should be called as follows:

```
Sound_Init(&PORTD, 3);
```

The *Sound_Play* function generates a square wave signal on the specified pin. The function requires two arguments (both type **unsigned int**): the required signal frequency (in Hertz) and the signal duration (in milliseconds). For example, the following code generates 1 KHz sound with the duration of 200 ms:

```
Sound_Play(1000, 200);
```

Example 4.5

Table 4.5 shows the frequencies of musical notes in middle octave. Write a program to generate all the notes with duration 250 and 50 ms gap between each note. Assume that a piezospeaker is connected to port pin RB7 of the microcontroller as shown in Figure 4.10. Use only the integer values of the frequencies.

Table 4.5: Frequencies of Musical Notes

Musical Note	Frequency
C_4	261.63
D_4	293.66
E_4	329.63
F_4	349.23
G_4	392.00
A_4	440.00
B_4	493.88
C_4	523.25

Solution 4.5

The program listing is shown in Figure 4.11 (program MUSIC.C). Array **Notes** stores the integer frequencies of the notes. At the beginning of the program, the analog inputs are configured as digital and PORT B direction is set as output. A **for** loop is then formed to send all the notes to the piezospeaker connected to port pin RB7.

4.6 ANSI C Library

ANSI C library consists of the following libraries. The functions have been implemented according to the ANSI C standard; however, some functions have been modified for reasons

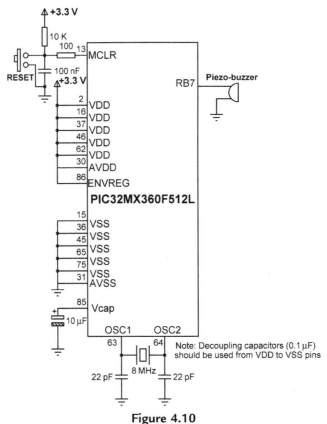

Figure 4.10
Connecting a piezospeaker to an I/O port pin.

of compatibility with the mikroC PRO for PIC32 language. Further details can be obtained from the language user manual:

- Ctype library
- Math library
- Stdlib library
- String library

4.6.1 Ctype Library

The functions in this library are mainly used for testing a character, or data conversion. Table 4.6 gives a list of the commonly used functions in this library. For example, function *isupper* is used to check whether or not a given character is upper case.

4.6.2 Math Library

The functions in this library are used for floating point trigonometric and mathematical operations. Table 4.7 gives a list of the commonly used functions in this library. For example,

```
/****************************************************************************

                    GENERATING MUSICAL NOTES
                    =========================

In this program a piezospeaker is connected to port pin RB7 of the microcontroller. The
program generates the musical notes of the middle octave. Each note is played for 250 ms
and 50 ms gap is inserted between each output.


Programmer:    Dogan Ibrahim
Date:          March, 2012
File:          MUSIC.C
****************************************************************************/

void main()
{
        unsigned int Notes[] = {261, 293, 329, 349, 392, 440, 493, 523};
        unsigned char i = 0;

        AD1PCFG = 0xFFFF;                          // Configure analog ports as digital
        TRISB = 0;                                 // Configure PORT B as output

        Sound_Init(&PORTB, 7);                     // Initialize sound library

        for(;;)                                    // DO FOREVER
        {

                Sound_Play(Notes[i], 250);         // Play a note
                Delay_Ms(50);                      // Wait 50 ms
                i++;                               // Next note
                if(i == 8)i = 0;                   // If end of notes
        }
}
```

Figure 4.11
Program listing of example 4.5.

Table 4.6: Commonly Used Ctype Library Functions

Function	Description
isalnum	Returns 1 if the specified character is alphanumeric (a–z, A–Z, and 0–9)
isalpha	Returns 1 if the specified character is alphabetic (a–z and A–Z)
isntrl	Returns 1 if the specified character is a control character (decimal 0–31, and 127)
isdigit	Returns 1 if the specified character is a digit (0–9)
islower	Returns 1 if the specified character is lower case
isprint	Returns 1 if the specified character is printable (decimal 32–126)
isupper	Returns 1 if the specified character is upper case
toupper	Convert a character to upper case
tolower	Convert a character to lower case

Table 4.7: Commonly Used Math Library Functions

Function	Description
acos	Returns in radians the arc cosine of its parameter
asin	Returns in radians the arc sine of its parameter
atan	Returns in radians the arc tangent of its parameter
atan2	Returns in radians the arc tangent of its parameter where the signs of both parameters are used to determine the quadrant of the result
cos	Returns the cosine of its parameter in radians
cosh	Returns the hyperbolic cosine of its parameter
exp	Returns the exponential of its parameter
fabs	Returns the absolute value of its parameter
log	Returns the natural logarithm of its parameter
log10	Returns the logarithm to base 10 of its parameter
pow	Returns the power of a number
sin	Returns the sine of its parameter in radians
sinh	Returns the hyperbolic sine of its parameter
sqrt	Returns the square root of its parameter
tan	Returns the tangent of its parameter in radians
tanh	Returns the hyperbolic sine of its parameter

function *tan* returns the trigonometric tangent of an angle where the angle must be specified in radians.

4.6.3 Stdlib Library

The functions in this library are standard library functions. Table 4.8 gives a list of the commonly used functions in this library. For example, function *abs* returns the absolute value of a variable.

Table 4.8: Commonly Used Stdlib Library Functions

Function	Description
Abs	Returns the absolute value
Atof	Converts ASCII character into floating point number
atoi	Converts ASCII character into integer number
atol	Converts ASCII character into long integer
max	Returns the greater of two integers
min	Returns the lower of two integers
rand	Returns a random number between 0 and 32767. Function srand must be called to obtain a different sequence of numbers
srand	Generates a seed for function rand so that a new sequence of numbers are generated
xtoi	Convert input string consisting of hexadecimal digits into integer

```
/***********************************************************

        TRIGONOMETRIC TANGENT OF ANGLES 0 to 45 DEGREES
        ==================================================

        This program calculates the trigonometric tangent of angles from 0° to
        45° in steps of 1°. The results are stored in an array called
        Trig_Tangent.

        Programmer:  Dogan Ibrahim
        File:        TANGENT.C
        Date:        May, 2012
***********************************************************/

void main()
{
        unsigned char j;
        float Trig_Tangent[46];
        float PI = 3.14159, rads, angle;

        for(j = 0; j <= 45; j++)
        {
                rads = j * PI/180.0;
                angle = tan(rad);
                Trig_Tangent[j] = angle;
        }
}
```

Figure 4.12
Calculating the tangent of angles $0°-45°$.

Example 4.6

Write a program to calculate the trigonometric tangent of the angles from $0°$ to $45°$ in steps of $1°$ and store the result in an array called **Trig_Tangent**.

Solution 4.6

The required program listing is shown in Figure 4.12 (program TANGENT.C). A loop is created using the **for** statement and inside this loop the tangent of the angles are calculated and stored in array **Trig_Tangent**. Note that the angles must be converted into radians before using in function **tan**.

Table 4.9: Commonly Used String Library Functions

Function	Description
memchr	Locate the first occurrence of a character
memcmp	Compare first n characters
memcpy	Copy first n characters
memmove	Copy first n characters
memset	Copy a character into many locations
strcat, strncat	Append two strings
strchr, strpbrk	Locate the first occurrence of a character in a string
strcmp, strncmp	Compare two strings
strcpy, strncpy	Copy one string into another one
strlen	Returns the length of a string

4.6.4 String Library

The functions in this library are used to perform string and memory manipulation operations. Table 4.9 gives a list of the commonly used functions in this library. For example, function *strlen* returns the length of a string. Further details on these functions can be obtained from the mikroC PRO for PIC32 user manual.

Example 4.7

Write a program to illustrate how the two strings "LAPTOP" and "COMPUTER" can be joined into a new string using string library functions.

```
/***********************************************************

                    JOINING TWO STRINGS
                    ===================


This program shows how two strings can be joined to obtain a new string.
mikroC Pro for PIC32 library function strcat is used to join the two strings pointed to
by p1 and p2 into a new string stored in character array New_String.



Programmer:  Dogan Ibrahim
File:        JOIN.C
Date:        May, 2012
***********************************************************/

void main()
{
        const char *p1 = "Laptop";            // First string
        const char *p2 = "COMPUTER";          // Second string
        char New_String[80];

        strcat(strcat(New_String, p1), p2);   // join the two strings
}
```

Figure 4.13
Joining two strings using function strcat.

Solution 4.7
The required program listing is shown in Figure 4.13 (program JOIN.C). mikroC PRO for PIC32 string library function **strcat** is used to join the two strings pointed to by p1 and p2 into a new string stored in character array called **New_String**.

4.7 Miscellaneous Library

Functions in the miscellaneous library consist of routines such as button operations, data conversions, time library, and so on. A list of the libraries inside the miscellaneous library is the following:

- Button library
- Conversion library

- PrintOut library
- SetJmp library
- Sprint library
- Time library
- Trigonometry library.

A description of the important library functions is described in this section.

4.7.1 Button Library

There is only one function in this library called *Button*. This function eliminates the influence of contact flickering (debouncing) when a button is pressed. The port pin where the button is connected to is tested just after the function call and then again after the debouncing period has expired. If the pin was in the active state in both cases, the function returns 255 (i.e. true).

The button function requires four arguments:

```
Port name where the button is connected to
Pin number of the port where the button is connected to
Debouncing period in milliseconds
Active state (0 or 1)
```

In the following example, it is assumed that a button is connected to port pin 0 of PORT D. The debouncing period is set to 2 ms, and the active button state is assumed to be logic 1:

```
Button(&PORTD, 0, 2, 1);
```

Example 4.8

A button is connected to port pin RD0 (PORT D, bit 0) of a microcontroller. In addition, an LED is connected to port pin RB7 (PORT B, bit 7). Write a program which will toggle the LED when the button is pressed and then released. Assume a 2 ms debouncing period. Figure 4.14 shows the circuit diagram for this example.

Solution 4.8

The required program listing is shown if Figure 4.15 (BUTTON.C). Port pin RD0 is normally at logic 0 and goes to logic 1 when the button is pressed. The program detects when the button is pressed and also when the button is released. The LED is toggled as soon as the button is released.

4.7.2 Conversion Library

The conversion library includes functions to convert different data types to strings, to hexadecimal numbers, and to trim variables. The library provides the functions listed in Table 4.10.

Some examples are given below to illustrate how these functions can be used in programs.

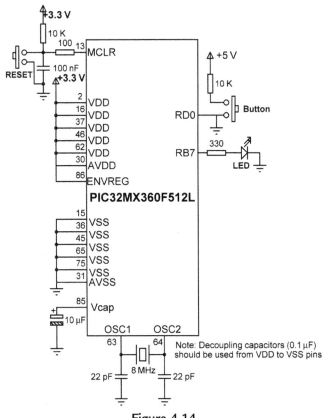

Figure 4.14
Circuit diagram of example 4.8.

Example 4.9

An LCD is connected to a microcontroller PORT B as shown in Figure 4.2. Write a program to count up starting from zero with 1 s delay and display the results on the first row of the LCD. Use function *ByteToStr* to convert the byte to be displayed into a string before sending it to the LCD.

Solution 4.9

The required program listing is shown in Figure 4.16 (UPCNT.C). At the beginning of the program, all analog ports are configured as digital. PORT B pins are then configured as output. Variable *Cnt* is used as the counter. A **for** loop is established where inside the loop *Cnt* is converted into a string called *Txt* and is sent to the LCD. *Cnt* is then incremented by one and the above process repeats after 1 s delay.

```
/***********************************************************

                USING THE BUTTON FUNCTION
                ========================

This program shows how the button function can be used. In this program a
button is connected to port pin RD0. In addition, an LED is connected to port
pin RB7. The LED is OFF when the program is started. The state of the LED
is toggled when the button is pressed and then released.

The button function eliminates the contact debounce problems associated with
mechanical buttons.


Programmer:  Dogan Ibrahim
File:        BUTTON.C
Date:        May, 2012
***********************************************************/

void main()
{
      unsigned char old = 0;

      ADPCFG = 0xFFFF;                // Make all ports digital
      PORTB = 0;                      // Make PORT B output
      TRISD = 1;                      // Make RD0 as input

      LATB = 0;                       // Turn OFF LED to start with

      do
      {
            if(Button(&PORTD, 0, 2, 1))        // Detect 0 to 1 (Pressed)
                  old = 1;
            if(old && Button(&PORTD, 0, 2, 0)) // Detect 1 to 0 (Released)
            {
                  old = 0;
                  LATB = ~LATB;
            }
      } while(1);
}
```

Figure 4.15
Program listing of example 4.8.

Example 4.10

Repeat Example 4.9, but display the number in hexadecimal format.

Solution 4.10

The required program listing is shown in Figure 4.17 (HEX.C). The number to be displayed is
converted into hexadecimal format using function *ByteToHex*.

Table 4.10: Functions in the Conversion Library

Function	Description
ByteToStr	Convert a byte into string
ShortToStr	Convert a short into string
WordToStr	Convert a word into string
IntToStr	Convert an integer into string
LongToStr	Convert a long into string
LongWordToStr	Convert a long word into string
FloatToStr	Convert a float into string
WordToStrWithZeros	Convert a word into string, fill with zeros
IntToStrWithZeros	Convert an integer into string, fill with zeros
LongWordToStrWithZeros	Convert a long word into string, fill with zeros
LongIntToStrWithZeros	Convert a long integer into string, fill with zeros
ByteToHex	Convert a byte into string in hexadecimal format
ShortTohex	Convert a short into string in hexadecimal format
WordToHex	Convert a word into string in hexadecimal format
IntToHex	Convert an integer into string in hexadecimal format
LongWordToHex	Convert a long word into string in hexadecimal format
LongIntToHex	Convert a long integer into string in hexadecimal format
Dec2Bcd	Convert a short integer into BCD
Bcd2Dec	Convert a BCD number into a short integer
Dec2Bcd16	Convert 16-bit decimal value into BCD
Bcd2Dec16	Convert a BCD number into 16-bit decimal value
Rtrim	Trim the trailing spaces from an array
Ltrim	Trim the leading spaces from an array

4.7.3 Time Library

The time library contains function for time calculations in the UNIX format, which counts the number of seconds since the "epoch". These functions are useful for programs that wish to use time intervals. An "epoch" is defined as the beginning of January 1, 1970 GMT. Time library makes use of a structure defined in file "_Time.h" which should be included at the beginning of a program.

The time library provides the following functions:

- Time_dateToEpoch
- Time_epochToDate
- Time_dateDiff.

Function *Time_dateToEpoch* returns the number of seconds since January 1, 1970 (UNIX epoch time) midnight. The function requires the time structure to be declared in its argument. In the following example, the epoch time is returned in long variable called **Ts**:

```
#include "_Time.h"
TimeStruct Tm;
```

```
/*****************************************************************************

                        COUNT UP ON LCD
                        ===============

In this program an LCD is connected to PORT B of a microcontroller. The program counts up
by one on the LCD with one second delay between each output. The data is displayed on the
first roe of the LCD.



Programmer:   Dogan Ibrahim
Date:         March, 2012
File:         UPCNT.C
*****************************************************************************/
//
// Define LCD interface
//
sbit LCD_RS at LATB2_bit;
sbit LCD_EN at LATB3_bit;
sbit LCD_D4 at LATB4_bit;
sbit LCD_D5 at LATB5_bit;
sbit LCD_D6 at LATB6_bit;
sbit LCD_D7 at LATB7_bit;
//
// LCD pin directions
//
sbit LCD_RS_Direction at TRISB2_bit;
sbit LCD_EN_Direction at TRISB3_bit;
sbit LCD_D4_Direction at TRISB4_bit;
sbit LCD_D5_Direction at TRISB5_bit;
sbit LCD_D6_Direction at TRISB6_bit;
sbit LCD_D7_Direction at TRISB7_bit;

void main()
{
        unsigned char Cnt = 0;
        unsigned char Txt[4];

        ADPCFG = 0xFFFF;                     // Configure all ports as digital
        TRISB = 0;                           // Configure PORT B as output
        Lcd_Init();                          // Initialize LCD

        for(;;)                              // DO FOREVER
        {
                ByteToStr(Cnt, Txt);         // Convert Cnt into string
                Lcd_Out(1, 1, Txt);          // Display on LCD
                Cnt++;                       // Increment Cnt
                Delay_Ms(1000);              // Wait 1 second
        }
}
```

Figure 4.16
Program listing of example 4.16.

```
/*******************************************************************************

                        COUNT UP ON LCD AND DISPLAY IN HEX
                        =================================
```

In this program an LCD is connected to PORT B of a microcontroller. The program counts up
by one on the LCD with one second delay between each output. The data is displayed on the
first row of the LCD in hexadecimal format.

```
Programmer:    Dogan Ibrahim
Date:          March, 2012
File:          HEXDISP.C
*******************************************************************************/
//
// Define LCD interface
//
sbit LCD_RS at LATB2_bit;
sbit LCD_EN at LATB3_bit;
sbit LCD_D4 at LATB4_bit;
sbit LCD_D5 at LATB5_bit;
sbit LCD_D6 at LATB6_bit;
sbit LCD_D7 at LATB7_bit;
//
// LCD pin directions
//
sbit LCD_RS_Direction at TRISB2_bit;
sbit LCD_EN_Direction at TRISB3_bit;
sbit LCD_D4_Direction at TRISB4_bit;
sbit LCD_D5_Direction at TRISB5_bit;
sbit LCD_D6_Direction at TRISB6_bit;
sbit LCD_D7_Direction at TRISB7_bit;

void main()
{
        unsigned char Cnt = 0;
        unsigned char Txt[3];

        ADPCFG = 0xFFFF;                    // Configure all ports as digital
        TRISB = 0;                          // Configure PORT B as output
        Lcd_Init();                         // Initialize LCD

        for(;;)                             // DO FOREVER
        {
                ByteToHex(Cnt, Txt);        // Convert Cnt into hex string
                Lcd_Out(1, 1, Txt);         // Display on LCD
                Cnt++;                      // Increment Cnt
                Delay_Ms(1000);             // Wait 1 second
        }
}
```

Figure 4.17
Program listing of example 4.10.

```
long Ts;
Ts = Time_dateToEpoch(&Tm);
```

Function *Time_epochToDate* converts UNIX epoch time to current date and time. The UNIX time and the current date and time structure should be passed as arguments to the function. In the following example, the epoch time "1234567890" is converted into current date and time and is stored in structure **Tm**:

```
#include "_Time.h"
TimeStruct Tm;
long Ts;
Ts = 1234567890;
Time_epochToDate(Ts, &Tm);
```

Function *Time_dateDiff* compares two dates and returns the time difference as a signed long. The function requires two time structures T1 and T2 to be passed as arguments during the call. The result is positive if T1 is before T2, null if the two dates are equal, and negative if T1 is after T2. The following example shows how the difference (in seconds) between two date structures T1 and T2 can be calculated and stored in a variable called **diff**:

```
#include "_Time.h"
TimeStructure T1, T2;
long diff;
diff = Time_dateDiff(&T1, &T2);
```

4.7.4 Trigonometry Library

The functions in the trigonometry library implement the fundamental trigonometry functions. These functions are implemented as look-up tables and the results are in integer format in order to save memory.

The functions in this library are the following:

- sinE3
- cosE3

Function *sinE3* calculates the sine of an angle, multiplied by 1000 and rounded to the nearest integer number. The angle must be passed as an argument in degrees. The following example shows how the sine of angle 45° can be calculated using this function. Since the sine of angle 45° is 0.707, the result returned by the function is 707.

```
int result;
result = sinE3(45);          // 707 is returned
```

Function *cosE3* is similar, but the cosine of the angle is multiplied by 1000, rounded to the nearest integer number and returned.

4.8 Summary

This chapter has explained the important topic of built-in library functions. It is shown that these functions can be very useful as they have already been tested by the software developers, and are readily available to the users. Built-in functions are particularly useful in peripheral-based programs as they simplify and speed up the program development time. Examples are given in this chapter to show how the important built-in functions can be used in user programs.

4.9 Exercises

1. A text-based LCD is connected to a PIC microcontroller in four-bit data mode. Write a program that will display a count from 0 to 255 on the LCD with 1 s interval between each count.

2. A text-based LCD is connected to a PIC microcontroller as in Exercise 1. Write a program to display the text "Exercise 2" on the first row of the LCD.

3. Repeat Exercise 2 but display the message on the first row, starting from column 3 of the LCD.

4. A two-row text-based LCD is connected to a microcontroller in four-bit data mode. Write a program to display text "COUNTS:" on row 1, and then to count repeatedly from 1 to 100 on row 2 with 2 s intervals.

5. Write a program to calculate the trigonometric cosine of angles from 0° to 45° in steps of 1° and store the results in a floating point array.

6. Write a function to calculate and return the length of the hypotenuse of a right-angle triangle, given its two sides. Show how you can use the function in a main program to calculate the hypotenuse of a right angle triangle whose two sides are 4.0 and 5.0 cm.

7. Write a program to configure port pin RB2 of a microcontroller as the RS232 serial output port. Send character "X" to this port at 4800 baud.

8. Port RB0 of a microcontroller is configured as the RS232 serial output port. Write a program to send out string "SERIAL" at 9600 baud.

9. Repeat Exercise 8 but use the hardware UART available on the microcontroller chip.

10. Explain the differences between the software-implemented serial data communication, and UART hardware-based serial communication.

11. Write a program to count up in hexadecimal format from 0 to 100 and display the result on an LCD. Insert a 2 s delay between each output.

12. Repeat Exercise 11 but remove the leading spaces and zeros.

13. Write a function to convert the string pointed to by p into lowercase or uppercase depending on the value of a mode parameter passed to the function. If the mode parameter is nonzero, then convert to lowercase, otherwise to uppercase. The function should return a pointer to the converted string.

14. Write a program to generate a table of the trigonometric functions sine and cosines, from 0° to 90°, in steps of 0.1°. Store the results in two arrays called SINS and COSINES.

15. A projectile is fired at an angle of θ degrees at an initial velocity of v meters per second. The distance travelled by the projectile (d), the initial flight time (t), and the maximum height reached (h) are given by the following formulas, respectively:

$$h = \frac{v^2 \sin(\theta)}{g} \qquad t = \frac{2v\sin(\theta)}{g} \qquad d = \frac{v^2 \sin(2\theta)}{g}$$

Write functions to calculate the height, flight time, and distance travelled. Assume that $g = 9.81$ m/s^2 and $\theta = 45°$, call the functions from a main program to calculate the three variables.

PIC32 Microcontroller Development Tools

The development of a microcontroller-based system is a complex process. Development tools are hardware and software tools which help programmers to develop and test systems in a relatively short time.

Developing software and hardware for microcontroller-based systems involves the use of editors, assemblers, compilers, debuggers, simulators, emulators, and device programmers. A typical development cycle starts with writing the application program using a text editor. The program is then translated into the executable code by using an assembler or a compiler. If the program consists of several modules, these are combined together into a single application program using a linker. At this stage, any syntax errors are detected by the assembler or the compiler and they have to be corrected before an executable code can be generated. In the next stage of the development cycle, a simulator can be used to test the application program without the actual hardware. Simulators can be useful to test the correctness of an algorithm or a program with limited or no input—outputs. Most of the errors can be removed during the simulation. When the programmer is happy and the program seems to be working, the next stage of the development cycle is to load the executable code to the target microcontroller chip using a device programmer and then test the overall system logic. During this cycle, software and hardware tools such as in-circuit debuggers or in-circuit emulators can be used to analyze the operation of the program, and to display the variables and registers in real time by setting breakpoints in the program.

Development tools for microcontrollers can be classified into two categories: software and hardware. There are many such tools and the discussion of all these tools is beyond the scope of this book. In this chapter, only the commonly used tools are reviewed briefly.

5.1 Software Development Tools

Software development tools are basically computer programs and they usually run on personal computers, helping the programmer (or system developer) to create and/or modify, or test applications programs. Some of the commonly used software development tools are as follows:

- Text editors
- Assemblers/compilers
- Simulators.

5.1.1 Text Editors

A text editor is a program that allows us to create or edit programs and text files. Windows operating system is distributed with a text editor program called *Notepad*. Using the Notepad, we can create a new program file, modify an existing file, or display or print the contents of a file. It is important to realize that programs used for word processing, such as the *Word*, cannot be used as a text editor. This is because word processing programs are not true text editors as they embed word formatting characters such as bold, italic, underline, and so on inside the text.

Most assemblers and compilers have built-in text editors. Using these editors, we can create our program and then assemble or compile it without having to exit from the editor environment. These editors also provide additional features, such as automatic keyword highlighting, syntax checking, parenthesis matching, comment line identification, and so on. Different parts of a program can be shown in different colors to make the program more readable. For example, comments can be shown in one color, keywords in another color, conditional statements in a different color, and so on. These features can speed up the program development process since most syntax errors can be eliminated during the programming stage.

5.1.2 Assemblers and Compilers

Assemblers generate executable code from assembly language programs. The generated code is usually loaded into the flash program memory of the target microcontroller.

Similarly, compilers generate executable code from high-level language programs. Some of the commonly used compilers for the PIC32 microcontrollers are BASIC, C, and PASCAL.

Assembly language is used in applications where the processing speed is very critical and the microcontroller is required to respond to external and internal events in the shortest possible time. The main disadvantage of assembly language is that it is difficult to develop complex programs using this language. Also, assembly language programs cannot be maintained easily. High-level languages on the other hand are easier to learn and complex programs can be developed and tested in a much shorter time. The maintenance of high-level programs is also much easier than the maintenance of assembly language programs.

In this book only programming using the C language will be described. There are a few versions of C language compilers available for developing PIC32 microcontroller-based programs. (Most microcontroller C compilers are for 8-, 16-, and 24-bit processors.) Some of the popular ones are the following:

- Hi-Tech PIC32 C compiler (http://htsoft.com)
- MPLAB C32 Compiler (http://www.microchip.com)
- mikroC Pro for PIC32 C compiler (http://www.mikroe.com).

Although most C compilers are essentially the same, each one has its own additions or modifications to the standard language. The C compiler used in this book is the mikroC Pro for PIC32, developed by mikroElektronika.

5.1.3 Simulators

A simulator is a computer program that runs on a PC without any microcontroller hardware and it simulates the behavior of the target microcontroller by interpreting the user program instructions using the target microcontroller instruction set. Simulators can display the contents of registers, memory, and the status of input—output ports of the target microcontroller as the user program is interpreted. The user can set breakpoints to stop execution of the program at desired locations and then examine the contents of various registers at the breakpoint. In addition, the user program can be executed in a single-step mode and the memory and registers can be examined as the program executes a single instruction each time a key is pressed. One problem associated with standard simulators is that they are only software tools and any hardware interface is not simulated.

Most microcontroller language development tools also incorporate some form of simulators.

5.1.4 High-Level Language Simulators

These are also known as source-level debuggers, and like simulators, they are programs which run on a PC. A source-level debugger allows us to find the errors in our high-level programs. We can set breakpoints in high-level statements, execute the program up to the breakpoint and then display the values of program variables, the contents of registers, and memory locations at the breakpoint. For example, we can stop a program execution and examine (or modify) the contents of an array.

A source-level debugger can also invoke hardware-based debugging activity using a hardware debugger device. For example, the user program on the target microcontroller can be stopped and the values of various variables and registers can be examined.

Some of the high-level language compilers also include built-in source-level debuggers. The following are some of the popular C compilers which also include source-level debuggers:

- MPLAB C32 C compiler
- Hi-Tech PIC32 C Compiler
- MikroC Pro for PIC32 C compiler.

5.1.5 Simulators with Hardware Simulation

Some simulators (e.g. Labcenter Electronics VSM, http://www.labcenter.com) incorporate hardware simulation options where various software-simulated hardware devices can be connected to microcontroller I/O pins. For example, an electromagnetic motor software module can be connected to an I/O port and its operation can be simulated in software. Although the hardware simulation does not simulate the actual device exactly, it is very useful during early project development cycle.

5.1.6 *Integrated Development Environment*

Integrated Development Environments (IDEs) are powerful PC-based programs which include everything to edit, assemble, compile, link, simulate, source-level debug, and download the generated executable code to the physical microcontroller chip (using a programmer device). These programs are in the form of graphical user interface where the user can select various options from the program without having to exit the program. IDEs can be extremely useful during the development phases of microcontroller-based systems. Most PIC32 high-level language compilers are in the form of an IDE, thus enabling the programmer to do most tasks within a single software development tool.

5.2 Hardware Development Tools

There are numerous hardware development tools available for the PIC32 microcontrollers. Some of these products are manufactured by Microchip Inc., and some others by third-party companies. The popular hardware development tools are the following:

- Development boards
- Device programmers
- In-circuit debuggers
- In-circuit emulators
- Breadboards.

5.2.1 *Development Boards*

The development boards are invaluable microcontroller development tools. Simple development boards contain just a microcontroller and the necessary clock circuitry. Some sophisticated development boards contain LEDs, LCD, push-buttons, serial ports, USB port, power supply circuit, device programming hardware, etc.

In this section, we shall be looking at the specifications of some of the commercially available PIC32 microcontroller development boards.

PIC32 Starter Kit

This board (Figure 5.1) is manufactured by Microchip Inc. and can be used in PIC32 microcontroller-based project development. The kit includes everything to write, compile, program, debug, and execute a program.

The kit contains the PIC32 Starter Kit board, and a USB Mini-B cable.

Figure 5.1
PIC32 starter kit.

The board contains the following:

- PIC32MX360F512L 32-bit microcontroller
- Regulated power supply (+3.3 V) for powering via the USB port
- Processor running at 72 MHz
- On-board debugging
- Debug and power on LEDs
- Three push-button switches for user inputs
- Three LED indicators
- Connectors for I/O ports
- Interface to the I/O expansion board.

Thirty-five example programs are provided with the kit. Users can download a free MPLAB C32 compiler with limited functionality from company's website and use the compiler in their projects.

The I/O expansion board (Figure 5.2) provides full access to all the microcontroller I/O signals. Additional daughter boards can be attached to the expansion board for added functionality.

Microstick II

Microstick II from Microchip Inc. is a small (the size of a stick of gum), low-cost development board (Figure 5.3), designed for small applications with a few I/O port requirements. The kit is supplied with the following:

- Microstick II board
- USB cable
- PIC32MX250F128 microcontroller (in addition, PIC24 and dsPIC33 microcontrollers are also included)

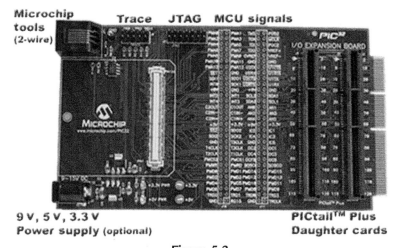

Figure 5.2
I/O expansion board.

Figure 5.3
Microstick II.

- Integrated USB programmer/debugger
- LED and reset button
- Pin headers for I/O access.

The board is distributed with free demo programs.

PIC32 USB Starter Kit II

The PIC32 USB Starter Kit II (Figure 5.4) is a low-cost PIC32 microcontroller development board with USB and Controller Area Network (CAN) functionality. Users can develop USB- and CAN-based applications easily using this kit.

The kit has the following features:

- PIC32MX795F512L 32-bit microcontroller
- On-board crystal

Figure 5.4
PIC32 USB starter kit II.

- USB for on-board programming/debugging
- Three push-button switches for user inputs
- Three LED indicators
- Debug and power LEDs
- Regulated power supply
- I/O connector for various expansion boards.

PIC32 Ethernet Starter Kit

The PIC32 Ethernet Starter Kit (Figure 5.5) is a low-cost 10/100 Ethernet development kit manufactured by Microchip Inc., using the PIC32 microcontroller.

The board has the following features:

- PIC32MX795F512L 32-bit microcontroller
- Thirty-two-bit microcontroller for on-board programming/debugging
- On-board crystal
- Ethernet oscillator
- Three push-button switches for user input
- Three LED indicators
- Debug and power supply LEDs
- RJ-45 Ethernet port
- Connector for various expansion boards.

Figure 5.5
PIC32 ethernet starter kit.

Cerebot MX3cK

The Cerebot MX3cK (Figure 5.6) is a 32-bit microcontroller development board based on PIC32MX320F128H, and is manufactured by Digilent (www.digilentinc.com). The kit is low cost and contains everything needed to start developing embedded applications based on 32-bit PIC microcontrollers using the MPIDE IDE. In order to use MLAB IDE, a programming/debugging device is required.

Figure 5.6
Cerebot MX3cK.

The kit has the following features:

- PIC32MX320F128H 32-bit microcontroller
- 80 MHz maximum operating frequency
- 42 I/O pins
- 12 analog inputs
- Programmed using MPIDE or MPLAB IDE
- Pmod headers for I/O signals
- I^{2}C connector
- Powered via USB port or using an external supply

Figure 5.7 shows the functional blocks of the Cerebot MX3cK development board.

Cerebot MX4cK

The Cerebot MX4cK (Figure 5.8) is a more advanced version of the Cerebot MK3cK with bigger printed circuit board (PCB) area and more functionality.

The kit has the following features:

- PIC32MX460F512L 32-bit microcontroller
- Pmod headers for I/O ports
- 2 × I^{2}C ports
- 1 × SPI port

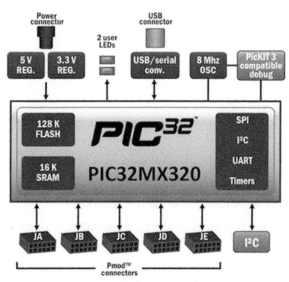

Figure 5.7
Cerebot MX3cK functional blocks.

Figure 5.8
Cerebot MX4cK.

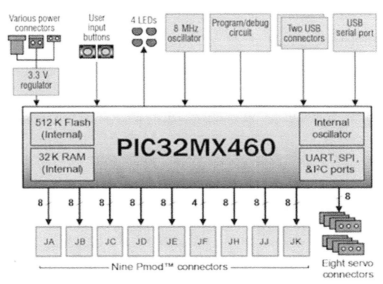

Figure 5.9
Cerebot MX4cK functional blocks.

- 8 × servo ports
- USB debug/programming port (for MPLAB IDE)
- USB port for debug/programming (for MPIDE IDE).

Figure 5.9 shows the functional blocks of the Cerebot MX4cK development board.

Cerebot MX7cK

The Cerebot MX7cK (Figure 5.10) is the most advanced version of the Digilent MX series of 32-bit development boards.

Figure 5.10
Cerebot MX7cK.

The kit has the following features:

- PIC32MX795F512L 32-bit microcontroller
- RJ-45 Ethernet port
- 2 × I^{2}C ports
- 1 × SPI port
- 2 × CAN ports
- 2 × SPI/UART ports
- 1 × USB UART port
- 2 × USB ports
- Pmod headers for I/O pins.

Figure 5.11 shows the functional blocks of the Cerebot MX4cK development board.

MINI-32 Board

This is a small development board (Figure 5.12) manufactured by mikroElektronika (www.mikroe.com) that contains a PIC32MX534F064H 32-bit microcontroller. The board operates with 3.3 V power supply and on-board regulator allows the board to be powered from a USB port.

The features of this board are the following:

- PIC32MX534F064H 32-bit microcontroller
- On-board crystal
- I/O pins at the edges
- Supports CAN communication

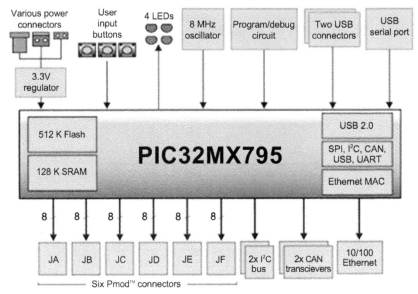

Figure 5.11
Cerebot MX4cK functional blocks.

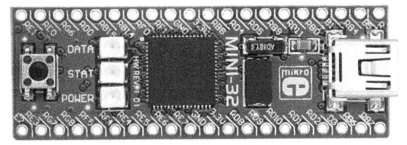

Figure 5.12
MINI-32 board.

- Comes with preprogrammed USB bootloader
- Fully supported by mikroElectronika mikroC Pro for PIC32 compiler.

EasyPIC Fusion V7

The EasyPIC Fusion V7 (Figure 5.13) combines support for three 16-bit and 32-bit different PIC microcontroller architectures: dsPIC33, PIC24, and PIC32 in a single development board.

The board has the following features:

- PIC32MX795F512L 32-bit microcontroller module
- On-board programmer (mikroProg)

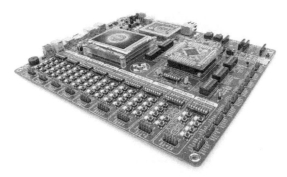

Figure 5.13
EasyPIC Fusion V7.

- On-board in-circuit-debugger (mikroICD)
- Sixty-eight push-button switches
- Sixty-eight LED indicators
- CAN support
- USB support
- Piezo-buzzer
- LM35/DS1820 temperature sensor sockets
- RJ-45 Ethernet connector
- I^2C EEPROM
- Serial flash memory
- Stereo MP3 Codec
- $2 \times$ mikroBUS sockets
- Audio in and out jack sockets
- mikroSD card slot
- TFT color display socket
- I/O port headers
- 3.3 V power supply (can be powered from USB or external supply)
- Reset button.

The EasyFusion V7 board accepts an external plug-in processor module on a small PCB. Figure 5.14 shows the processor module for the PIC32MX460F512L-type processor.

Mikromedia for PIC32

The mikromedia for PIC32 board (Figure 5.15) is a small board with an integrated touch screen TFT color display. In addition, the board contains a stereo MP3 Codec chip, and a microSD card slot. The device can be powered from an external USB port or from an external battery. A preprogrammed bootloader program enables the microcontroller chip to be programmed. The device is reset using a reset button.

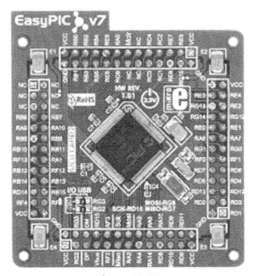

Figure 5.14
PIC32MX460F512L processor module.

Figure 5.15
Mikromedia for PIC32 board.

Multimedia for PIC32MX7

This development board includes a PIC32MX795F512L 32-bit microcontroller and is used for multimedia-based applications. A large, touch screen, color TFT display is provided (Figure 5.16) with on-board push-button switches for game applications. In addition, an Ethernet interface and a microSD card slot are available to store images or data.

Olimex PIC32 Development Board

This is a low-cost 32-bit microcontroller development board (Figure 5.17) with a high-performance PIC32MX460F512L microcontroller (http://www.olimex.com).

The board offers the following features:

- PIC32MX460F512L microcontroller
- Audio input and output
- USB interface
- SD card slot
- JTAG connector
- 84 × 84-pixel LCD
- On-board crystal
- Joystick
- Reset button

Figure 5.16
Multimedia for PIC32MX7 board.

Figure 5.17
Olimex PIC32 development board.

- 3.3 V voltage regulator
- I/O pins on connectors
- Development PCB area.

PIC32-MAXI-WEB Development Board

This board (Figure 5.18) from Olimex features a PIC32 microcontroller with embedded 100 Mbit Ethernet module. A large 240 × 320 TFT touch-screen LCD is provided with the board for graphical applications.

Figure 5.18
PIC32-MAXI-WEB board.

This board has the following features:

- PIC32MX795F512L 32-bit microcontroller
- 320 × 240 LCD
- 2 × opto-isolated digital inputs
- 2 × CAN interface
- Accelerometer sensor
- Temperature sensor
- microSD card slot
- 2 × relays
- RS232 interface
- 3 × LED indicators
- Reset button
- 3.3 V voltage regulator.

LV-32MX V6

The LV-32MX V6 (Figure 5.19) is a PIC32 development system manufactured by mikroElektronika (www.mikroe.com), and is equipped with many on-board modules, including multimedia peripherals which give great power and flexibility for system development. This development board is fully compatible with the mikroC Pro for PIC32 compiler.

The board offers the following features:

- PIC32MX460F512L
- Eighty-five push-button switches

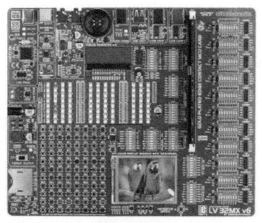

Figure 5.19
LV-32MX V6.

- Eighty-five LEDs
- SD card slot
- Reset button
- Power supply regulator
- Color TFT display with touch screen
- CAN support
- On-board programmer and debugger
- Serial EEPROM
- Serial flash memory
- Stereo codec chip
- Chip-on-Glass LCD
- 2 × UART connectors
- DS1820 temperature sensor socket
- I/O headers.

5.2.2 Device Programmers

After writing and translating the program into executable code, the resulting HEX file should be loaded to the target microcontroller program memory. Device programmers are used to load the program memory of the actual microcontroller chip. The type of device programmer to be used depends on the type of microcontroller to be programmed. For example, some device programmers can only program PIC16 series, some can program both PIC16 and PIC18 series, and some are used to program different models of microcontrollers (e.g. Intel 8051 series).

As we have seen in the previous section, some microcontroller development kits include on-board device programmers and thus there is no need to remove the microcontroller chip and insert into the programming device. In this section, some of the popular device programmers that can be used to program PIC32 series of microcontrollers are described.

mikroProg

mikroProg (Figure 5.20) is a small handheld programmer manufactured by mikroElektronika and supports all PIC microcontrollers from PIC10, PIC12−PIC16, PIC18, dsPIC, PIC24, and PIC32.

mikroProg programmer is supported by all the compilers of the company. The device is connected to a PC via a USB cable, and to the target development system. The microcontroller in the target system is programmed by first compiling and sending the HEX code to the programmer device.

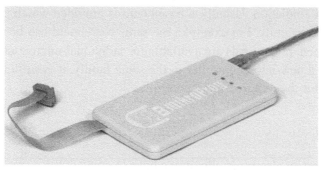

Figure 5.20
mikroProg device programmer.

5.2.3 In-Circuit Debuggers

An in-circuit debugger is a hardware connected between a PC and the target microcontroller test system and is used to debug real-time applications faster and easier. With in-circuit debugging, a monitor program runs in the PIC microcontroller in the test circuit. The programmer can set breakpoints on the PIC, run code, single step the program, examine variables and registers on the real device and if required change their values. An in-circuit debugger uses some memory and I/O pins of the target PIC microcontroller during the debugging operations. With some in-circuit debuggers only, the assembly language programs can be debugged. Some more powerful debuggers enable high-level language programs to be debugged.

In-circuit debuggers also include programming functions that enable the target microcontroller to be programmed. Some of the popular in-circuit debuggers are PicKit 3, ICD3 and Real Ice from Microchip (www.microchip.com), and mikroProg from mikroElektronika. These devices can be used with all types of PIC microcontrollers.

5.2.4 In-Circuit Emulators

The in-circuit emulator is one of the oldest and the most powerful method of debugging a microcontroller system. In fact it is the only tool that substitutes its own internal processor for the one in your target system. Like all in-circuit debuggers, the emulator's most fundamental resource is target access—the ability to examine and change the contents of registers, memory and I/O. However, since the ICE replaces the CPU, it generally does not require working CPU on the target system to provide this capability. This makes the in-circuit emulator by far the best tool for troubleshooting new or defective systems. Usually every microcontroller family has its own set of in-circuit emulator. For example, an in-circuit emulator for the PIC16 microcontrollers cannot be used for the PIC18 microcontrollers. Because of this, in order to lower the costs, emulator manufacturers provide a multiboard solution to in-circuit emulation. Usually, a base board is provided which is common to most microcontrollers in the family. For example, the same base board can be used by all PIC microcontrollers. Then, probe cards are available for individual microcontrollers. When it is required to emulate a new microcontroller in the same family, it is sufficient to purchase just the probe card for the required microcontroller.

5.2.5 Breadboard

When we are building an electronic circuit, we have to connect the components as shown in the given circuit diagram. This task can usually be carried out on a strip-board or a PCB by soldering the components together. The PCB approach is used for circuits which have been tested and which function as desired and also when the circuit is to be made permanent. It is not economical to make a PCB design for one or only a few applications.

During the development stage of an electronic circuit, it may not be known in advance whether or not the circuit will function correctly when assembled. A solderless breadboard is then usually used to assemble the circuit components together. A typical breadboard is shown in Figure 5.21. The board consists of rows and columns of holes which are spaced so that integrated circuits and other components can be fitted inside them. The holes have spring actions so that the component leads can be hold tight inside the holes. There are various types and sizes of breadboards depending on the complexity of the circuit to be built. The boards can be stacked together to make larger boards for very complex circuits. Figure 5.22 shows the internal connection layout of the breadboard given in Figure 5.21.

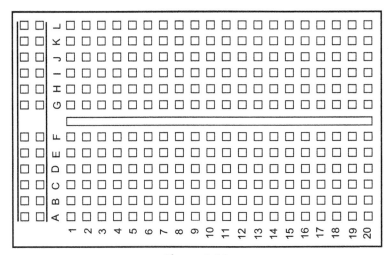

Figure 5.21
A typical breadboard layout.

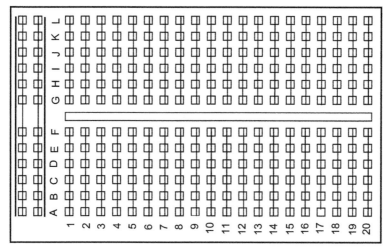

Figure 5.22
Internal wiring of the breadboard in Figure 5.21.

The top and bottom half parts of the breadboard are separate with no connection between them. Columns 1–20 in rows A–F are connected to each other on a column basis. Similarly, rows G–L in columns 1–20 are connected to each other on a column basis. Integrated circuits are placed such that the legs on one side are on the top half of the breadboard, and the legs on the other side of the circuit are on the bottom half of the breadboard. First two columns on the left of the board are usually reserved for the power and earth connections. Connections between the components are usually carried out by using stranded (or solid) wires plugged inside the holes to be connected.

Figure 5.23 shows the picture of a breadboard with two integrated circuits and a number of resistors and capacitors placed on it.

The nice thing about the breadboard design is that the circuit can be modified very easily and quickly and different ideas can be tested without having to solder any components. The components can easily be removed and the breadboard can be used for other projects after the circuit has been tested and working satisfactorily.

5.3 mikroC Pro for PIC32 IDE

In this book, we shall be using the mikroC Pro for PIC32 compiler developed by mikroElektronika (www.mikroe.com). Before using the compiler, we need to know how the compiler IDE is organized and how to write, compile, and simulate a program written in this language. In this section we shall be looking at the operation of the mikroC Pro for PIC32 IDE in detail.

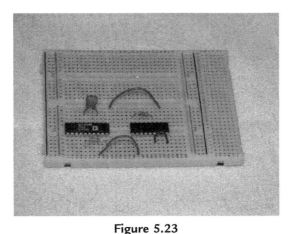

Figure 5.23
Picture of a breadboard with some components.
(For color version of this figure, the reader is referred to the online version of this book.)

A free 2 K program size-limited version of the mikroC Pro for PIC32 IDE is available from mikroElektronika. You could download and use this limited version for most small to medium sized applications. Alternatively, you can purchase a license and turn the limited version into a fully working unlimited IDE and use it for projects of any size and complexity.

After installing the mikroC Pro for PIC32 IDE, you should by default have a new icon installed in your Desktop. Double-click this icon to start the IDE.

5.3.1 mikroC Pro for PIC32 IDE Screen

After double-clicking the compiler icon, the IDE starts and the form shown in Figure 5.24 is displayed by default.

The screen is basically in four sections: top-left section, bottom section, middle section, and top-right section.

Top-Left Section

This section contains the Project Settings and the Code Explorer. The Project Settings is shown in Figure 5.25 and this section is about the type of microcontroller used, the oscillator frequency, and build type. Users can change any of the settings in this section if they wish.

The Code Explorer section gives a view of every declared item in the source code. In the example shown in Figure 5.26, **main** is listed in the **Functions** section.

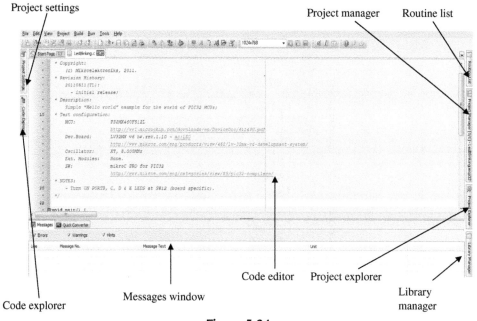

Figure 5.24
mikroC IDE screen. (For color version of this figure, the reader is referred to the online version of this book.)

Figure 5.25
Project settings form.

Figure 5.26
Code explorer form.

Bottom Section

This section is also called the Message Window and it consists of two tabs: Messages tab and Quick Converter tab. Compilation errors and warnings are reported inside the Messages tab. Double-clicking a message line will highlight the line where the error has encountered. A HEX file is generated only if the source file contains no errors. Figure 5.27 shows the results of a successful compilation listed in the Message Window.

The Quick Converter tab can be used for number conversions from decimal to binary or hexadecimal, or floating point number conversions. The conversion can be in 8, 16, or 32 bits.

Line	Message No.	Message Text
0	1144	Static RAM (bytes): 0 Dynamic RAM (bytes): 32765
0	1144	Used ROM (bytes): 560 (0%) Free ROM (bytes): 523728 (100%)
0	125	Project Linked Successfully
0	128	Linked in 203 ms
0	129	Project 'LedBlinking.mcp32' completed: 515 ms
0	103	Finished successfully: 06 Oct 2012, 09:45:12

Figure 5.27
Display of a successful compilation.

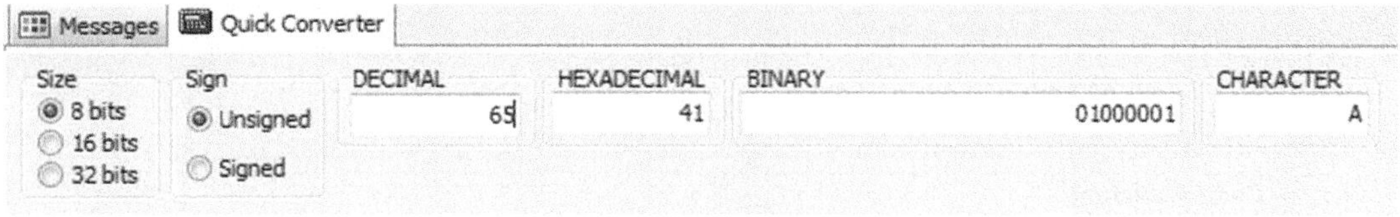

Figure 5.28
Converting decimal number 65.

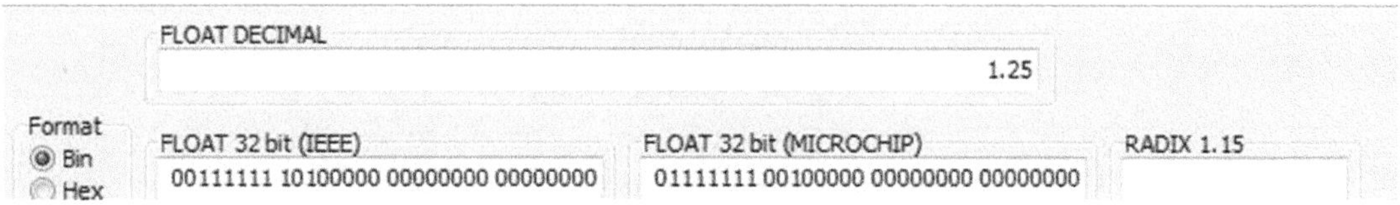

Figure 5.29
Representing floating point number 1.25.

Figure 5.28 shows an example where the decimal number 65 is converted to both hexadecimal and its ASCII character representation.

As another example, Figure 5.29 shows the floating point number 1.25 converted into its binary representation in different standards.

Middle Section

This section is the Code Editor, which is an advanced text editor. Programs are written in this section of the screen. The Code Editor supports the following features:

• Code Assistant
• Parameter Assistant
• Code Template
• Auto Correct
• Bookmarks.

The Code Assistant is useful when writing a program. Type the first few letters of an identifier and then press CTRL + SPACE keys to list all valid identifiers beginning with those letters. In Figure 5.30, for example, to locate identifier *Delay_ms*, the letters *Del* are typed and CTRL + SPACE is pressed. *Delay_ms* can be selected from the displayed list of matching valid words by using keyboard arrows and pressing ENTER.

The Parameter Assistant is invoked when a parenthesis is opened after a function or a procedure name. The expected parameters are listed in a small window just above the parenthesis. In Figure 5.31, function *Strlen* has been entered, and *unsigned char *s* appears in a small window when a parenthesis is opened.

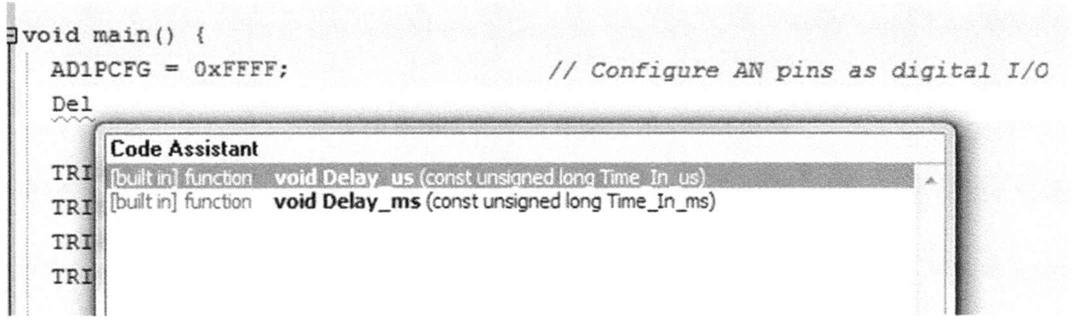

```
void main() {
  AD1PCFG = 0xFFFF;                    // Configure AN pins as digital I/O
  Del
```

Code Assistant
[built in] function **void Delay_us** (const unsigned long Time_In_us)
[built in] function **void Delay_ms** (const unsigned long Time_In_ms)

Figure 5.30
Using the Code Assistant.

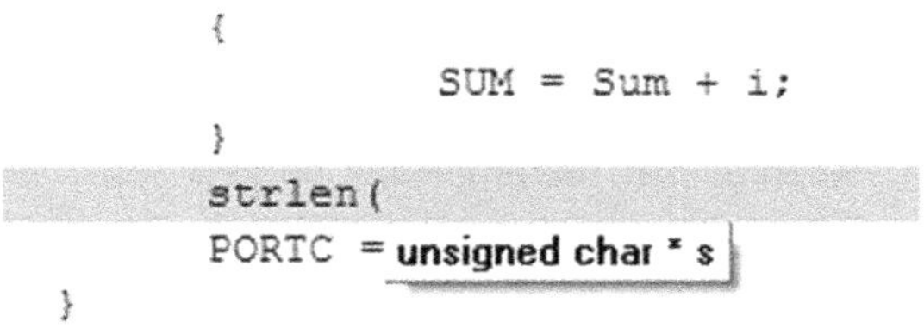

```
  {
          SUM = Sum + i;
  }
  strlen(
  PORTC = unsigned char * s
}
```

Figure 5.31
Using the Parameter Assistant.

Code Template is used to generate code in the program. For example, as shown in Figure 5.32, typing *switch* and pressing CTRL + J automatically generates code for the *switch* statement. We can add our own templates by selecting *Tools → Options → Auto Complete*. Some of the available templates are *array*, *switch*, *for*, and *if*.

Auto Correct corrects typing mistakes automatically. A new list of recognized words can be added by selecting *Tools → Options → Auto Correct Tab*.

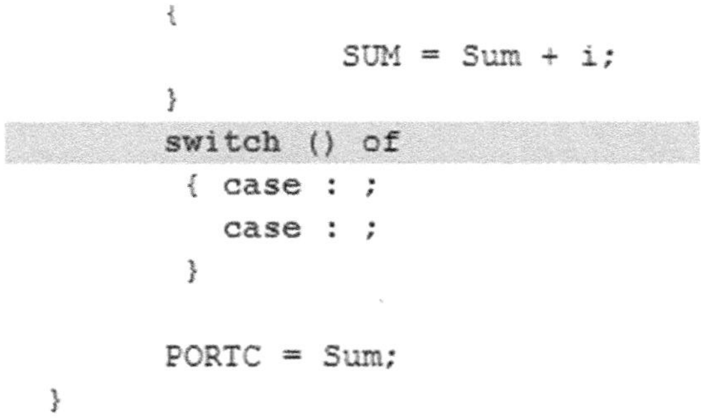

```
  {
          SUM = Sum + i;
  }
  switch () of
    { case : ;
      case : ;
    }

  PORTC = Sum;
}
```

Figure 5.32
Using the Code Template. (For color version of this figure, the reader is
referred to the online version of this book.)

Bookmarks make the navigation easier in large code. We can set bookmarks by entering CTRL + SHIFT + number, and then jump to the bookmark by pressing CTRL + number where number is the bookmark number.

Top-Right Section

The top-right section of the IDE consists of the following:

- Routine List
- Project Manager
- Project Explorer
- Library Manager.

The Routine List displays list of routines used in the program, and enables filtering of routines by name. It is possible to jump to a desired routine by double-clicking on it, or by pressing the ENTER button. In addition, routine names can be sorted by size of by address. Figure 5.33 shows a snapshot from this window.

The Project Manager displays a list of the source files (such as C files and header files) used in the program, image files, and output files created by the program (such as the HEX file, assembler file, list file, and log file). Figure 5.34 shows a typical Project Manager window display.

The Project Explorer provides a list of useful files and sample programs related to the microcontroller type being used for our project. Sometimes it may be hard to locate these files, and using the Project Explorer window we can select and display various useful project-related files and demonstration programs on various topics. Figure 5.35 shows the typical contents of this window.

The Library Manager enables various library files to be selected for our project. When selected this window lists all available libraries (extension .emcl) which are stored in the compiler. Desired libraries should be added to the project by selecting the check boxes next

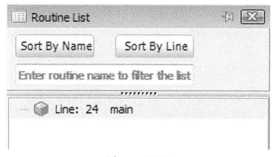

Figure 5.33
Routine List window.

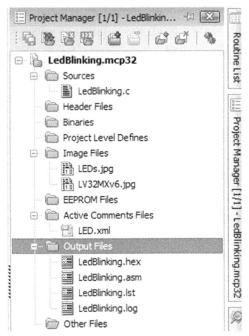

Figure 5.34
Project Manager window.

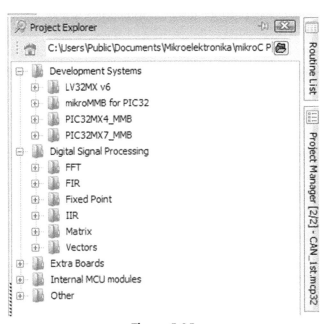

Figure 5.35
Project Explorer window.

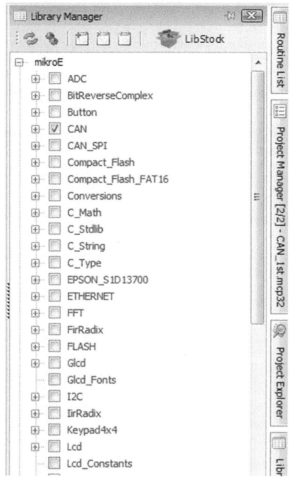

Figure 5.36
Library Manager window.

to the library names. Figure 5.36 shows part of the Library Manager window with the CAN library being selected for our project.

5.3.2 Creating and Compiling a New File

mikroC Pro for PIC32 files are organized in projects and all related files of a project are stored in the same folder. By default a project file has the extension ".mcp32". A project file contains the project name, target microcontroller device, device configuration flags, device clock, and list of source files with their paths. C source files have the extension **".c"**.

The example below illustrates step by step how a program source file can be created and compiled successfully.

Example 5.1

It is required to write a C program to calculate the sum of integer numbers from 1 to 10 and then send the result to PORT B of a PIC32MX360F512L type 32-bit microcontroller. Assume that eight LEDs are connected to PORT C of the microcontroller via current-limiting resistors. Draw the circuit diagram and show the steps involved in creating and compiling the program.

Solution 5.1

Figure 5.37 shows the circuit diagram of the project. The LEDs are connected to PORT B of the microcontroller using 390 Ω current-limiting resistors. The microcontroller is operated from an 8 MHz crystal.

The program is created and compiled as follows:

Step 1
Double-click the mikroC Pro for PIC32 icon to start the IDE.
Step 2
Create a new project called EXAMPLE. Click *Project → New Project.* You should see the new project creation wizard as in Figure 5.38.

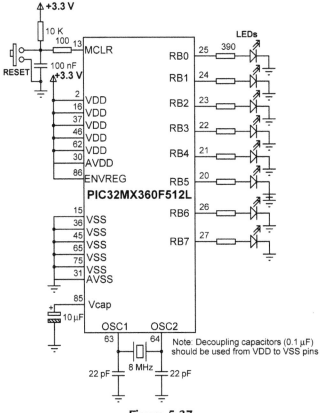

Figure 5.37
Circuit diagram of the project.

Figure 5.38
Project creation wizard.

Press *Next* to continue. Enter the project name, choose a project folder, select the device type, and enter the crystal frequency as shown in Figure 5.39.

Press *Next* to add any files to the project. As we are not adding any files, just press *Next* again. Select to include all library files as in Figure 5.40.

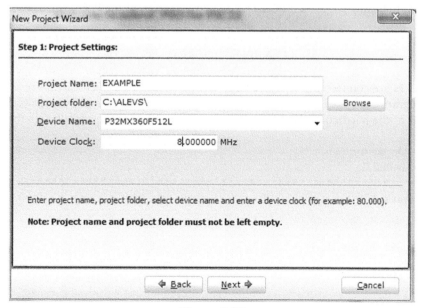

Figure 5.39
Enter project details.

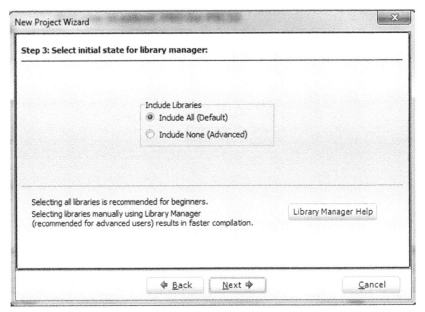

Figure 5.40
Include all library files.

Press *Next* and then press *Finish* to complete the new project creation steps.
You should now get a blank form where you can write your program.
Step 3
Enter the following program into the Code Editor section of the IDE:

```
/***********************************************************
                    EXAMPLE PROGRAM
                    ============
Eight LEDs are connected to a PIC32MX360F512L type 32-bit microcontroller.
This program calculates the sum of integer numbers from 1 to 10 and then sends the sum on
PORT B of the microcontroller. A for loop is used in the program to calculate the sum of
numbers.
Author: Dogan Ibrahim
File: EXAMPLE.C.
***********************************************************/
void main()
{

    unsigned int Sum, i;

    AD1PCFG = 0xFFFF;              // Configure all PORTB pins as digital
    TRISB = 0;                     // Configure PORT B pins as outputs

    Sum = 0;                       // Sum = 0 to start with
```

```
    for(i = 1; i <= 10; i++)              // Do 10 times
    {
        Sum = Sum + i ;                    // Calculate Sum
    }

    PORTB = Sum;                           // Send the result to PORT B
}
```

Part of the program is shown in Figure 5.41.

Step 4

Save the program with the name EXAMPLE by clicking *File → Save As*. The program will be saved with the name EXAMPLE.C.

Step 5

Compile the project by pressing CTRL + F9 or by clicking the Build Project button (Figure 5.42):

Step 6

If the compilation is successful, you should see the Success message in the Message Window as shown in Figure 5.43. Any errors in the program will be generated in the Message Window and you should correct these errors before proceeding further.

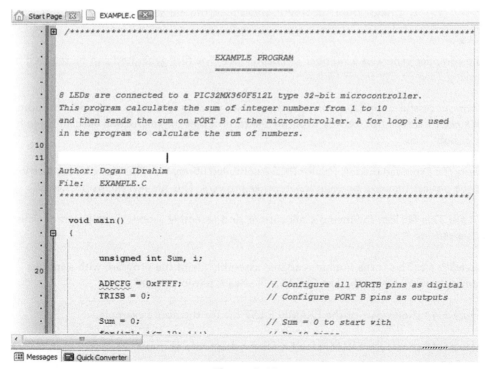

Figure 5.41
Example program.

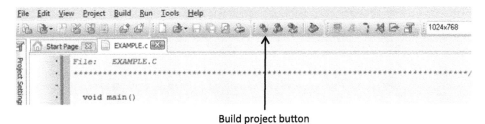

Build project button

Figure 5.42
Build project button.

Line	Message No.	Message Text
0	1144	Static RAM (bytes): 0 Dynamic RAM (bytes): 32765
0	1144	Used ROM (bytes): 560 (0%) Free ROM (bytes): 523728 (100%)
0	125	Project Linked Successfully
0	128	Linked in 93 ms
0	129	Project 'LedBlinking.mcp32' completed: 249 ms
0	103	Finished successfully: 06 Oct 2012, 11:51:19

Figure 5.43
Compilation successful message in the message window.

The compiler generates a number of output files. The files generated can be selected by clicking *Tools → Options → Output*.

Upon successful compilation of a program, the following files are generated by the compiler:
HEX file: Intel type hex records. Use this file to program the microcontroller. This file has extension ".hex".

Binary file: Extended mikroC Pro for PIC32 compiled library file. This is the binary distribution of the project that can be included in other projects. This file has extension ".emcl".

List file: This file lists the memory allocations and assembly listing of the program. This file has extension ".lst".

Assembler file: This is the human readable assembly file of the program with symbolic names, extracted from the List file. This file has extension ".asm".

Figure 5.44 shows part of the EXAMPLE.LST file for the above example.

Figure 5.45 shows the EXAMPLE.HEX file for the above example.

```
|; LST file generated by mikroListExporter - v.2.0
; Date/Time: 06/10/2012 11:51:19
;-------------------------------------------------

;Address Opcode          ASM
____SysVT:
0x9FC01180        0x0B400000   J       ____GenExcept
0x9FC01184        0x70000000   NOP
; end of ____SysVT
____BootVT:
0xBFC00380        0x3C1E9D00   LUI     R30, 40192
0xBFC00384        0x37DE0018   ORI     R30, R30, 24
0xBFC00388        0x03C00008   JR      R30
0xBFC0038C        0x70000000   NOP
; end of ____BootVT
_main:
;LedBlinking.c, 23 ::                 void main() {
;LedBlinking.c, 24 ::                 AD1PCFG = 0xFFFF;           // Conf
0x9D000030        0x3402FFFF   ORI     R2, R0, 65535
0x9D000034        0x3C1EBF81   LUI     R30, 49025
0x9D000038        0xAFC29060   SW      R2, -28576(R30)
;LedBlinking.c, 27 ::                 TRISB = 0;           // Initialize PORTE
0x9D00003C        0x3C1EBF88   LUI     R30, 49032
0x9D000040        0xAFC06040   SW      R0, 24640(R30)
;LedBlinking.c, 28 ::                 TRISC = 0;           // Initialize PORTC
0x9D000044        0x3C1EBF88   LUI     R30, 49032
0x9D000048        0xAFC06080   SW      R0, 24704(R30)
```

Figure 5.44
EXAMPLE.LST.

5.3.3 Other Features of the mikroC Pro for PIC32 IDE

In this section, we shall be looking at some other useful and important features of the mikroC Pro for PIC32 IDE.

Statistics Window

The memory and register usage statistics can be displayed after a successful compilation. *Click View → Statistics* from the main top menu of the IDE. To display a summary of the data and program memory usage click *Summary*. The memory usage for our example program in the previous section is shown in Figure 5.46.

Edit Project Window

The project details, such as the microcontroller type, clock frequency, and the values of configuration registers can be modified by selecting *Project → Edit Project* from the top menu of the IDE. Figure 5.47 shows the default project values for the program created in the previous section.

Programming the Target Microcontroller

The IDE provides an option for programming mikroC Pro for PIC32-compatible microcontrollers. Connect the compatible hardware to USB port of the PC and Select *Tools → mE Programmer* from the top menu of the IDE.

```
:020000041FC01B
:08118000080000400B00000070A4
:10038000009D1E3C3800DE370800C00300000070EE
:020000041D00DD
:10005000FFFF023481BF1E3C6090C2AF88BF1E3CD0
:100060004060C0AF0A20000001000334FFFF62308F
:10007000B00422C060040100000007021108300D8D
:10008000FFFF4430010062241B00400BFFFF4330A0
:10009000FFFF823088BF1E3C5060C2AF2700400B7C
:0400A00000000070EC
:10000000FCFFBD2700001E830000FEA20100F726B2
:10001000FCFFF616010018270800E0030400BD27C6
:00002000E0
:020000041FC01B
:10000000FCFFBD270000000700000007000000070C1
:100010000000000070000000070000000700000007020
:100020000000000070000000070000000700000007010
:100030000000000070000000070000000700000007000
:100040000000000070000000070000000700000007F0
:100050000000000070000A01D3CFC7FBD3700A0013CEB
:100060000008021340261E400AE0C003801EDB7F56
:1000700084497E7F02609E400008C14102609C402E
:1000800021A4023C830542340AF0400000809E40D7
:100090000A1000000AF0400000489E40FFFF023CAA
:1000A000FFFF42340AF0400000589E40C09F023CCF
:1000B00001042340AF0400001789E4020000234D3
:1000C0000AF0400001609E400004023C0AF040003B
:1000D00002609E400A1000000AF0400003609E404B
:1000E0001000023C0AF0400000609E4030000234E4
:1000F00088BF1E3C0040C2AF0010023488BF1E3CC7
:10010000010C2AF8000023C0AF0400000689E4030
:1001100009D1E3C5000DE370800C0030000007048
:08012000800E0030400BD2704
:020000041D00DD
:10003800FCFFBD270F00400B000000700400BD2727
:080048001800004200000070E6
:10002000FCFFBD270900400B000000700400BD2745
:080030001800004200000070FE
:020000041FC01B
:102FF0000000000051000000A3C594000BF00F1169
:00000001FF
```

Figure 5.45

EXAMPLE.HEX.

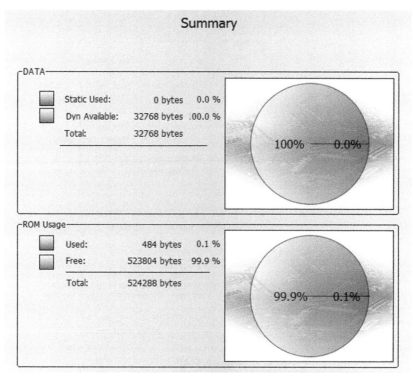

Figure 5.46
Memory usage summary.

ASCII Chart

The ASCII chart can be displayed by selecting *Tool → ASCII Chart* from the top menu of the IDE. Figure 5.48 shows the display.

GLCD Bitmap Editor

The compiler enables the user to load a bitmap image and then to generate code automatically corresponding to this image for the required compiler. Select *Tools → GLCD Bitmap Editor* to generate bitmap code.

HID Terminal

This option is useful during USB-based applications development. This option enables the user to send and receive data from HID-type USB devices. Select *Tools → HID Terminal* for this option.

Interrupt Assistant

This option helps the user to configure interrupt-based operations. Select *Tools → Interrupt Assistant* for this option.

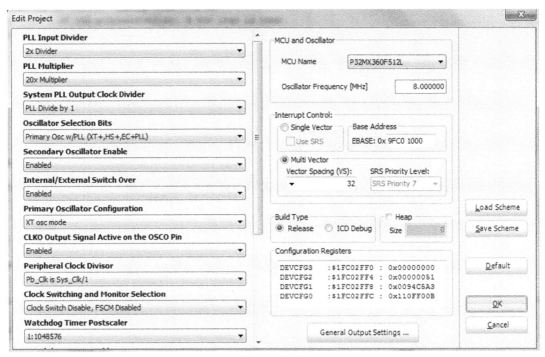

Figure 5.47
Edit project window.

LCD Custom Character Generation

This option enables the user to generate code for the standard LCDs. The user can create the required image on a pixel-based display and then generate code for the required type of compiler. In Figure 5.49 an up array image is created on a 5 × 7-pixel LCD display, and then the required code is generated by clicking the *Generate Code* button. The generated code can directly be inserted into our C programs so that the required image can easily be generated.

Seven-Segment Editor

This option enables the user to generate code for seven-segment displays. The code can be generated both in decimal and in hexadecimal. Figure 5.50 shows the code generated for displaying number "2" on the display (hexadecimal 0xA4 in this example).

UDP Terminal

The UDP Terminal is useful when it is required to develop UDP-based TCP/IP network applications. Select *Tools → UDP Terminal* for this option.

Ascii Chart

	0	1	2	3	4	5	6	7	8	9	A	B	C	D	E	F
0	NUL	SOH	STX	ETX	EOT	ENQ	ACK	BEL	BS	HT	LF	VT	FF	CR	SO	SI
	0	1	2	3	4	5	6	7	8	9	10	11	12	13	14	15
1	DLE	DC1	DC2	DC3	DC4	NAK	SYN	ETB	CAN	EM	SUB	ESC	FS	GS	RS	US
	16	17	18	19	20	21	22	23	24	25	26	27	28	29	30	31
2	SPC	!	"	#	$	%	&	'	(	)	*	+	,	-	.	/
	32	33	34	35	36	37	38	39	40	41	42	43	44	45	46	47
3	0	1	2	3	4	5	6	7	8	9	:	;	<	=	>	?
	48	49	50	51	52	53	54	55	56	57	58	59	60	61	62	63
4	@	A	B	C	D	E	F	G	H	I	J	K	L	M	N	O
	64	65	66	67	68	69	70	71	72	73	74	75	76	77	78	79
5	P	Q	R	S	T	U	V	W	X	Y	Z	[	\	]	^	_
	80	81	82	83	84	85	86	87	88	89	90	91	92	93	94	95
6	`	a	b	c	d	e	f	g	h	i	j	k	l	m	n	o
	96	97	98	99	100	101	102	103	104	105	106	107	108	109	110	111
7	p	q	r	s	t	u	v	w	x	y	z	{	|	}	~	DEL
	112	113	114	115	116	117	118	119	120	121	122	123	124	125	126	127
8	€		‚	ƒ	„	…	†	‡	^	‰	Š	‹	Œ		Ž	
	128	129	130	131	132	133	134	135	136	137	138	139	140	141	142	143
9		'	'	"	"	•	–	—	~	™	š	›	œ		ž	Ÿ
	144	145	146	147	148	149	150	151	152	153	154	155	156	157	158	159
A		¡	¢	£	¤	¥	¦	§	¨	©	ª	«	¬	-	®	¯
	160	161	162	163	164	165	166	167	168	169	170	171	172	173	174	175
B	°	±	²	³	´	µ	¶	·	¸	¹	º	»	¼	½	¾	¿
	176	177	178	179	180	181	182	183	184	185	186	187	188	189	190	191
C	À	Á	Â	Ã	Ä	Å	Æ	Ç	È	É	Ê	Ë	Ì	Í	Î	Ï
	192	193	194	195	196	197	198	199	200	201	202	203	204	205	206	207
D	Ð	Ñ	Ò	Ó	Ô	Õ	Ö	×	Ø	Ù	Ú	Û	Ü	Ý	Þ	ß
	208	209	210	211	212	213	214	215	216	217	218	219	220	221	222	223
E	à	á	â	ã	ä	å	æ	ç	è	é	ê	ë	ì	í	î	ï
	224	225	226	227	228	229	230	231	232	233	234	235	236	237	238	239
F	ð	ñ	ò	ó	ô	õ	ö	÷	ø	ù	ú	û	ü	ý	þ	ÿ
	240	241	242	243	244	245	246	247	248	249	250	251	252	253	254	255

Figure 5.48
Displaying the ASCII chart.

USART Terminal

This option is very useful while developing RS232-based serial communication programs. The option enables the user to send and receive serial data from RS232-based external equipment. The user can select the communication parameters (baud rate, data bits, parity and stop bits) and the communication ports. Notice that the PC must be equipped with a suitable

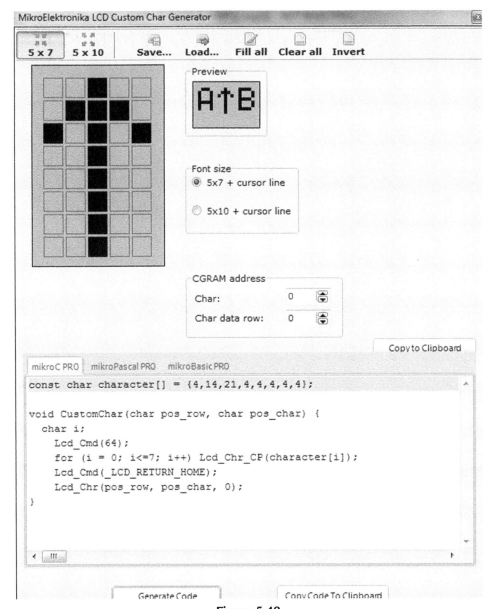

Figure 5.49
Using the LCD custom character generator.

serial communication port, or an external USB-to-RS232 converter device should be used before this tool can be invoked.

Help

The Help option, selected from the top menu of the IDE, can be very useful as it provides information on various aspects of the compiler and IDE operation. In addition,

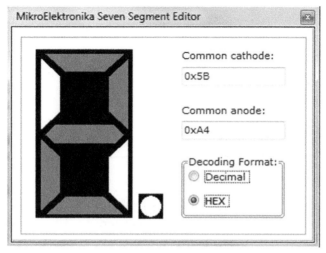

Figure 5.50
Generating code for the seven-segment display.

information about the syntax and use of mikroC language concepts can be obtained from the help menu.

5.3.4 Using the Software Simulator

The program developed in Section 5.3.2 is simulated as shown by the steps in the following example, using the simulator in software (release mode). That is, no hardware is used in this simulation.

Example 5.2

Describe the steps for simulating the program developed in Example 5.1. Remove integer variables *Sum* and *i* from inside the *main* program and place them as global variables before the *main* so that they are accessible to the debugger. Display the values of various variables and PORT B during the simulation while single-stepping the program. What is the final value displayed on PORT B?

Solution 5.2
The steps are as follows:

Step 1
Start the mikroC Pro for PIC32 IDE and make sure the program developed in Example 5.1 is displayed in the Code Editor window.

Step 2
From the drop-down menu, select *Run* → *Start Debugger*, as shown in Figure 5.51, to start the software simulator (or the software debugger).

Figure 5.51
The software simulator.

Step 3
Now select the variables we want to see during the simulation. Assuming that we want to display the values of variables *Sum, i*, and *PORT B*:

- Click on *Select from variable list* and then find and click on variable name *Sum*
- Click *Add* to add this variable to the Watch list
- Repeat these steps for variable *i* and *PORT B*

   The debugger watch window should now look like Figure 5.52.

Step 4
We can now single step the program and see the variables changing.

Press the F8 key on your keyboard. You should see a blue line to move down. This shows the line where the program is currently executing. Keep pressing F8 until you are inside the loop and you will see that variables *Sum* and *i* have become 1, as shown in Figure 5.53. Recently changed items are shown in red color. Double-clicking an item in the watch window opens the Edit Value window, where you can change the value of a variable or a register, or display the value in other bases such as decimal, hexadecimal, binary, or as floating point or character.

Step 5
Keep pressing F8 until the program comes out of the *for loop* and executes the line that sends data to PORT B. At this point, as shown in Figure 5.54, $i = 11$ *and Sum* $= 55$.

Step 6
Press F8 again to send the value of variable *Sum* to PORT B. As shown in Figure 5.55, PORT B will have the decimal value 55, which is the sum of numbers from 1 to 10.

This is the end of the simulation. Select from drop-down menu, *Run* → *Stop Debugger*.

In the above simulation example, we single-stepped through the program until we came to the end and then we could see the final value of PORT B. The next example shows how to set breakpoints in the program and then execute up to a breakpoint.

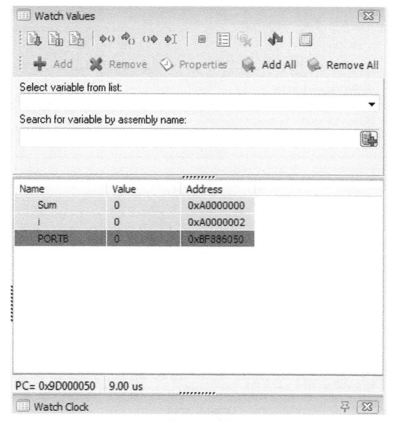

Figure 5.52
Selecting variables to be displayed.

Example 5.3

Describe the steps involved in simulating the program developed in Example 5.1. Set a breakpoint at the end of the program and run the debugger up to this breakpoint. Display the values of various variables and PORT B at this point. What is the final value displayed on PORT B?

Solution 5.3
The steps are as follows:

Step 1
Start the mikroC Pro for PIC32 IDE and make sure the program developed in Example 5.1 is displayed in the Code Editor window.

Step 2
From the drop-down menu select, *Run → Start Debugger* to start the software simulator as before.

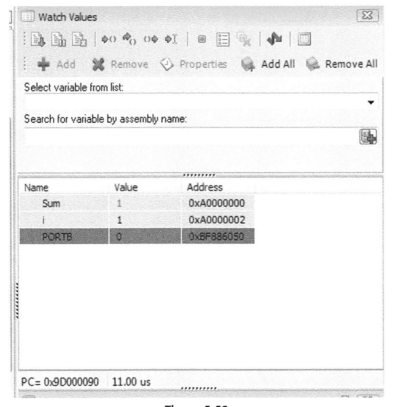

Figure 5.53
Single stepping through the program.

Step 3
Select variables *Sum, i*, and *PORT B* from the Watch window as described in Example 5.2.

Step 4
To set a breakpoint at the end of the program, click the mouse at the last closing bracket of the program, which is line 29, and press F5. As shown in Figure 5.56, you should see a red line at the breakpoint and a little marker in the left column of the Code Editor window.

Step 5
Now, press the F6 key to run the program. The program will run and then stop at the breakpoint, displaying variables as shown in Figure 5.55.

This is the end of the simulation. Select from drop-down menu, *Run* → *Stop Debugger*.

To clear a breakpoint, move the cursor over the line where the breakpoint is and then press F5. To clear all breakpoints in a program, you should press the SHIFT + CTRL + F5 keys. To display the breakpoints in a program, you can press SHIFT + F4 keys.

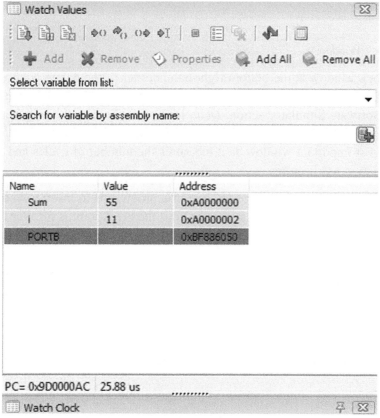

```
Author: Dogan Ibrahim
File:   EXAMPLE.C
*******************************************************************/
unsigned int Sum, i;

    void main()
    {
        AD1PCFG = 0xFFFF;          // Configure all PORTB pins as digital
        TRISB = 0;                 // Configure PORT B pins as outputs

        Sum = 0;                   // Sum = 0 to start with
        for(i=1; i<= 10; i++)      // Do 10 times
        {
            Sum = Sum + i;         // Calculate Sum
        }

        PORTB = Sum;               // Send the reslt to PORT B
    }
```

Figure 5.54
Single stepping through the program.

Watch Values

Select variable from list:

Search for variable by assembly name:

Name	Value	Address
Sum	55	0xA0000000
i	11	0xA0000002
PORTB		0xBF836050

PC= 0x9D0000AC   25.88 us

Watch Clock

Figure 5.55
PORT B has the value 55.

```
Author: Dogan Ibrahim
File:    EXAMPLE.C
*************************************************************************/
unsigned int Sum, i;

  void main()
{

    AD1PCFG = 0xFFFF;              // Configure all PORTB pins as digital
    TRISB = 0;                     // Configure PORT B pins as outputs

    Sum = 0;                       // Sum = 0 to start with
    for(i=1; i<= 10; i++)          // Do 10 times
    {
            Sum = Sum + i;         // Calculate Sum
    }

    PORTB = Sum;                   // Send the reslt to PORT B
}
```

Figure 5.56
Setting a breakpoint at line 29.

The Watch Clock Window

The Watch Clock window at the bottom right-hand corner of the debug window measures the execution time of a program. The *Current Count* measures the cycle count and time since the last Software Simulator action. *Delta* represents the number of cycles and time between the lines where the Software Simulator action has started and stopped. *Stopwatch* is probably the most important window as it measures the number of cycles and the time since the simulation has been reset.

The easiest way to measure the execution time of a program is to insert a breakpoint at the end of the program and then run the simulator until the breakpoint is hit. For the example given in this section, the total number of cycles and the execution time were 135 and 16.88 μs, respectively, as shown in Figure 5.57.

The following are some other useful debugger commands:

Step Into [F7]

Executes the current instruction and then halts. If the instruction is a call to a routine, the program enters the routine and halts at the first instruction.

Step Over [F8]

Executes the current instruction and then halts. If the instruction is a call to a routine, it skips it and halts at the first instruction following the call.

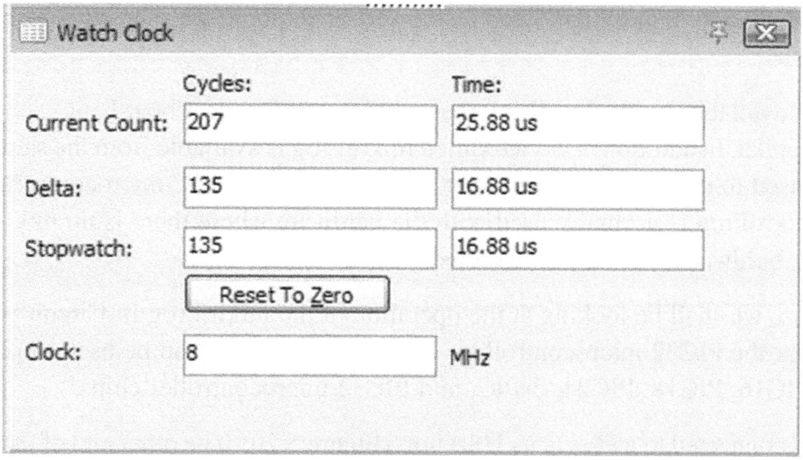

Figure 5.57
The Watch Clock window.

Step Out [CTRL + F8]

Executes the current instruction and then halts. If the instruction is within a routine, executes the instruction and halts at the first instruction following the call.

Run To Cursor [F4]

Executes all instructions between the current instruction and the cursor position.

Jump To Interrupt [F2]

Jumps to the interrupt service routine address and executes the procedure located at that address (a window appears to select the required interrupt source).

Toggle Breakpoint

Toggles the breakpoint where the cursor is placed. To remove a breakpoint, place the cursor over the red line and select *Run → Toggle Breakpoint* from the drop-down menu of the IDE (or simply press F5). To place a new breakpoint, place the cursor at the required place in the program and select *Run → Toggle Breakpoint* (or press F5).

5.3.5 Using the mikroICD In-Circuit Debugger

mikroICD is a hardware In-Circuit debugger developed by mikroElektronika. An In-Circuit debugger is extremely useful as it enables us to simulate a program on the target microcontroller as we have been doing using a software simulator. We can observe and modify registers and memory data, and also insert breakpoints on the actual target microcontroller.

mikroICD is available nearly on all microcontroller development boards of mikroElektronika. In addition, a device called mikroProg is available from the same company that can be used to program and In-Circuit debug all types of PIC microcontrollers in hardware. mikroProg is normally used with the hardware where there is no mikroICD, for example, the hardware that we put together.

In this section, we shall be looking at the operation of the mikroProg In-Circuit programmer when used for the PIC32 microcontrollers. The device actually can be used to program and debug PIC16, PIC18, PIC24, dsPIC, and PIC32 microcontroller chips.

mikroProg is connected to a PC via its USB port (Figure 5.20). The other end of the device is a 10-way IDC connector that is connected to the target microcontroller. The mikroProg is connected to the development circuit using the following pins of the microcontroller:

- MCLR
- RB6
- RB7
- +5 V
- GND.

The connection between the mikroProg and the microcontroller pins are as follows:

mikroProg	Microcontroller
MCU-VCC	VCC
MCU-PGC	RB6
MCU-PGD	RB7
MCU-MCLR	MCLR
MCU-GND	GND

In-Circuit Debugging Example

After building the hardware, we are ready to program the microcontroller and test the system's operation with the in-circuit debugger. The steps are as follows:

Step 1
Start the mikroC Pro for PIC32 IDE as before and make sure the program developed in Example 5.1 is displayed in the Code Editor window.

Step 2
Click Project Settings window and make the following selections for the *Build/Debugger Type*

```
Build Type:  ICD Debug
Debugger:    mikroICD
```

Step 3

Click Build Project icon to compile the program with the debugger, and make sure that successful build message is displayed in the Message Window.

Step 4

Make sure the mikroProg programmer/debugger device is connected as described earlier and program the target microcontroller using the mikroProg software supplied with the mikroProg device (Figure 5.58). Select *Load* to load the HEX file to mikroProg and then select *Write* to program the target microcontroller.

Step 5

From the drop-down menu, select *Run → Debugger*

Figure 5.58
Programming the target microcontroller using mikroProg.

Figure 5.59
Showing decimal number 55 by LEDs.

Step 6

Select variables to monitor *Sum, i,* and *PORT B* as described in Example 5.2

Step 7

Single-step through the program by pressing the F8 key. You should see the values of variables changing. At the end of the program, decimal value 55 will be sent to PORT B and you should see LEDs 0, 1, 2, 4 and 5 to turn ON as shown in Figure 5.59, corresponding to this number.

Step 8

Stop the debugger.

Using *Step into [F7]* and *Step over [F8]* commands in routines containing delays can take too long, and functions *Run to cursor [F4]* and breakpoints should be used in such routines.

5.3.6 Using a Development Board

In this section we shall see how easy it is to develop 32-bit microcontroller-based applications using a development board. This section explains how to use the development board LV-32MX V6, described earlier in this chapter. The use of the development board will be described with a very simple example. In this example, PORT B LEDs will count up in binary with 1 s delay between each count.

However, before using the development board, we need to know in detail how the LV-32MX is organized and how to use the various devices and indicators on the board.

LV-32MX V6 Development Board

Figure 5.60 shows the LV-32MX V6 development board with the functions of various devices identified with arrows (further details can be obtained from the LV-32MX V6 User Guide). The board can be powered either from an external power supply (8−16 V AC/DC) or from the USB port of a computer, using a jumper. In this application, the board will be powered from the USB port.

Push-Button Switches

The push-button switches at the bottom-left corner of the board can be used as inputs to the microcontroller. Jumper J15 (near the reset button) is used to determine the logic state

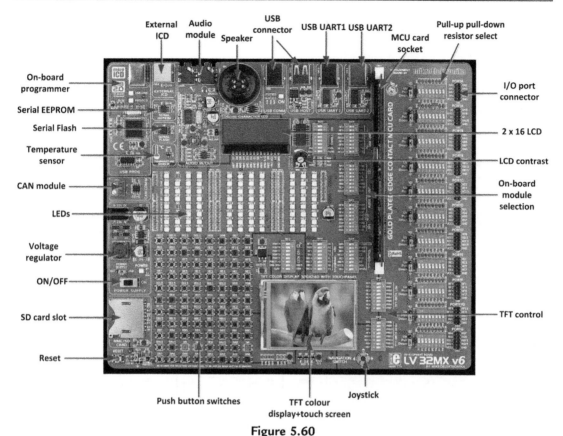

Figure 5.60

LV-32MX V6 development board. (For color version of this figure,
the reader is referred to the online version of this book.)

to be applied to a microcontroller input when a push-button switch is pressed. A 220 Ω
protective resistor is used to limit the maximum current, thus preventing the development
system and peripheral modules from being damaged, in case a short circuit occurs. This
resistor can be shorted by jumper J12 if desired. Figure 5.61 shows how the push-button
switches are configured.

LEDs

The LEDs are placed at the middle part of the board. A common LED voltage is about 2.5 V,
while the current varies from 1 to 20 mA depending on the type of LED used and the
required brightness. Resistors of 4.7 K are used on the board to limit the LED currents to
1 mA. There are 85 LEDs on the board that visually indicate the state of each microcontroller
pin. As shown in Figure 5.62, it is necessary to connect ports to LEDs using DIP switch
SW12.

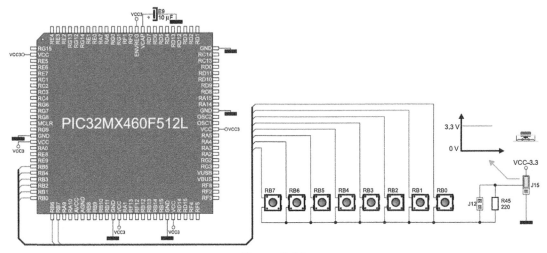

Figure 5.61
Push-button switches.

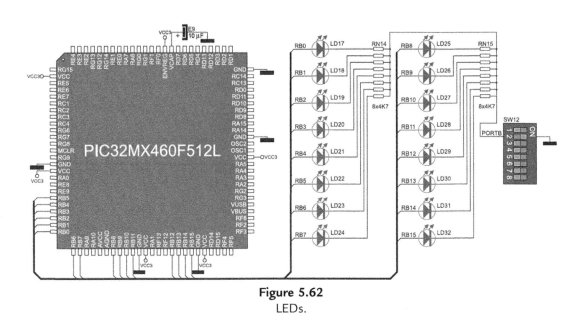

Figure 5.62
LEDs.

Input/Output Ports

The microcontroller I/O ports are available through 10-way connectors at the right-hand side of the board. DIP switches SW1–SW11 enable each connector pin to be connected to one pull-up/pull-down resistor. Jumpers J1–J11 determine whether the pull-up or the pull-down resistors are to be selected. Figure 5.63 shows the I/O connection in detail.

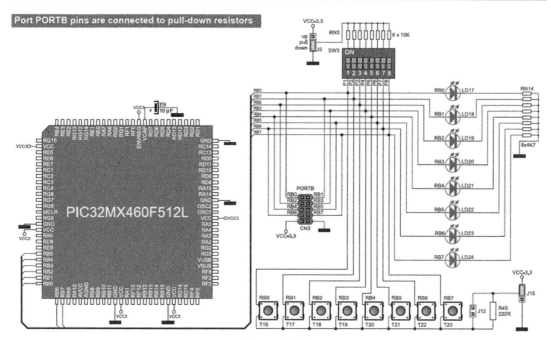

Figure 5.63
I/O port connections.

The MCU Card

The LV-32MX V6 board provides a DIMM-168P connector to place an MCU card into. There are several MCU cards available but the board is distributed with an MCU card with a PIC32MX460F512L 100-pin TQFP package installed on it. In addition to the microcontroller chip, the MCU cards contain a crystal oscillator and 102 soldering pads connected to the microcontroller pins. Figure 5.64 shows the board with an MCU card installed.

Figure 5.64
The board with an MCU card installed.

The Power Supply

The LV-32MX V6 development board can be powered from +5 V supply through the USB programming cable, or an external AC/DC power source can be connected to the board through its power connector. Jumper J16 is used to select the power supply source (the USB source is normally selected during normal project development). The development system is turned on/off by the Power Supply switch. The board contains an on-board voltage regulator to reduce the power supply voltage from +5 to +3.3 V. The 3.3 V is then used for powering the microcontroller and most other modules on the board. The voltage regulator is capable of supplying up to 800 mA.

USB UART Modules

The board is equipped with two USB UART modules that enable the development system to be connected to a PC or to some other compatible serial communications port via a USB connector. DIP switch SW15 controls the USB UART connections on the board. In order to connect USB UART 1 module and the microcontroller, turn ON switches 1−4 on SW15. For USB UART 2 connections, turn on switches 5−8 on SW15. UART 1 uses microcontroller I/O pins RF2, RF8, RD14 and RD15. Similarly, UART 2 uses microcontroller I/O pins RF4, RF5, RF12 and RF13. As shown in Figure 5.65, the USB/RS232 interface is established using FT232RL-type chips.

CAN Communication Module

The CAN communication module is mainly used in automotive applications. A CAN interface with an MCP2551 CAN controller chip is provided on the LV-32MX V6 board through a two-way screw connector (CN16) at the middle left-hand part of the board. DIP switch SW16 controls the CAN to microcontroller interface. Set switches 1 and 3 of SW16 to ON position (to use I/O ports RF0 and RF1), or switches 2 and 4 to ON (to use I/O ports RG0 and RG1). Figures 5.66 and 5.67 show the CAN module connector and circuit diagram, respectively.

SD Card Connector

The SD card connector enables a standard SD card to be used with the development board. DIP switch SW13, positions 4−18 should be ON to establish connection between the microcontroller and the SD card slot. Figures 5.68 and 5.69 show the SD card connector and the interface circuit diagram, respectively.

The Temperature Sensor

The board is equipped with the MCP9700A temperature sensor chip which can measure the temperature from −40 to +125 °C with an accuracy of ±2 °C. The MCP9700A

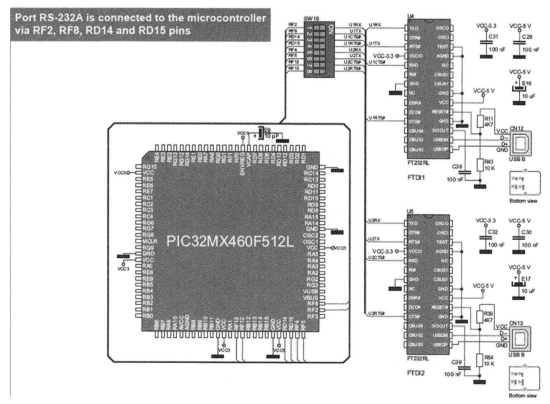

Figure 5.65
USB UART modules.

converts the temperature into an analog voltage signal which is then transferred to the RB8 microcontroller pin via DIP switch SW12. Figures 5.70 and 5.71 show the temperature sensor and the MCP9700A interface circuit, respectively.

The Joystick

The small joystick on the board (at the bottom right-hand corner of the TFT display) is a movable stick that can be moved in four directions, and also has a push-button action. The joystick is normally used in menu-based applications and games applications. Microcontroller I/O pins RB0-RB3 sense the direction of the joystick, while RA10 senses when the joystick is pressed. Figures 5.72 and 5.73 show the joystick and the interface circuit, respectively.

Flash Module and EEPROM Module

The board is equipped with a M25P80 chip that provides additional 8 Mbit flash memory for storing programs. This chip is controlled with the SPI signals. In addition, a 24AA01 chip

Figure 5.66
CAN module connector.

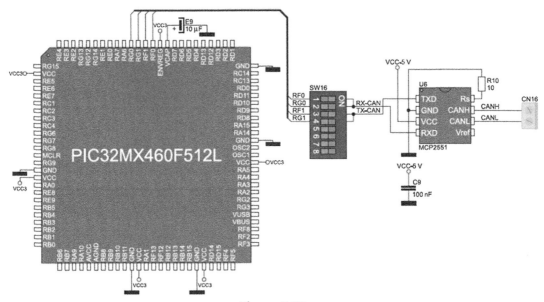

Figure 5.67
CAN module circuit diagram.

is provided that can store up to 1 Kbit EEPROM data. This chip is controlled with the I^2C signals.

Audio Module

The LV-32MX V6 board contains an audio module that enables the board to be connected to an external microphone and headphone. A stereo CODEC is provided to convert an

Figure 5.68
SD Card connector.

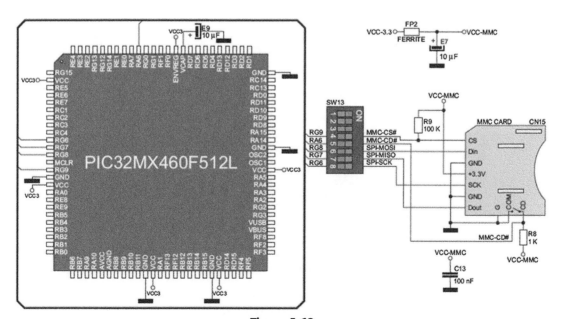

Figure 5.69
SD card interface circuit.

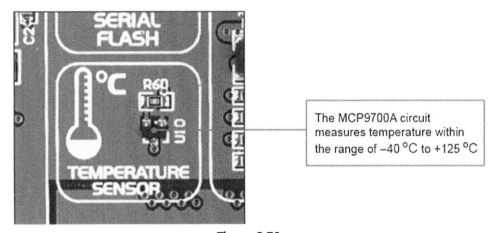

Figure 5.70
Temperature sensor.

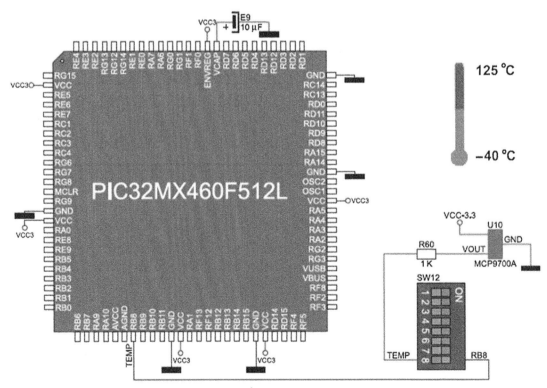

Figure 5.71
MCP9700A interface.

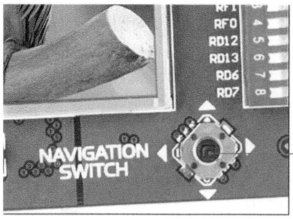

In addition to navigation function, the joystick can serve as a push button

Figure 5.72
Joystick.

analog signal from the interfaced microphone to a digital value, and then to transfer it to the microcontroller. The board can also generate an audio signal. An on-board speaker is also available on the board. Interested readers should refer to the LV-32MX V6 User Guide for further details and the circuit diagram of the audio module.

2 × 16 LCD

A 2 × 16 LCD is available on the board. A transceiver chip (74LVCC3245) is provided to provide the interface between the +3.3 V microcontroller side and the +5 V LCD side. SW20 pins 1−6 should be ON to connect the LCD to the microcontroller. PORT B pins RB2−RB7 are used for the LCD interface. Figures 5.74 and 5.75 show the LCD and the LCD interface circuit, respectively.

The TFT Display

The LV-32MX V6 board provides a 320 × 240 pixel color TFT display for graphical applications. In addition, a touch panel is provided to enable the user create interactive applications. Further details about the TFT display interface can be obtained from the LV-32MX V6 User Guide.

Example 5.4

A very simple example is given in this section to show how the LV-32MX V6 development board can be used. In this example, the PORT B LEDs count up by one with 1 s delay between each count.

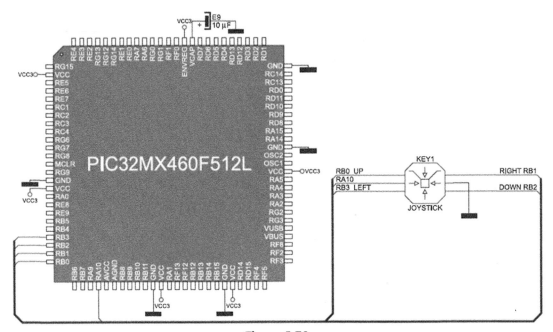

Figure 5.73
Joystick interface circuit.

Figure 5.74
The LCD.

Solution 5.4

The steps in developing the application for this example are as follows:

Step 1

Start the mikroC Pro for PIC32 compiler. Create a project as described in Section 5.3.2 earlier, and enter the following program into the Code Editor section of the IDE:

```
/******************************************************************
                        COUNTER PROGRAM
                        ==============
```

Eight LEDs are connected to a PIC32MX460F512L type 32-bit microcontroller.
This program counts up by one on PORT B LEDs with 1 s delay between each count.

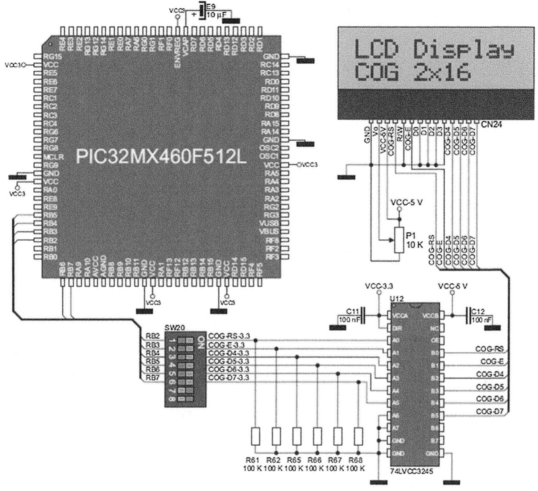

Figure 5.75
The LCD interface circuit.

```
Author: Dogan Ibrahim
File: COUNTER.C
*****************************************************************************/
void main()
{

    unsigned int Count = 0;

    AD1PCFG = 0xFFFF;              // Configure all PORTB pins as digital
    TRISB = 0;                     // Configure PORT B pins as outputs
```

```
    for(;;)                        // Do FOREVER
    {
        PORTB = Count;             // Send Count to PORT B
        Count++;                   // Increment Count
        Delay_Ms(1000);            // Wait 1 second
    }
}
```

Step 2
Select the following in the Project Settings:

Device	PIC32MX460F512L
MCU clock	80 MHz
Build type	Release
Debugger	Software

Compile the program by clicking *Build* icon on the drop-down menu and make sure there are no errors.

Step 3
Install the MCU card (PIC32MX460F512L) onto the slot on the LV-32MX V6 development board. Connect the development board to the PC via the supplied USB cable, and turn ON the Power Supply on the board.

Step 4
Download the program HEX code to the program memory of the target microcontroller by clicking *Tools → mE Programmer* on the drop-down menu of the mikroC Pro for PIC32 compiler. You should see the blue programming light (PRG/ICD) to flash on the development board during this process.

Step 5
Set DIP switch SW12, PORT B to ON to enable the PORT B LEDs.

Step 6
Press the Reset button. You should see the PORT B LEDs counting up with 1 s delay between each count.

5.4 Summary

This chapter has described the PIC microcontroller software and hardware development tools. It is shown that software tools such as text editors, assemblers, compilers, and simulators may be useful tools during microcontroller-based system development. The required useful hardware tools include development boards/kits, programming devices, in-circuit debuggers, or in-circuit emulators. The required useful software tools include assemblers, compilers, simulators, device programming software, and in-circuit debugger software. In this book, the mikroC Pro for PIC32 compiler is used in the examples and projects.

Steps in developing and testing a mikroC Pro for PIC32-based C program are given in this chapter with and without a hardware in-circuit debugger. In addition, an example use of the highly popular LV-32MX V6 32-bit PIC microcontroller development board is shown step by step using a simple example.

5.5 Exercises

1. Describe various phases of the microcontroller-based system development cycle.

2. Give brief description of the microcontroller development tools.

3. Explain the advantages and disadvantages of assemblers and compilers.

4. Explain why a simulator can be a useful tool during the development of a microcontroller-based product.

5. Explain in detail what a device programmer is. Give an example of device programmers for the PIC32 series of microcontrollers.

6. Describe briefly the differences between in-circuit debuggers and in-circuit emulators. List the advantages and disadvantages of each type of debugging tool.

7. Enter the following program into the mikroC Pro for PIC32 IDE and compile the program, correcting any syntax errors you might have. Then, using the software ICD, simulate the operation of the program by single stepping through the code and observe the values of various variables during the simulation.

    ```
    /*===========================================
    A SIMPLE LED PROJECT
    This program flashes the 8 LEDs connected to PORT B of a PIC32MX460F512L
    microcontroller.
    ============================================*/
    void main()
    {
    AD1PCFG = 0xFFFF;            // Configure all PORTB pins as digital
    TRISB = 0;                   // PORT B is output

    do
    {
       PORTB = 0xFF;             // Turn ON 8 LEDs on PORT B
       PORTC = 0;                // Turn OFF 8 LEDs on PORT B
    } while(1);                  // Endless loop

    }
    ```

8. Describe the steps necessary to use the mikroICD in-circuit debugger.

9. The following C program contains some deliberately introduced errors. Compile the program to find these errors and correct the errors.

```
void main()
{
        unsigned char i,j,k
        i = 10;
        j = i + 1;

        for(i = 0; i < 10; i++)
        {
            Sum = Sum + i;
            j++
        }
        }
}
```

10. The following C program contains some deliberately introduced errors. Compile the program to find these errors and correct the errors.

```
int add(int a, int b)
{
        result = a + b
}

void main()
{
        int p,q;
        p = 12;
        q = 10;
        z = add(p, q)
        z++;
        for(i = 0; i < z; i++)p++
}
}
```

11. Describe the steps necessary to create a program and to download it to the LV-32MX V6 development board.

12. Explain the functions of DIP switch SW12 on the LV-32MX V6 development board.

13. Describe how the pull-up/pull-down jumpers can be used on the LV-32MX V6 development board.

14. Explain the functions of DIP switches SW1−SW11 on the LV-32MX V6 development board

CHAPTER 6

Microcontroller Program Development

Chapter Outline

Before writing a program, it is always helpful to first derive the program's algorithm. Although simple programs can easily be developed by writing the code without any prior preparation, the development of complex programs has almost always become easier if the algorithm is first derived. Once the algorithm is ready, writing of the actual program code is not a difficult task.

A program's algorithm can be described in a variety of graphic- and text-based methods, such as flowchart, structure chart, data flow diagram, program description language, unified modeling language (UML) activity diagrams, and so on. The problem with graphical techniques is that it will be very time consuming to draw shapes with text inside them. Also, it is a tedious task to modify an algorithm described using graphical techniques.

Flowcharts can be very useful to describe the flow of control and data in small programs where there are only a handful of diagrams, usually not extending beyond a page or two. Flowcharts are the earliest software development tools, introduced in 1920s. There are now commercially available flowchart design programs (e.g. SmartDraw, RFFlow, Edraw, and so on) that help the user to create and manage large and complex flowcharts. One of the problems with flowcharts is that the code based on a flowchart tends to be rather unstructured with lots of branches all over the place, and it is difficult to maintain such code. Flowcharts are not used nowadays during the development of large programs. Interested readers can find many examples of flowcharts on the Internet.

Designing Embedded Systems with 32-Bit PIC Microcontrollers and MikroC.

The program description language (PDL, sometimes called Program Design Language) can be useful to describe the flow of control and data in small- to medium-size programs. The main advantage of the PDL description is that it is very easy to modify a given PDL since it consists of only text. In addition, the code generated from a PDL-based program is normally structured and is easy to maintain.

In this book, we will mainly be using the program description language, but the flowcharts will also be given where it is felt to be useful. The following sections briefly describe the basic building blocks of the program description language and flowcharts. It is left to the readers to decide which methods to use during the development of their programs.

6.1 Using the Program Description Language and Flowcharts

Program Description Language (PDL) is a free-format English-like text which describes the flow of control and data in a program. It is important to realize that PDL is not a programming language, but it is a collection of some keywords which enable a programmer to describe the operation of a program in a stepwise and logical manner. In this section, we will look at the basic PDL statements and their flowchart equivalents. The superiority of the PDL over flowcharts will become obvious when we have to develop medium- to large-size programs.

6.1.1 BEGIN—END

Every PDL program description should start with a BEGIN and end with an END statement. The keywords in a PDL description should be highlighted (e.g. in bold) to make the reading easier. The program statements should be indented and described between the PDL keywords. An example is shown in Figure 6.1 together with the equivalent flow diagram. Notice that the flowchart also uses BEGIN and END keywords.

6.1.2 Sequencing

For normal sequencing, the program statements should be written in English text and describe the operations performed one after the other. An example is shown in Figure 6.2 together

Figure 6.1
BEGIN—END statement and equivalent flowchart.

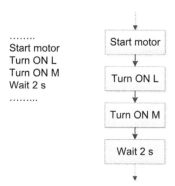

Figure 6.2
Sequencing and equivalent flowchart.

with the equivalent flowchart. It is clear from this example how much easier it is to describe the sequence using PDL.

6.1.3 IF–THEN–ELSE–ENDIF

IF, THEN, ELSE, and ENDIF should be used to conditionally change the flow of control in a program. Every IF line should be terminated with a THEN statement, and every IF block should be terminated with an ENDIF statement. Use of the ELSE statement is optional and depends on the application. Figure 6.3 shows an example of using IF–THEN–ENDIF, while Figure 6.4 shows the use of IF–THEN–ELSE–ENDIF statements in a program and their equivalent flowcharts. Again, the simplicity of using PDL is apparent from this example.

6.1.4 DO–ENDDO

The DO–ENDDO statements should be used when it is required to create iterations, or conditional or unconditional loops in programs. Every DO statement should be terminated

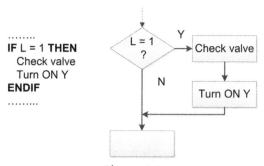

Figure 6.3
Using IF–THEN–ENDIF statements.

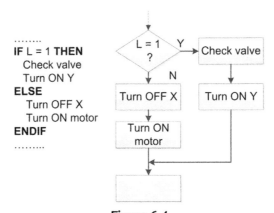

Figure 6.4
Using IF—THEN—ELSE—ENDIF statements.

with an ENDDO. Other keywords, such as FOREVER or WHILE can be used after the DO statement to indicate an endless loop or a conditional loop, respectively. Figure 6.5 shows an example of a DO—ENDDO loop executed 10 times. Figure 6.6 shows an endless loop created using the FOREVER statement. The flowchart equivalents are also shown in the figures.

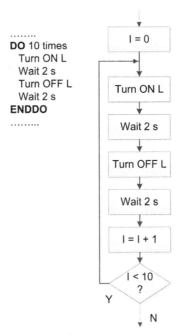

Figure 6.5
Using DO—ENDDO statements.

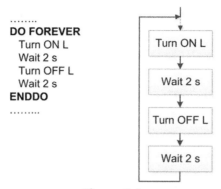

Figure 6.6
Using DO—FOREVER statements.

6.1.5 REPEAT—UNTIL

REPEAT—UNTIL is similar to DO—WHILE but here the statements enclosed by the REPEAT—UNTIL block are executed at least once, while the statements enclosed by DO—WHILE may not execute at all if the condition is not satisfied just before entering the DO statement. An example is shown in Figure 6.7, with the equivalent flowchart.

6.1.6 Calling Subprograms

In some applications, a program consists of a main program and a number of subprograms (or functions). Subprogram activation in PDL should be shown by adding the CALL

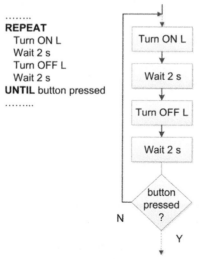

Figure 6.7
Using REPEAT—UNTIL statements.

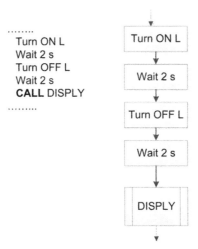

Figure 6.8
Calling a subprogram.

statement before the name of the subprogram. In flowcharts, a rectangle with vertical lines at each side should be used to indicate the invocation of a subprogram. An example call to a subprogram is shown in Figure 6.8 for both a PDL description and a flowchart.

6.1.7 Subprogram Structure

A subprogram should begin and end with the keywords BEGIN/*name* and END/*name*, respectively, where *name* is the name of the subprogram. In flowchart representation, a horizontal line should be drawn inside the BEGIN box and the name of the subprogram should be written at the lower half of the box. An example subprogram structure is shown in Figure 6.9 for both a PDL description and a flowchart.

6.2 Examples

Some examples are given in this section to show how the PDL and flowcharts can be used in program development.

Example 6.1

It is required to a write a program to convert hexadecimal numbers "A" to "F" into decimal. Show the algorithm using a PDL and also draw the flowchart. Assume that the number to be converted is called HEX_NUM and the output number is called DEC_NUM.

Solution 6.1

The required PDL is:

```
BEGIN
    IF HEX_NUM = "A" THEN
        DEC_NUM = 10
    ELSE IF HEX_NUM = "B" THEN
        DEC_NUM = 11
    ELSE IF HEX_NUM = "C" THEN
        DEC_NUM = 12
    ELSE IF HEX_NUM = "D" THEN
        DEC_NUM = 13
    ELSE IF HEX_NUM = "E" THEN
        DEC_NUM = 14
    ELSE IF HEX_NUM = "F" THEN
        DEC_NUM = 15
    ENDIF
END
```

The required flowchart is shown in Figure 6.10. Notice that it is much easier to write the PDL statements than drawing the flowchart shapes and writing text inside them.

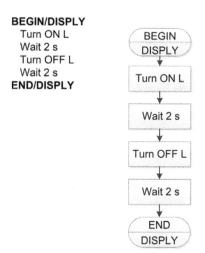

BEGIN/DISPLY
Turn ON L
Wait 2 s
Turn OFF L
Wait 2 s
END/DISPLY

Figure 6.9
Subprogram structure.

Example 6.2

The PDL of part of a program is given as follows:

```
J = 0
M = 0
DO WHILE J < 10
```

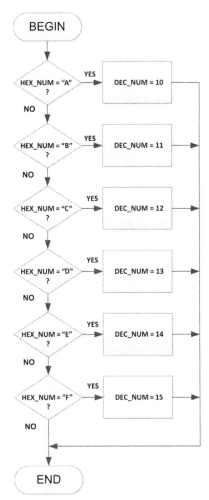

Figure 6.10
Flowchart solution.

```
DO WHILE M < 20
      Flash the LED
      Increment M
ENDDO
   Increment J
ENDDO
```

Show how this PDL can be implemented by drawing a flowchart.

Solution 6.2
The required flowchart is shown in Figure 6.11. Here again notice how complicated the flow-chart can be even for a simple nested DO–WHILE loop.

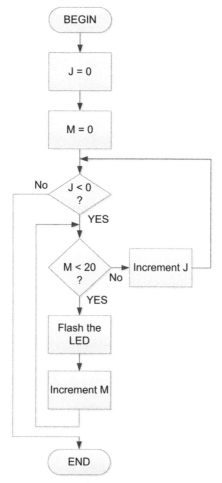

Figure 6.11
Flowchart solution.

Example 6.3

It is required to write a program to calculate the sum of integer numbers between 1 and 100. Show the algorithm using a PDL and also draw the flowchart. Assume that the sum will be stored in a variable called SUM.

Solution 6.3

The required PDL is:

```
BEGIN
    SUM = 0
    I = 1
    DO 100 TIMES
```

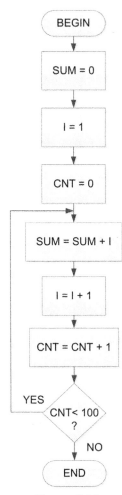

Figure 6.12
Flowchart solution.

```
            SUM = SUM + I
            Increment I
        ENDDO
    END
```

The required flowchart is shown in Figure 6.12.

Example 6.4

It is required to write a program to calculate the sum of all the even numbers between 1 and 10 inclusive of 10. Show the algorithm using a PDL and also draw the flowchart. Assume that the sum will be stored in a variable called SUM.

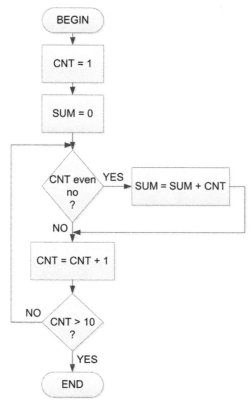

Figure 6.13
Flowchart solution.

Solution 6.4
The required PDL is:

```
BEGIN
     SUM = 0;
     CNT = 1
     REPEAT
          IF CNT is even number THEN
               SUM = SUM + CNT
          ENDIF
          INCREMENT CNT
     UNTIL CNT > 10
END
```

The required flowchart is shown in Figure 6.13. Notice how complicated the flowchart can be for a very simple problem such as this.

6.3 Representing for Loops in Flowcharts

Most programs include some form of iteration or looping. One of the easiest ways to create a loop in a C program is by using the *for* statement. This section shows how a *for* loop can be represented in a flowchart. As shown below, there are several methods of representing a *for* loop in a flowchart.

Suppose that we have a *for* loop as below and we wish to draw an equivalent flowchart.

```
for(m = 0; m < 10; m++)
{
        Cnt = Cnt + 2*m;
}
```

Method 1
Figure 6.14 shows one of the methods for representing the above *for* loop as with a flowchart. Here, the flowchart is drawn using the basic primitive components.
Method 2
Figure 6.15 shows the second method for representing the *for* loop with a flowchart. Here, a hexagon-shaped flowchart symbol is used to represent the *for* loop and the complete *for* loop statement is written inside this symbol.

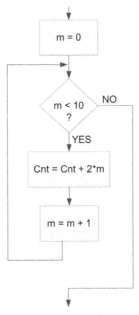

Figure 6.14
Method 1 for representing a *for* loop.

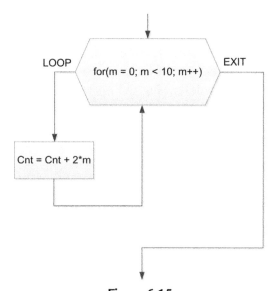

Figure 6.15
Method 2 for representing a *for* loop.

Method 3
Figure 6.16 shows the third method for representing the *for* loop with a flowchart. Here again a hexagon-shaped flowchart symbol is used to represent the *for* loop and the symbol is divided into three to represent the initial condition, the increment, and the terminating condition.

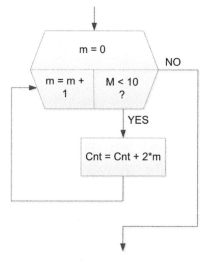

Figure 6.16
Method 3 for representing a *for* loop.

6.4 Summary

This chapter has described the program development process using the program description language (PDL) and flowcharts as tools. The PDL is commonly used as it is a simple and convenient method of describing the operation of a program. The PDL consists of several English-like keywords. Although the flowchart is also a useful tool, it can be very tedious in large programs to draw and modify shapes and write text inside them.

6.5 Exercises

1. Describe the various shapes used in drawing flowcharts.

2. Describe how the various keywords used in PDL can be used to describe the operation of a program.

3. What are the advantages and disadvantages of flowcharts?

4. It is required to write a program to calculate the sum of numbers from 1 to 10. Draw a flowchart to show the algorithm for this program.

5. Write the PDL statements for question in (4) above.

6. It is required to write a program to calculate the roots of a quadratic equation, given the coefficients. Draw a flowchart to show the algorithm for this program.

7. Write the PDL statements for the question in (6).

8. Draw the equivalent flowchart for the following PDL statements:
    ```
    DO WHILE count < 10
      Increment J
      Increment count
    ENDDO
    ```

9. It is required to write a function to calculate the sum of numbers from 1 to 10. Draw a flowchart to show how the function subprogram and the main program can be implemented.

10. Write the PDL statements for the question (9) above.

11. It is required to write a function to calculate the cube of a given integer number and then call this function from a main program. Draw a flowchart to show how the function subprogram and the main program can be implemented.

12. Write the PDL statements for the question (8) above.

13. Draw the equivalent flowchart for the following PDL statements:

```
J = 0
K = 0
REPEAT
      Flash LED A
      Increment J
      REPEAT
            Flash LED B
            Increment K
      UNTIL K = 10
UNTIL J > 15
```

CHAPTER 7

Simple PIC32 Microcontroller Projects

Chapter Outline

In this chapter, we shall be looking at the design of simple PIC32 32-bit microcontroller-based projects, with the idea of becoming familiar with basic interfacing techniques and learning how to use the various microcontroller peripheral registers. We will look at the design of projects using light-emitting diodes (LEDs), push-button switches, keyboards, LED arrays, sound devices and so on, and we will develop programs in C language using the mikroC Pro for PIC32 compiler. Most of the hardware for the projects is designed around the popular LV-32MX V6 development board (with an on-board PIC32MX460F512L 32-bit microcontroller). A breadboard is used where necessary to include any additional components required for the projects. We will start with very simple projects and proceed to more complex ones. It is recommended that the reader moves through the projects in their given order. The following are provided for each project:

- Description of the project
- Block diagram of the project
- Circuit diagram of the project
- Description of the hardware
- Algorithm description (in program description language (PDL))
- Program listing
- Photos of the project (where applicable)
- Suggestions for further development.

In this book, we will be using the PDL for all the projects.

7.1 Project 7.1—LED DICE

7.1.1 Project Description

This is a simple dice project based on LEDs, a push-button switch, and a PIC32MX460F512L microcontroller operating with 8 MHz crystal. Notice that in all of the projects in this book although the crystal frequency is 8 MHz, the microcontroller is operated at a clock rate of 80 MHz b configuring the built-in clock phase locked loop (PLL) module and the clock circuitry. This is done by clicking *Project → Edit Project* in the drop-down menu of the integrated development enviornment (IDE) and then selecting (the following are the default settings):

```
PLL input divider:                   2
PLL multiplier:                      20
System PLL output clock divider:     PLL divide by 1
Oscillator selection bits:           Primary Osc w/PLL
Primary oscillator configuration:    XT osc mode
```

The block diagram of the project is shown in Figure 7.1.

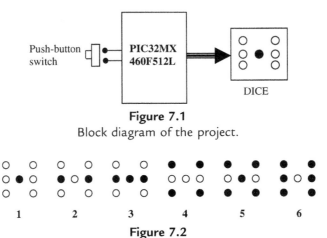

Figure 7.1
Block diagram of the project.

O   O   O   O   O   O   ●   ●   ●   ●   ●   ●
O ● O   ● O ●   ● ● ●   O O O   O ● O   ● O ●
O   O   O   O   O   O   ●   ●   ●   ●   ●   ●
  1   2   3   4   5   6

Figure 7.2
LED dice.

As shown in Figure 7.2, the LEDs are organized such that when they turn ON, they indicate the numbers as in a real dice. Operation of the project is as follows: normally, the LEDs are all OFF to indicate that the system is ready to generate a new number. Pressing the switch generates a random dice number between 1 and 6 and displays on the LEDs for 3 s. After 3 s, the LEDs turn OFF again.

An LED can be connected to a microcontroller output port in two different modes: *current-sinking* mode and *current-sourcing* mode.

7.1.2 Current-Sinking

As shown in Figure 7.3, in current-sinking mode, the anode leg of the LED is connected to the *Vdd* supply, and the cathode leg is connected to the microcontroller output port through a current-limiting resistor.

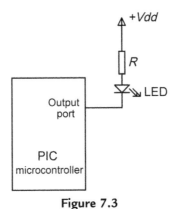

Figure 7.3
LED connected in current-sinking mode.

The voltage drop across an LED is around 2 V. The brightness of the LED depends on the current through the LED and this current can vary between 8 and 16 mA, with a typical value of 10 mA.

The LED is turned ON when the output of the microcontroller is at logic 0, so that current flows through the LED. We can calculate the value of the required resistor as follows:

$$R = \frac{Vdd - V_{\mathrm{LED}}}{I_{\mathrm{LED}}}$$

where,

Vdd is the supply voltage (3.3 V);

V_{LED} is the voltage drop across the LED (2 V);

I_{LED} is the current through the LED (10 mA).

substituting the values into the equation, we get $R = \frac{3.3-2}{10} = 130\ \Omega$. The nearest physical resistor is 120 Ω.

7.1.3 Current-Sourcing

As shown in Figure 7.4, in current-sourcing mode, the anode leg of the LED is connected to the microcontroller output port and the cathode leg is connected to the ground through a current-limiting resistor.

In this mode, the LED is turned ON when the microcontroller output port is at logic 1, i.e. *Vdd*. In practice, the output voltage is about 3.3 V and the value of the resistor can be determined as follows:

$$R = \frac{Vdd - V_{\mathrm{LED}}}{I_{\mathrm{LED}}}$$

Which gives the same resistor value of 120 Ω.

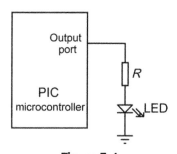

Figure 7.4
LED connected in current-sourcing mode.

7.1.4 Project Hardware

The circuit diagram of the project is shown in Figure 7.5. Seven LEDs representing the faces of a dice are connected to PORT B of a PIC32MX460F512L microcontroller in current-sourcing mode using 120 Ω current-limiting resistors. A push-button switch is connected to bit 1 of PORT C (RC1) using a pull-up resistor. The microcontroller is operated from 8 MHz crystal connected between its crystal oscillator pins. The microcontroller is powered from a +3.3 V power supply.

In this project the LV-32MX V6 development board is used. The following jumpers should be configured on the board to enable PORT B LEDs, and also to configure the push-button switch RC1 so that its output goes from logic 1 to 0 when the switch is pressed:

```
DIP switch SW12 position PORTB, set to ON
Jumper J15, set to Ground
DIP switch SW5 position 1, set to ON
Jumper J5, set to Pull-Up
```

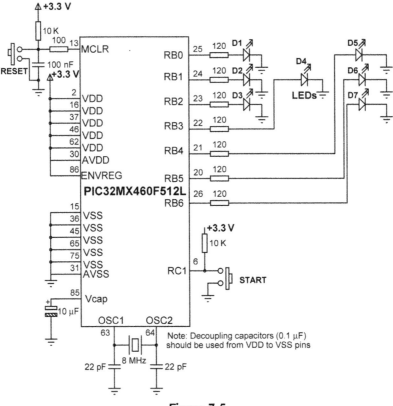

Figure 7.5
Circuit diagram of the project.

```
BEGIN
        Create DICE table
        Configure PORT B as outputs
        Configure RC1 as input
        Set J = 1
        DO FOREVER
                IF button pressed THEN
                        Get LED pattern from DICE table
                        Turn ON required LEDs
                        Wait 3 seconds
                        Set J = 0
                        Turn OFF all LEDs
                ENDIF
                Increment J
                IF J = 7 THEN
                        Set J = 1
                ENDIF
        ENDDO
END
```

Figure 7.6
PDL of the project.

7.1.5 Project PDL

The operation of the project is described in the PDL given in Figure 7.6. At the beginning of the program PORT B pins are configured as outputs, and bit 1 of PORT C (RC1) is configured as input. The program then executes in a loop continuously and increments a variable between 1 and 6. The state of the push-button switch is checked and when the switch is pressed (switch output at logic 0), the current number is sent to the LEDs. A simple array is used to find out the LEDs to be turned ON corresponding to the dice number.

Table 7.1 gives the relationship between a dice number and the corresponding LEDs to be turned ON to imitate the faces of a real dice. For example, to display number 1 (i.e. only the middle LED is ON), we have to turn on D4. Similarly, to display number 4, we have to turn ON D1, D3, D5 and D7.

Table 7.1: Dice Number and LEDs to be Turned ON

Required Number	LEDs to be Turned On
1	D4
2	D2, D6
3	D2, D4, D6
4	D1, D3, D5, D7
5	D1, D3, D4, D5, D7
6	D1, D2, D3, D5, D6, D7

Table 7.2: Required Number and PORT B Data

Required Number	PORT B Data (Hex)
1	0x08
2	0x22
3	0x2A
4	0x55
5	0x5D
6	0x77

The relationship between the required number and the data to be sent to PORT B to turn on the correct LEDs is given in Table 7.2. For example, to display dice number 2, we have to send hexadecimal 0x22 to PORT B. Similarly, to display number 5, we have to send hexadecimal 0x5D to PORT B and so on.

7.1.6 Project Program

The program is called DICE.C and the program listing is given in Figure 7.7. At the beginning of the program, **Switch** is defined as bit 1 of PORT C, and **Pressed** is defined as 0. The relationship between the dice numbers and the LEDs to be turned on are stored in an array called **DICE**. PORT B is configured as digital output by clearing TRISC and also by setting AD1PCFG bits. Bit 1 of PORT C is configured as digital input by setting TRISC register to 2. All the LEDs are turned OFF to start with.

Variable **J** is used as the dice number. Variable **Pattern** is the data sent to the LEDs. The program then enters an endless **for** loop where the value of variable **J** is incremented very fast between 1 and 6. When the push-button switch is pressed (*Switch* equals to *Pressed*) the LED pattern corresponding to the current value of J is read from the array and is sent to the LEDs. The LEDs remain at this state for 3 s (using function Delay_ms with argument set to 3000 ms) and after this time they all turn OFF to indicate that the system is ready to generate a new dice number.

7.1.7 Using a Random Number Generator

In the above project the value of variable **J** changes very fast between 1 and 6 and when the push-button switch is pressed the current value of this variable is taken and is used as the dice number. Because the values of **J** are changing very fast we can say that the numbers generated are random, i.e. new numbers do not depend on the previous numbers.

In this section we shall see how a pseudo random number generator function can be used to generate the dice numbers. The modified program listing is shown in Figure 7.8 (Program DICERAND.C). In this program a function called **Number** generates the dice numbers.

```
/*****************************************************************************
                    SIMPLE DICE
                    ===========

In this project 7 LEDs are connected to PORT B of a PIC32MX460F512L microcontroller
and the microcontroller is operated from 8 MHz crystal. The LEDs are organised
as the faces of a real dice. When a push-button switch connected to RC1 is
pressed a dice pattern is displayed on the LEDs. The display remains in this
state for 3 s  and after this period the LEDs all turn OFF to indicate that
the system is ready for the button to be pressed again.


Author: Dogan Ibrahim
Date:   July 2012
File:   DICE.C
*****************************************************************************/

#define Switch PORTCbits.RC1
#define Pressed 0

void main()
{
    unsigned char J = 1;
    unsigned char Pattern;
    unsigned char DICE[] = {0,0x08,0x22,0x2A,0x55,0x5D,0x77};

    TRISB = 0;                      // PORT B outputs
    TRISC = 2;                      // RC1 input
    PORTB = 0;                      // Turn OFF all LEDs
    AD1PCFG = 0xFF;                 // PORT B digital

    for(;;)                         // Endless loop
    {
        if(Switch == Pressed)       // Is switch pressed ?
        {
            Pattern = DICE[J];      // Get LED pattern
            PORTB = Pattern;        // Turn on LEDs
            Delay_ms(3000);         // Delay 3 second
            PORTB = 0;              // Turn OFF all LEDs
            J = 0;                  // Initialise J
        }
        J++;                        // Increment J
        if(J == 7) J = 1;           // Back to 1 if > 6
    }
}
```

Figure 7.7

Program listing.

The function receives the upper limit of the numbers to be generated (6 in this example), and also a seed value which defines the number set to be generated. In this example, the seed is set to 1. Every time the function is called, a number will be generated between 1 and 6.

The operation of the program is basically same as in Figure 7.7. When the push-button switch is pressed function **Number** is called to generate a new dice number between 1 and 6

```
/********************************************************************************
                        SIMPLE DICE
                        ===========

In this project 7 LEDs are connected to PORT B of a PIC32MX460F512L microcontroller
and the microcontroller is operated from 8 MHz crystal. The LEDs are organised
as the faces of a real dice. When a push-button switch connected to RC1 is
pressed a dice pattern is displayed on the LEDs. The display remains in this
state for 3 s and after this period the LEDs all turn OFF to indicate that
the system is ready for the button to be pressed again.


Author: Dogan Ibrahim
Date:   July 2012
File:   DICERAND.C
********************************************************************************/

#define Switch PORTCbits.RC1
#define Pressed 0

//
// This function generates a pseudo random integer Number between 1 and Lim
//
unsigned char Number(int Lim, int Y)
{
   unsigned char Result;
   static unsigned int Y;

   Y = (Y * 32719 + 3) % 32749;
   Result = ((Y % Lim) + 1);
   return Result;
}


void main()
{
   unsigned char J, Pattern, Seed = 1;
   unsigned char DICE[] = {0,0x08,0x22,0x2A,0x55,0x5D,0x77};

   TRISB = 0;                      // PORT B outputs
   TRISC = 2;                      // RC1 input
   PORTB = 0;                      // Turn OFF all LEDs
   AD1PCFG = 0xFF;                 // PORT B digital

   for(;;)                         // Endless loop
   {
      if(Switch == Pressed)        // Is switch pressed ?
      {
        J = Number(6, Seed);       // Generate a Number 1 to 6
        Pattern = DICE[J];         // Get LED pattern
        PORTB = Pattern;           // Turn on LEDs

        Delay_ms(3000);            // Delay 3 second
        PORTB = 0;                 // Turn OFF all LEDs
      }
   }
}
```

Figure 7.8
Dice program using a pseudo random number generator.

and this number is used as an index in array **DICE** in order to find the bit pattern to be sent to the LEDs.

7.2 Project 7.2—Liquid-Crystal Display Event Counting

7.2.1 Project Description

This is a simple project that shows how external events can be counted and the count displayed on a liquid-crystal display (LCD). The block diagram of the project is shown in Figure 7.9. In this project, an external event is said to occur when bit 1 of PORT C (RC1) goes from logic 1 to logic 0. A 2 × 16 LCD is connected to PORT B of the microcontroller. First row, starting from column 5 of the LCD displays text "COUNT", while the actual count is displayed in the second row as, for example, count 100 will be displayed as follows:

```
    COUNT
100
```

7.2.2 Project Hardware

The circuit diagram of the project is shown in Figure 7.10. The project is based on a PIC32MX460F512L 32-bit microcontroller. A 2 × 16 LCD is connected to PORT B, and external events are applied to bit 1 of PORT C (RC1). An event is recognized by the microcontroller and the event count is incremented and displayed in real-time on the LCD. The microcontroller is powered from a 3.3 V supply, and an 8 MHz crystal is used to provide the clock pulses. The LCD operates with +5 V supply and the microcontroller operates with +3.3 V supply. As a result of this, the microcontroller output pins cannot drive

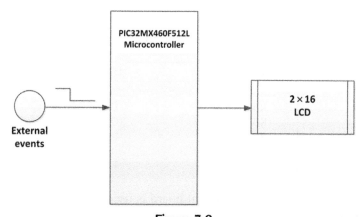

Figure 7.9
Block diagram of the project.

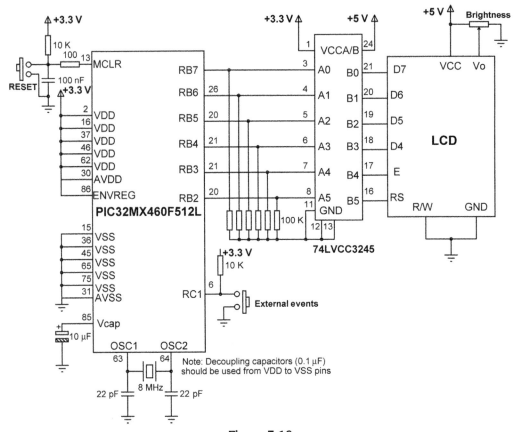

Figure 7.10
Circuit diagram of the project.

the LCD. A voltage translator chip (74VCC3245) is used between the LCD and the microcontroller I/O pins.

The interface between the microcontroller and the LCD is as follows:

PORT B	LCD
RB7	D7
RB6	D6
RB5	D5
RB4	D4
RB3	E
RB2	RS

In this project the LV-32MX V6 development board is used. External events are simulated by pressing push-button switch RC1 (notice that because of switch contact bouncing you may get more than one count when you press RC1 to simulate an external event).

The following jumpers should be configured on the board to enable the on-board LCD, and also to configure port pin RC1, so that changes from logic 1 to logic 0 can be detected on this port:

```
DIP switch SW20 positions 1–6, set to ON
Jumper J15, set to Ground
DIP switch SW5 position 1, set to ON
Jumper J5, set to Pull-Up
```

7.2.3 Project PDL

The operation of the project is described in the PDL given in Figure 7.11. At the beginning of the program PORT B pins where the LCD is connected are configured as digital and bit 1 of PORT C (RC1) are configured as input. Variable *Count* is cleared to zero. The program waits until pin RC1 goes from logic 1 to logic 0, and then increments and displays the value of *Count*. RC1 should go back to logic 1 before any other input can be accepted.

7.2.4 Project Program

The program is called EVENTS.C and the program listing is given in Figure 7.12. At the beginning of the program the interface between the LCD and the microcontroller is defined using a set of **sbit** statements. Then, variable *Count* is cleared to zero and a 4-byte text array called *Txt* is declared to store the value of *Count* as a string before it is sent to the LCD. PORT B pins are configured as digital and PORT C pin RC1 is configured as an input. The LCD is then initialized, cursor is turned OFF so that it is not visible, and the display is cleared before text COUNT is displayed on row 1, column 5.

```
BEGIN
        Define LCD interface
        Clear Count to zero
        Configure PORT B as digital
        Configure RC1 as input
        Initialize LCD
        Set cursor OFF
        Clear LCD
        Display "COUNT" at row 1, column 5 of LCD
        DO FOREVER
                Wait for an event to occur on RC1
                Increment Count
                Convert Count to string
                Display Count on second row of LCD
                Wait until event is removed
        ENDDO
END
```

Figure 7.11
PDL of the project.

```
/**************************************************************************
                        LCD EVENT COUNTER
                        =================
```

In this project an LCD is connected to PORT B of the microcontroller as follows:

```
PORT B    LCD
RB7       D7
RB6       D6
RB5       D5
RB4       D4
RB3       E
RB2       RS
```

In addition, port pin RC1 is used as the event input. An event is said to
occur if the event inpu changes from logic 1 to logic 0. In this project,
the push-button switch connected to RC1 is used to simulate an event occurring.

When an event is detected a counter is incremented and the total count is
displayed on the LCD. Row 1, starting from column 5 of the LCD displays heading
"COUNT". The actual count is displayed in the second row of the LCD. A byte is
used to store the event count. i.e. it is assumed that the event count is no
more than 255.

```
Author:        Dogan Ibrahim
Date:          July 2012
File:          EVENTS.C
**************************************************************************/
#define Event PORTCbits.RC1
#define Occured 0
#define NotOccured 1

// LCD module connections
sbit LCD_RS at LATB2_bit;
sbit LCD_EN at LATB3_bit;
sbit LCD_D4 at LATB4_bit;
sbit LCD_D5 at LATB5_bit;
sbit LCD_D6 at LATB6_bit;
sbit LCD_D7 at LATB7_bit;

sbit LCD_RS_Direction at TRISB2_bit;
sbit LCD_EN_Direction at TRISB3_bit;
sbit LCD_D4_Direction at TRISB4_bit;
sbit LCD_D5_Direction at TRISB5_bit;
sbit LCD_D6_Direction at TRISB6_bit;
sbit LCD_D7_Direction at TRISB7_bit;
// End LCD module connections


void main()
```

Figure 7.12

(Continued on next page)

```
{
    unsigned char Count = 0;
    unsigned char Txt[4];

    TRISC = 2;                              // RC1 input
    AD1PCFG = 0xFFFF;                       // PORT B digital

    Lcd_Init();                             // Initialize LCD
    Lcd_Cmd(_LCD_CURSOR_OFF);               // Turn OFF cursor
    Lcd_Cmd(_LCD_CLEAR);                    // Clear LCD
    Lcd_Out(1, 5, "COUNT");                 // Display "COUNT" on first row

    for(;;)                                 // Endless loop
    {
        while(Event == NotOccured);         // No event occurred
                                            // We are here so an Event occurred
        Count++;                            // Increment Count
        ByteToStr(Count, Txt);              // Convert Count to string
        Lcd_Out(2, 1, Txt);                 // Display Count
        while(Event == Occured);            // Wait until event is removed
    }
}
```

Figure 7.12
Program listing.

The main part of the program is executed inside a **for** loop which never terminates. Inside this loop the program waits until an event occurs, and then increments *Count*, converts it into a string in array *Txt* using library function **ByteToStr**, and then displays the value of *Count* on the second row of the LCD. The program then waits until pin RC1 returns back to 0, i.e. waits until the event has been removed. The above process is repeated forever.

Figure 7.13 shows an event count displayed on the LCD.

Figure 7.13
Displaying an event count.

Modifying the Display Format

It may be desirable to modify the display format such that the word COUNT and the actual event count number are displayed on the same row of the LCD. For example, we may want to display the event count as follows:

```
EVENT COUNT
Count = nnn
```

The modified program (named EVENTS2.C) to implement these changes is shown in Figure 7.14. Notice here that the word "Count = " is stored in a character array called *Head*. The event count is converted into a string and stored in character array *Txt*. The leading spaces in the converted string are removed using library function **Ltrim**. The event count is displayed at the current cursor position using LCD function **Lcd_Out_Cp**.

Figure 7.15 shows an event count displayed on the LCD with the new format.

7.3 Project 7.3—Creating a Custom LCD Character

7.3.1 Project Description

In this project we will see how to create a custom character or symbol and then display it on the LCD. The block diagram of the project is shown in Figure 7.16.

7.3.2 Project Hardware

The circuit diagram of the project is shown in Figure 7.17. The project is based on a PIC32MX460F512L 32-bit microcontroller. A 2 × 16 LCD is connected to PORT B as in the previous project. The microcontroller is powered from a 3.3 V supply, and an 8 MHz crystal is used to provide the clock pulses. The LCD operates with +5 V supply and the microcontroller operates with +3.3 V supply. As a result of this the microcontroller output pins cannot drive the LCD. A voltage translator chip (74VCC3245) is used between the LCD and the microcontroller I/O pins.

The interface between the microcontroller and the LCD is as in Project 7.2.

In this project the LV-32MX V6 development board is used. The following jumpers should be configured on the board to enable the on-board LCD:

```
DIP switch SW20 positions 1–6, set to ON
```

7.3.3 Project PDL

The operation of the project is described in the PDL given in Figure 7.18. At the beginning of the program PORT B pins where the LCD is connected to are configured as digital. Then the created new character is displayed on the LCD.

```
/****************************************************************************
                    LCD EVENT COUNTER
                    =================
```

In this project an LCD is connected to PORT B of the microcontroller as follows:

```
PORT B     LCD
RB7        D7
RB6        D6
RB5        D5
RB4        D4
RB3        E
RB2        RS
```

In addition, port pin RC1 is used as the event input. An event is said to occur if the event input changes from logic 1 to logic 0. In this project, the push-button switch connected to RC1 is used to simulate an event occurring.

When an event is detected a counter is incremented and the total count is displayed on the LCD. In this program the display format is as follows:

```
EVENT COUNT
Count = nnn
```

A byte is used to store the event count. i.e. it is assumed that the event count is no more than 255.

```
Author:      Dogan Ibrahim
Date:        July 2012
File:        EVENTS2.C
*****************************************************************************/
#define Event PORTCbits.RC1
#define Occured 0
#define NotOccured 1

// LCD module connections
sbit LCD_RS at LATB2_bit;
sbit LCD_EN at LATB3_bit;
sbit LCD_D4 at LATB4_bit;
sbit LCD_D5 at LATB5_bit;
sbit LCD_D6 at LATB6_bit;
sbit LCD_D7 at LATB7_bit;

sbit LCD_RS_Direction at TRISB2_bit;
sbit LCD_EN_Direction at TRISB3_bit;
sbit LCD_D4_Direction at TRISB4_bit;
sbit LCD_D5_Direction at TRISB5_bit;
sbit LCD_D6_Direction at TRISB6_bit;
sbit LCD_D7_Direction at TRISB7_bit;
// End LCD module connections
```

Figure 7.14
(Continued on next page)

```
void main()
{
   unsigned char Count = 0;
   unsigned char Txt[4];
   unsigned char Head[] = "Count = ";

   TRISC = 2;                              // RC1 input
   AD1PCFG = 0xFFFF;                       // PORT B digital

   Lcd_Init();                             // Initialize LCD
   Lcd_Cmd(_LCD_CURSOR_OFF);               // Turn OFF cursor
   Lcd_Cmd(_LCD_CLEAR);                    // Clear LCD
   Lcd_Out(1, 1, "EVENT COUNT");           // Display "COUNT" on first row

   for(;;)                                 // Endless loop
   {
      while(Event == NotOccured);          // No event occurred
                                           // We are here so an Event occurred
      Count++;                             // Increment Count
      ByteToStr(Count, Txt);               // Convert Count to string
      Lcd_Out(2, 1, Head);                 // Display Count
      Lcd_Out_Cp(Ltrim(Txt));              // Remove leading spaces
      while(Event == Occured);             // Wait until event is removed
   }
}
```

Figure 7.14
Modified event count display format.

Figure 7.15
Displaying an event count.

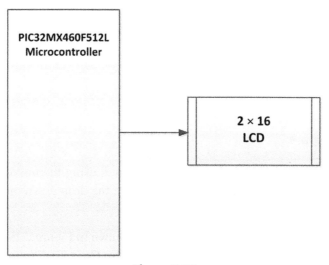

Figure 7.16
Block diagram of the project.

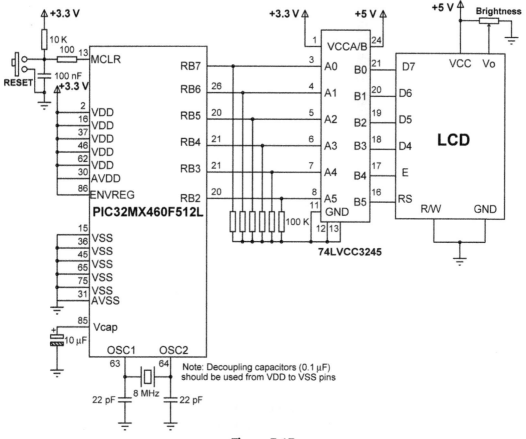

Figure 7.17
Circuit diagram of the project.

BEGIN
    Define LCD interface
    Insert the new character code
    Display the new character
END

Figure 7.18
PDL of the project.

7.3.4 Project Program

Custom LCD characters or symbols can easily be created using the mikroC PRO for PIC32 compiler LCD Custom Character option, selected from the drop down menu by clicking *Tools → LCD Custom Character*.

In this project, a simple up arrow has been created as shown in Figure 7.19. Select mikroC as the compiler and then click *Generate Code*, followed by *Copy To Clipboard*. The compiler will generate the required code as shown in Figure 7.20 and store it in the Clipboard. You

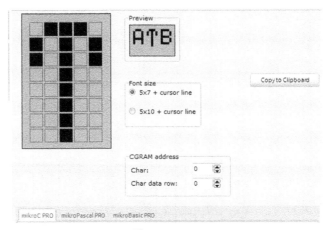

Figure 7.19
Created symbol.

```
mikroC PRO   mikroPascal PRO   mikroBasic PRO
const char character[] = {14,21,21,4,4,4,4,4};

void CustomChar(char pos_row, char pos_char) {
  char i;
    Lcd_Cmd(64);
    for (i = 0; i<=7; i++) Lcd_Chr_CP(character[i]);
    Lcd_Cmd(_LCD_RETURN_HOME);
    Lcd_Chr(pos_row, pos_char, 0);
}
```

Figure 7.20
Created code for the new symbol.

should paste this code at the beginning of your program before the **main**. The new character or symbol can be displayed by calling to function *CustomChar* created by the compiler and specifying the required row and column positions.

The program is called NEWSYMBOL.C and the program listing is given in Figure 7.21. At the beginning of the program the interface between the LCD and the microcontroller is defined using a set of **sbit** statements as in the previous project. Then the code created by the compiler is inserted. In this project the main program is very small where PORT B is configured as digital, the cursor is turned OFF, display is cleared and the new symbol is displayed by calling to function *CustomChar*. In this example, as shown in Figure 7.22, the symbol is displayed at row 1, column 1 of the LCD.

7.4 Project 7.4—LCD Progress Bar

7.4.1 Project Description

This project shows how to create a progress bar on the LCD. A function is written to display the percentage as well as a horizontal bar chart in the form of a progress bar. This function is called from a main program to show how progress bar can be created and displayed. The block diagram of the project is same as in Figure 7.16.

7.4.2 Project Hardware

The circuit diagram of the project is as shown in Figure 7.17.

7.4.3 Project PDL

The operation of the project is described in the PDL shown in Figure 7.23. Function *Progress_Bar* displays the progress bar on two rows of the LCD. The function has an argument which is the percentage to be displayed on the LCD. The top row displays the percentage as a number, while the second row displays it as a horizontal bar chart. The main program calls the function to display progress bars from 0% to 100% in steps of 10%, and with 3 s delay between each display.

7.4.4 Project Program

The program listing (called PBAR.C) is shown in Figure 7.24. The main program configures PORT B as digital I/O. Then the LCD is initialized and cursor is turned OFF. The main part of the program is executed in an endless **for** loop. Inside this loop, a variable is incremented from 0 to 100 in steps of 10, and then function *Progress_Bar* is called to display the progress bar, with 3 s delay between each output. Function *Progress_Bar* first converts the

```
/**************************************************************************
                     CREATE A NEW LCD CHARACTER/SYMBOL
                     =================================

In this project an LCD is connected to PORT B of the microcontroller as follows:

    PORT B    LCD
    RB7       D7
    RB6       D6
    RB5       D5
    RB4       D4
    RB3       E
    RB2       RS

A new symbol (an up arrow) is created and displayed at row 1, column 1 of the LCD.

Author:       Dogan Ibrahim
Date:         July 2012
File:         NEWSYMBOL.C
**************************************************************************/
// LCD module connections
sbit LCD_RS at LATB2_bit;
sbit LCD_EN at LATB3_bit;
sbit LCD_D4 at LATB4_bit;
sbit LCD_D5 at LATB5_bit;
sbit LCD_D6 at LATB6_bit;
sbit LCD_D7 at LATB7_bit;

sbit LCD_RS_Direction at TRISB2_bit;
sbit LCD_EN_Direction at TRISB3_bit;
sbit LCD_D4_Direction at TRISB4_bit;
sbit LCD_D5_Direction at TRISB5_bit;
sbit LCD_D6_Direction at TRISB6_bit;
sbit LCD_D7_Direction at TRISB7_bit;
// End LCD module connections

const char character[] = {14,21,21,4,4,4,4,4};

void CustomChar(char pos_row, char pos_char) {
 char i;
  Lcd_Cmd(64);
  for (i = 0; i <= 7; i++) Lcd_Chr_CP(character[i]);
  Lcd_Cmd(_LCD_RETURN_HOME);
  Lcd_Chr(pos_row, pos_char, 0);
}


void main()
{
   AD1PCFG = 0xFFFF;                  // PORT B digital
   Lcd_Init();                       // Initialize LCD
   Lcd_Cmd(_LCD_CURSOR_OFF);          // Turn OFF cursor
   Lcd_Cmd(_LCD_CLEAR);              // Clear LCD
   CustomChar(1,1);                  // Display the created symbol
}
```

Figure 7.21
Program listing.

Figure 7.22
Displaying the created symbol.

```
BEGIN/Main
        Define LCD interface
        Configure PORT B as digital
        Initialize LCD
        Turn OFF cursor
        DO FOREVER
                CALL Progress_Bar to display from 0% to 100% with 3
                        seconds between each display
        ENDDO
END/Main


BEGIN/Progress_Bar
        Display percentage as a number in row 1
        Display percentage as horizontal bar chart in row 2
END/Progress_Bar
```

Figure 7.23
PDL of the project.

percentage from numeric format into string using built-in function **ByteToStr** and sends it to the LCD. Then the percentage character "%" is displayed just next to the number. The LCD has 16 columns. Zero percent corresponds to no display and 100% corresponds to filling all 16 columns. Therefore, given a percentage n%, we can calculate how many columns to fill from the following equation:

```
LCD columns to fill = n * 16/100
```

In this project only integer arithmetic is considered for simplicity. Thus, for example, 50% corresponds to 50 * 16/100 = 8 columns to be filled. The ASCII character 0xFF is used as the fill character as this character fills a column with a black pattern.

```
/****************************************************************
                    PROGRESS BAR PROGRAM
                    =======================
```

In this project an LCD is connected to PORT B of the microcontroller as follows:

```
PORT B    LCD
RB7       D7
RB6       D6
RB5       D5
RB4       D4
RB3       E
RB2       RS
```

The program consists of a function called Progress_Bar. This function receives
the percentage to be displayed as its argument and then displays a progress bar.
Top row of the LCD shows the progress in numeric form, while the bottom row shows
it as a horizontal bar chart.

The main program calls the function to display progress bars from 0% to 100% in
steps of 10%, with 3 seconds delay between each display.


```
Author:      Dogan Ibrahim
Date:        July 2012
File:        PBAR.C
****************************************************************/
// LCD module connections
sbit LCD_RS at LATB2_bit;
sbit LCD_EN at LATB3_bit;
sbit LCD_D4 at LATB4_bit;
sbit LCD_D5 at LATB5_bit;
sbit LCD_D6 at LATB6_bit;
sbit LCD_D7 at LATB7_bit;

sbit LCD_RS_Direction at TRISB2_bit;
sbit LCD_EN_Direction at TRISB3_bit;
sbit LCD_D4_Direction at TRISB4_bit;
sbit LCD_D5_Direction at TRISB5_bit;
sbit LCD_D6_Direction at TRISB6_bit;
sbit LCD_D7_Direction at TRISB7_bit;
// End LCD module connections

//
// This function displays the progress bar on the LCD
// 100% corresponds to 16 LCD columns. Therefore, n% progress corresponds to
// n * 16 /100 LCD columns.
//
void Progress_Bar(unsigned char Percentage)
{
    unsigned char First, k, p, Txt[4];
```

Figure 7.24

(Continued on next page)

```
            ByteToStr(Percentage, Txt);
            Lcd_Out(1, 8, Ltrim(Txt));
            Lcd_Chr_Cp('%');
            p = Percentage * 16 / 100;

            First = 1;
            for(k = 0; k < p; k++)
            {
                if(First == 1)
                {
                     Lcd_Chr(2, 1, 0xFF);
                     First = 0;
                }
                Lcd_Chr_Cp(0xFF);
            }
}


void main()
{
    unsigned char i;

    AD1PCFG = 0xFFFF;                           // PORT B digital

    Lcd_Init();                                 // Initialize LCD
    Lcd_Cmd(_LCD_CURSOR_OFF);                   // Turn OFF cursor
    i = 0;

    for(;;)                                     // DO FOREVER
    {
      Progress_Bar(i);                          // Display progress bar
      i = i + 10;                               // Increment progress
      if(i > 100) i = 0;                        // If 100%, make it 0%
      Delay_Ms(3000);                           // Wait 3 second
      Lcd_Cmd(_LCD_CLEAR);                      // Clear LCD
    }
}
```

Figure 7.24
Program listing.

Figure 7.25 shows a typical display of the LCD. First row displays the percentage as a number, while the second row displays it as a horizontal bar.

7.5 Project 7.5—Shifting Text on LCD

7.5.1 Project Description

This project shows how text can be shifted left and right on the LCD. Some text is displayed on both rows of the LCD and then this text is shifted left and right continuously. The block diagram of the project is same as in Figure 7.16.

Figure 7.25
Displaying a progress bar.

7.5.2 Project Hardware

The circuit diagram of the project is as shown in Figure 7.17.

7.5.3 Project PDL

The operation of the project is described in the PDL shown in Figure 7.26. The program is in an endless **for** loop where a text displayed on the LCD is shifted to left and right continuously with some delay.

7.5.4 Project Program

The program listing (called SHIFT.C) is shown in Figure 7.27. The program consists of the main program and two functions named *Shift_Right* and *Shift_Left*. The functions require two arguments: the number of digits to shift, and the delay between each shift operation.

The main program configures PORT B as digital I/O. Then the LCD is initialized and cursor is turned OFF. Texts "ABC" and "DEF" are displayed at row 1 and row 2 of the display, respectively. The main part of the program is executed in an endless **for** loop. Inside this loop, function *Shift_Right* is called to shift the texts right five digits, with 1 s between each shift. Then, function *Shift_Left* is called to shift the characters left three digits with 2 s delay between each shift. Finally, function *Shift_Left* is called again to shift texts left two digits with 500 ms between each shift. Built-in function Vdelay_Ms is used to create the required delay.

```
BEGIN/Main
       Define LCD interface
       Configure PORT B as digital
       Initialize LCD
       Turn OFF cursor
       Display "ABC" at row 1
       Display " DEF" at row 2
       DO FOREVER
              CALL  Shift_Right to shift right 5 digits with 1 s delay
              CALL Shift_Left to shift left 3 digits with 2 s delay
              CALL Shift_Left to shift left 2 digits with 500 ms delay
       ENDDO
END/Main


BEGIN/Shift_Right
       Shift right display required number of times with the required delay
END/Shift_Right


BEGIN/Shift_Left
       Shift left display required number of times with the required delay
END/Shift_Left
```

Figure 7.26
PDL of the project.

Notice that the parameter of function **VDelay_Ms** must be a variable, whereas the parameter of function **Delay_Ms** must be an integer number. The above process is repeated forever.

Figure 7.28 shows a typical display.

Modifying the Program

The program given in Figure 7.27 can be simplified by combining the two shift functions and then introducing another parameter to define the shift direction. The new program (called SHIFT2.C) listing is shown in Figure 7.29, where the new function is called *Shift_Left_Right*.

The program can be made more readable by using enumerated data type to declare left and right shift operations. In the program listing given in Figure 7.30 (SHIFT3.C), left and right are assigned 0 and 1, respectively.

7.6 Project 7.6—External Interrupt-Based Event Counting Using LCD

7.6.1 Project Description

This project shows how to use the external interrupt pin INT0 (Port D, bit 0) of the PIC32MX460F512L microcontroller to generate interrupts to count events occurring at this pin. In this project interrupts are defined as the HIGH to LOW transition of the INT0 input. The event count is displayed on an LCD as in Project 7.2. The block diagram of the project is same as in Figure 7.9.

```
/*******************************************************************************
                             SHIFT TEXT ON LCD
                             ================
```

In this project an LCD is connected to PORT B of the microcontroller as follows:

```
PORT B    LCD
RB7       D7
RB6       D6
RB5       D5
RB4       D4
RB3       E
RB2       RS
```

In this program text "ABC" and "DEF" are displayed on row 1 and row 2 of the LCD. The program operates in a for loop, where inside this loop these texts are shifted left and write. The shifting speed depends on the delay inserted inside the loop. In this project the texts are shifted as follows:

> Shift right 5 digits with 1 s between each shift
> Shift left 3 digits with 2 s between each shift
> Shift left 2 digits with 500 ms between each digit

```
Author:       Dogan Ibrahim
Date:         July 2012
File:         SHIFT.C
*******************************************************************************/
// LCD module connections
sbit LCD_RS at LATB2_bit;
sbit LCD_EN at LATB3_bit;
sbit LCD_D4 at LATB4_bit;
sbit LCD_D5 at LATB5_bit;
sbit LCD_D6 at LATB6_bit;
sbit LCD_D7 at LATB7_bit;

sbit LCD_RS_Direction at TRISB2_bit;
sbit LCD_EN_Direction at TRISB3_bit;
sbit LCD_D4_Direction at TRISB4_bit;
sbit LCD_D5_Direction at TRISB5_bit;
sbit LCD_D6_Direction at TRISB6_bit;
sbit LCD_D7_Direction at TRISB7_bit;
// End LCD module connections


//
// This function shifts the characters to right. Parameter N is the number of
// to shift the display, and Del is the delay between each shift
//
void Shift_Right(unsigned char N, unsigned int Del)
{
```

Figure 7.27
(Continued on next page)

```
   unsigned char i;

   for(i = 0; i < N; i++)
   {
       Lcd_Cmd(_LCD_SHIFT_RIGHT);                    // Shift right
       VDelay_Ms(Del);                              // Wait Del milliseconds
   }
}


//
// This function shifts the characters to left. Parameter N is the number of
// to shift the display, and Del is the delay to be inserted
//
void Shift_Left(unsigned char N, unsigned int Del)
{
   unsigned char i;

   for(i = 0; i < N; i++)
   {
       Lcd_Cmd(_LCD_SHIFT_LEFT);                     // Shift right
       VDelay_Ms(Del);                              // Wait Del milliseconds
   }
}

//
// Start of main program
//
void main()
{
   unsigned char i;

   AD1PCFG = 0xFFFF;                                 // PORT B digital

   Lcd_Init();                                       // Initialize LCD
   Lcd_Cmd(_LCD_CURSOR_OFF);                         // Turn OFF cursor

   Lcd_Out(1, 1, "ABC");                             // Display "ABC" at row 1
   Lcd_Out(2, 1, "DEF");                             // Display "DEF" at row 2

   for(;;)                                           // DO FOREVER
   {
     Shift_Right(5, 1000);                           // Shift right with 1 s delay
     Shift_Left(3, 2000);                            // Shift left with 2 s delay
     Shift_Left(2, 500);                             // Shift left with 500 ms delay
   }
}
```

Figure 7.27
Program listing.

Figure 7.28
Shifting text left and right.

7.6.2 Project Hardware

The circuit diagram of the project is as shown in Figure 7.31. A 2 × 16 LCD is connected to PORT B as in Project 7.2, and external events are applied to pin INT0 as external interrupts. The microcontroller is powered from a 3.3 V supply, and an 8 MHz crystal is used to provide the clock pulses. A voltage translator chip (74VCC3245) is used as in Project 7.2 to interface the LCD to the microcontroller.

In this project the LV-32MX V6 development board is used. External interrupts are simulated by pressing push-button switch RD0. (Notice that you may get more than one count when you press the switch. This is because of switch contact bouncing as we shall see in a later project.) The following jumpers should be configured on the board to enable the on-board LCD, and also to configure port pin RD0 so that changes from logic 1 to logic 0 can be detected on this port:

```
DIP switch SW20 positions 1–6, set to ON
Jumper J15, set to Ground
DIP switch SW6 position 1, set to ON
Jumper J6, set to Pull-Up
```

7.6.3 Project PDL

The operation of the project is described in the PDL shown in Figure 7.32. Inside the main program the LCD is initialized and cursor is turned OFF. Variable *Count* is cleared to zero, the LCD is cleared and word COUNT is displayed at the first row.

```
/*********************************************************************
                        SHIFT TEXT ON LCD
                        ================
```

In this project an LCD is connected to PORT B of the microcontroller as follows:

```
PORT B    LCD
RB7       D7
RB6       D6
RB5       D5
RB4       D4
RB3       E
RB2       RS
```

In this program text "ABC" and " DEF" are displayed on row 1 and row 2 of the LCD.
The program operates in a for loop, where inside this loop these texts are shifted left
and write. The shifting speed depends on the delay inserted inside the loop. In this project
the texts are shifted as follows:

> Shift right 5 digits with 1 s between each shift
> Shift left 3 digits with 2 s between each shift
> Shift left 2 digits with 500 ms between each digit

In this program the left and right shift functions are combined into a single function.

```
Author:        Dogan Ibrahim
Date:          July 2012
File:          SHIFT2.C
*********************************************************************/
// LCD module connections
sbit LCD_RS at LATB2_bit;
sbit LCD_EN at LATB3_bit;
sbit LCD_D4 at LATB4_bit;
sbit LCD_D5 at LATB5_bit;
sbit LCD_D6 at LATB6_bit;
sbit LCD_D7 at LATB7_bit;

sbit LCD_RS_Direction at TRISB2_bit;
sbit LCD_EN_Direction at TRISB3_bit;
sbit LCD_D4_Direction at TRISB4_bit;
sbit LCD_D5_Direction at TRISB5_bit;
sbit LCD_D6_Direction at TRISB6_bit;
sbit LCD_D7_Direction at TRISB7_bit;
// End LCD module connections


//
// This function shifts the characters to left or right. Parameter N is the number
// of digits to shift the display, and Del is the delay between each shift. Parameter
// mode defines the shift direction. 0 shifts left, while 1 shifts right.
```

Figure 7.29

(Continued on next page)

```
//
void Shift_Left_Right(unsigned char N, unsigned int Del, unsigned mode)
{
  unsigned char i;

  for(i = 0; i < N; i++)
  {
    if(mode == 1)
      Lcd_Cmd(_LCD_SHIFT_RIGHT);              // Shift right
    else
      Lcd_Cmd(_LCD_SHIFT_LEFT);               // Shift left
      VDelay_Ms(Del);                         // Wait Del milliseconds
  }
}


//
// Start of main program
//
void main()
{
  unsigned char i;

  AD1PCFG = 0xFFFF;                           // PORT B digital

  Lcd_Init();                                 // Initialize LCD
  Lcd_Cmd(_LCD_CURSOR_OFF);                   // Turn OFF cursor

  Lcd_Out(1, 1, "ABC");                       // Display "ABC" at row 1
  Lcd_Out(2, 1, " DEF");                      // Display "DEF" at row 2

  for(;;)                                     // DO FOREVER
  {
    Shift_Left_Right(5, 1000, 1);             // Shift right with 1 s delay
    Shift_Left_Right(3, 2000, 0);             // Shift left with 2 s delay
    Shift_Left_Right(2, 500, 0);              // Shift left with 500 ms delay
  }
}
```

Figure 7.29
Modified program.

The main program then configures and enables external interrupts on pin INT0. The rest of the program is handled by the interrupt service routine (called EVENTS) where the *Count* is incremented and displayed on the LCD.

7.6.4 Project Program

The program is called EXTINT.C and the program listing is given in Figure 7.33. At the beginning of the program the interface between the LCD and the microcontroller is defined as before. Then, variable *Count* is cleared to zero and a 4-byte text array called *Txt* is declared to store the value of *Count* as a string before it is sent to the LCD.

```
/*****************************************************************************
                          SHIFT TEXT ON LCD
                          =================
```

In this project an LCD is connected to PORT B of the microcontroller as follows:

```
PORT B    LCD
RB7       D7
RB6       D6
RB5       D5
RB4       D4
RB3       E
RB2       RS
```

In this program text "ABC" and " DEF" are displayed on row 1 and row 2 of the LCD. The program operates in a for loop, where inside this loop these texts are shifted left and write. The shifting speed depends on the delay inserted inside the loop. In this project the texts are shifted as follows:

> Shift right 5 digits with 1 s between each shift
> Shift left 3 digits with 2 s between each shift
> Shift left 2 digits with 500 ms between each digit

In this program the left and right shift functions are combined into a single function, and Left and Right re declared as enumarated variables to simplify the program.

```
Author:       Dogan Ibrahim
Date:         July 2012
File:         SHIFT3.C
*****************************************************************************/
// LCD module connections
sbit LCD_RS at LATB2_bit;
sbit LCD_EN at LATB3_bit;
sbit LCD_D4 at LATB4_bit;
sbit LCD_D5 at LATB5_bit;
sbit LCD_D6 at LATB6_bit;
sbit LCD_D7 at LATB7_bit;

sbit LCD_RS_Direction at TRISB2_bit;
sbit LCD_EN_Direction at TRISB3_bit;
sbit LCD_D4_Direction at TRISB4_bit;
sbit LCD_D5_Direction at TRISB5_bit;
sbit LCD_D6_Direction at TRISB6_bit;
sbit LCD_D7_Direction at TRISB7_bit;
// End LCD module connections

    enum { Left, Right};
//
// This function shifts the characters to left or right. Parameter N is the number
// of digits to shift the display, and Del is the delay between each shift. Parameter
```

Figure 7.30

(Continued on next page)

```
// mode defines the shift direction. 0 shifts left, while 1 shifts right.
//
void Shift_Left_Right(unsigned char N, unsigned int Del, unsigned mode)
{
   unsigned char i;

   for(i = 0; i < N; i++)
   {
     if(mode == Right)
       Lcd_Cmd(_LCD_SHIFT_RIGHT);                          // Shift right
     else
       Lcd_Cmd(_LCD_SHIFT_LEFT);                           // Shift left
       VDelay_Ms(Del);                                     // Wait Del milliseconds
   }
}


//
// Start of main program
//
void main()
{
   unsigned char i;

   AD1PCFG = 0xFFFF;                                       // PORT B digital
   Lcd_Init();                                             // Initialize LCD
   Lcd_Cmd(_LCD_CURSOR_OFF);                               // Turn OFF cursor

   Lcd_Out(1, 1, "ABC");                                   // Display "ABC" at row 1
   Lcd_Out(2, 1, " DEF");                                  // Display "DEF" at row 2

   for(;;)                                                 // DO FOREVER
   {
     Shift_Left_Right(5, 1000, Right);                     // Shift right with 1 s delay
     Shift_Left_Right(3, 2000, Left);                      // Shift left with 2 s delay
     Shift_Left_Right(2, 500, Left);                       // Shift left with 500 ms delay
   }
}
```

Figure 7.30
Modified program.

Inside the main program, INT0 pin is configured as an input, and PORT B pins are configured as digital I/O. The LCD is initialized, cursor turned OFF, LCD cleared and word "COUNT" is displayed at row 1, column 5 of the LCD.

The remaining code in the main program configures external interrupts on pin INT0 (RD0) so that HIGH to LOW transition on this pin generates an interrupt. The following bits need to be configured for the microcontroller to accept external interrupts at pin INT0:

```
IFS0bits.INT0IF = 0;         // Clear external interrupt INT0 flag
INTCONbits.INT0EP = 0;       // External interrupt on HIGH to LOW transition
IPC0bits.INT0IP = 1;         // Set priority to 1
```

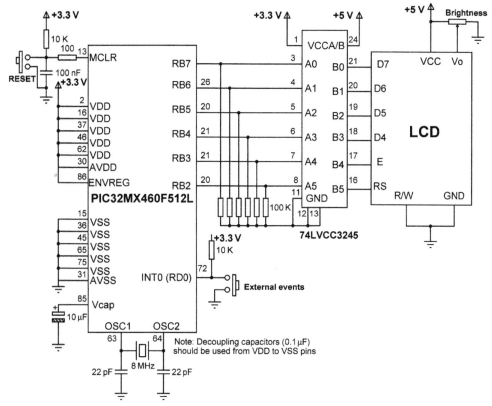

Figure 7.31
Circuit diagram of the project.

```
IECObits.INT0IE = 1;              // Enable external interrupts on INT0
EnableInterrupts();               // Enable interrupts
```

The main program then waits in a **for** loop forever. When an external interrupt is detected at pin INT0, the program jumps to the interrupt service routine, named EVENTS.

The interrupt service routine has been created using the Interrupt Assistant of the mikroC Pro for PIC32 compiler IDE, as described in Chapter 2. Here, the routine is given the name EVENTS, External_0 has been selected as the interrupt source, interrupt priority level set to 1, and context switching has been set to occur in software. The interrupt service routine code is as follows. Here, *Count* is incremented by 1, then converted into a string in character array *Txt* and is displayed in row 2 of the LCD. Notice that the external interrupt flag INT0IF must be cleared so that further interrupts can be accepted from pin INT0:

```
void EVENTS() iv IVT_EXTERNAL_0 ilevel 1 ics ICS_SOFT
{
  Count++;                        // Increment Count
  ByteToStr(Count, Txt);          // Convert Count to string
```

BEGIN
        Define LCD interface
        Clear Count to zero
        Configure PORT B as digital
        Initialize LCD
        Set cursor OFF
        Clear LCD
        Display "COUNT" at row 1, column 5 of LCD
        Configure external interrupts on pin INT0
        Enable interrupts
        DO FOREVER
                Wait for an external interrupt to occur on INT0
                Jump to Interrupt Service Routine, EVENTS
        ENDDO
END

BEGIN/EVENTS
        Increment Count
        Convert Count to string
        Display Count on second row of LCD
        Clear external interrupt INT0 flag
END/EVENTS

Figure 7.32
PDL of the project.

```
    Lcd_Out(2, 1, Txt);              // Display Count
    IFS0.INT0IF = 0;                 // Clear external interrupt INT0 flag
 }
```

The event count displayed on the LCD is same as in Figure 7.13.

7.7 Project 7.7—Switch Contact Debouncing

7.7.1 Project Description

Whenever a push-button switch is pressed and released, we get the effect known as "switch bouncing" where the switch contacts bounce and generate noise, causing undesirable logic levels to be sent to the connected hardware. In this project, we will demonstrate the switch bouncing effect by a project and then show how this effect can be removed in software.

Before going into the design of the project, it is worthwhile to learn a bit more about switch bouncing and see how this effect can be removed in hardware.

Switch Bouncing and Removing it in Hardware

Switch bouncing is a real problem when we want to detect a switch being pressed. Figure 7.34 shows what happens when a switch is pressed and released.

Switch bouncing problem can be solved in hardware by several ways. Perhaps the simplest way is to use a capacitor to delay the contact bouncing until it settles down. The circuit in

```
/*******************************************************************************
              EXTERNAL INTERRUPT BASED LCD EVENT COUNTER
              ===========================================
```

In this project an LCD is connected to PORT B of the microcontroller as follows:

```
PORT B    LCD
RB7       D7
RB6       D6
RB5       D5
RB4       D4
RB3       E
RB2       RS
```

The main program configures the I/O ports, initializes the LCD, turns OFF the cursor, and enables external interrupts on pin INT0 (RD0).

External interrupt INT0 pin (RD0) is used as the event input. An event is said to occur if the event input changes from logic 1 to logic 0. In this project, the push button switch connected to RD0 is used to simulate an external event occurring.

When an event is detected the program jumps to the Interrupt Service Routine (called EVENTS). Here, a counter is incremented and the total count is displayed on the LCD. In addition, external interrupt INT0 flag is cleared so that further interrupts can be detected by the microcontroller. In this project, the interrupt priority is set to 1.

The heading "COUNT" is displayed starting from row 1, column 5 of the LCD. A byte is used to store the event count. i.e. it is assumed that the event count is no more than 255.

```
Author:      Dogan Ibrahim
Date:        July 2012
File:        EXINT.C
*******************************************************************************/
// LCD module connections
sbit LCD_RS at LATB2_bit;
sbit LCD_EN at LATB3_bit;
sbit LCD_D4 at LATB4_bit;
sbit LCD_D5 at LATB5_bit;
sbit LCD_D6 at LATB6_bit;
sbit LCD_D7 at LATB7_bit;

sbit LCD_RS_Direction at TRISB2_bit;
sbit LCD_EN_Direction at TRISB3_bit;
sbit LCD_D4_Direction at TRISB4_bit;
sbit LCD_D5_Direction at TRISB5_bit;
sbit LCD_D6_Direction at TRISB6_bit;
sbit LCD_D7_Direction at TRISB7_bit;
// End LCD module connections

   unsigned char Count = 0;
```

Figure 7.33
(Continued on next page)

```
                unsigned char Txt[4];

    //
    // This is the Interrupts Service Routine. Whenever an event occurs on external
    // interrupt pin INT0, the program jumps here automatically. The priority level
    // of INT0 is set to 1. Notice that the interrupt flag of the interrupting device
    // must be cleared before exiting from the interrupt service routine, so that further
    // interrupts can be accepted from the same source.
    //
    void EVENTS() iv IVT_EXTERNAL_0 ilevel 1 ics ICS_SOFT
    {
        Count++;                          // Increment Count
        ByteToStr(Count, Txt);            // Convert Count to string
        Lcd_Out(2, 1, Txt);               // Display Count
        IFS0.INT0IF = 0;                  // Clear external interrupt INT0 flag
    }


    //
    // Start of main program
    //
    void main()
    {
        TRISD0_bit = 1;                   // Configure INT0 interrupt pin as input
        AD1PCFG = 0xFFFF;                 // Configure PORT B as digital

        Lcd_Init();                       // Initialize LCD
        Lcd_Cmd(_LCD_CURSOR_OFF);         // Turn OFF cursor
        Lcd_Cmd(_LCD_CLEAR);              // Clear LCD
        Lcd_Out(1, 5, "COUNT");           // Display "COUNT" on first row
    //
    // Configure external interrupts on pin INT0
    //
        IFS0bits.INT0IF = 0;              // Clear external interrupt INT0 flag
        INTCONbits.INT0EP = 0;            // External interrupt on HIGH to LOW transition
        IPC0bits.INT0IP = 1;              // Set priority to 1
        IEC0bits.INT0IE = 1;              // Enable external interrupts on INT0
        EnableInterrupts();               // Enable interrupts

        for(;;)                           // Wait for an external interrupt
        {
        }
    }
```

Figure 7.33
Program listing.

Figure 7.35 can be used for this purpose. Here, we choose values for the R and C such that the time-constant of the circuit (R × C) is longer than the contact bounce period, e.g. about 50—100 ms is typical. The delay should not be too large, otherwise we have to wait before we press the switch again. The buffer (e.g. Schmitt trigger gate) used after the RC circuit makes sure that we get a nice signal edge.

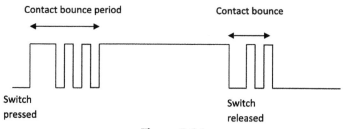

Figure 7.34
Switch contact bouncing.

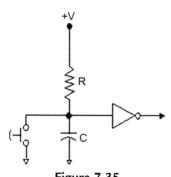

Figure 7.35
Hardware to eliminate switch contact bouncing.

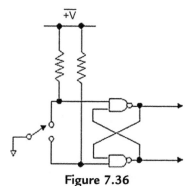

Figure 7.36
Eliminating switch contact bouncing using a flip-flop circuit.

Another more commonly used switch debouncing circuit is a cross-coupled flip-flop as shown in Figure 7.36. This circuit of this circuit is that a nice sharp signal edge is obtained and there is no need to use a buffer amplifier. Also, the circuit has no delay and we can press the button again without having to wait.

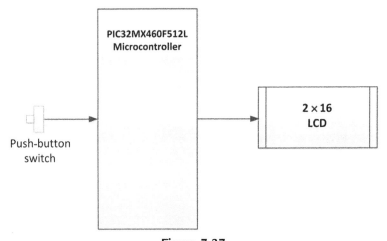

Figure 7.37
Block diagram of the project.

We are now ready to design our project. In this project, we will first write a program and show the effects of switch bouncing. A simple binary counter will be designed with a push-button switch and an LCD connected to the microcontroller. Every time the switch is pressed we expect the counter to count up by 1. We will see that because of switch bouncing effects the counting is not regular and more than one count may be produced when the button is pressed. We will then introduce switch debouncing into our program and see how the switch bouncing can be eliminated in software.

The block diagram of the project is shown in Figure 7.37.

7.7.2 Project Hardware

The circuit diagram of the project is shown in Figure 7.38. A 2 × 16 LCD is connected to PORT B as in Project 7.2, and a push-button switch is connected to bit 1 of PORT C (RC1). The microcontroller is powered from a 3.3 V supply, and an 8 MHz crystal is used to provide the clock pulses. A voltage translator chip (74VCC3245) is used as in Project 7.2 to interface the LCD to the microcontroller.

In this project, the LV-32MX V6 development board is used. The following jumpers should be configured on the board to enable the on-board LCD, and also to configure port pin RD0 so that changes from logic 1 to logic 0 can be detected on this port:

```
DIP switch SW20 positions 1-6, set to ON
Jumper J15, set to Ground
DIP switch SW5 position 1, set to ON
Jumper J5, set to Pull-Up
```

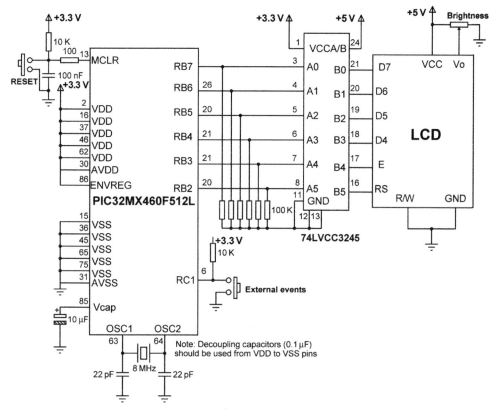

Figure 7.38
Circuit diagram of the project.

7.7.3 Project PDL

The PDL of the project is shown in Figure 7.39. At the beginning of the program PORT B pins where the LCD is connected are configured as digital, and bit 1 of PORT C (RC1) is configured as input. Variable *Count* is cleared to zero. The program waits until push button RC1 is pressed (i.e. button output goes from logic 1 to logic 0), and then increments and displays the value of *Count*, and then waits until the button is released.

7.7.4 Project Program

The program is called BOUNCE.C and the program listing is given in Figure 7.40. At the beginning of the program, the interface between the LCD and the microcontroller is defined, variable *Count* is cleared to zero, and a 4-byte text array called *Txt* is declared to store the value of *Count* as a string before it is sent to the LCD. PORT B pins are configured as digital and PORT C pin RC1 is configured as an input. The LCD is then initialized, cursor is turned OFF so that it is not visible, and the display is cleared.

```
BEGIN
        Define LCD interface
        Clear Count to zero
        Configure PORT B as digital
        Configure RC1 as input
        Initialize LCD
        Set cursor OFF
        Clear LCD
        DO FOREVER
                Wait for button RC1 to be pressed
                Increment Count
                Convert Count to string
                Display Count on first row of LCD
                Wait until button is released
        ENDDO
END
```

Figure 7.39
PDL of the project.

The main part of the program is executed inside a **for** loop which never terminates. Inside this loop the program waits until an event occurs, and then increments *Count*, converts it into a string in array *Txt*, and then displays the value of *Count* on the first row of the LCD. The program then waits until the button is released. The above process is repeated forever.

You should notice that when the button is pressed the counting is not one by one, but it jumps because of the switch bouncing.

Switch Contact Debouncing Code

We shall see now how we can eliminate the switch contact bouncing in software. mikroC PRO for PIC32 language has a "Button" library that contains a function called "Button" developed just for eliminating switch contact bouncing problems.

Function Button has the following arguments:

Port, Pin, Time, Active_State

Port and Pin specify which I/O port and which pin we will be debouncing. In our project here, we will be debouncing PORTC, pin 1.

Time is the debouncing period in milliseconds, where a few milliseconds are usually all that is required to eliminate the bouncing effects. In our project we will be using 2 ms.

Active_State specifies if the port pin is active from zero-to-one or from one-to-zero. In our project, the Active_State is from one-to-zero.

The new program listing (DEBOUNCE.C) is shown in Figure 7.41. If we press the button now, we will see that the button action is clean and the count goes up by one each time the button is pressed.

```
/****************************************************************************
                        CONTACT BOUNCING
                        ================
```

In this project an LCD is connected to PORT B of the microcontroller as follows:

```
PORT B    LCD
RB7       D7
RB6       D6
RB5       D5
RB4       D4
RB3       E
RB2       RS
```

In addition, a push-button switch is connected to port pin 1 of PORT C (RC1).
When the button is pressed the LCD is supposed to count up by one. But, because
of switch contact bouncing effect, the count normally jumps.

```
Author:      Dogan Ibrahim
Date:        July 2012
File:        BOUNCE.C
****************************************************************************/
// LCD module connections
sbit LCD_RS at LATB2_bit;
sbit LCD_EN at LATB3_bit;
sbit LCD_D4 at LATB4_bit;
sbit LCD_D5 at LATB5_bit;
sbit LCD_D6 at LATB6_bit;
sbit LCD_D7 at LATB7_bit;

sbit LCD_RS_Direction at TRISB2_bit;
sbit LCD_EN_Direction at TRISB3_bit;
sbit LCD_D4_Direction at TRISB4_bit;
sbit LCD_D5_Direction at TRISB5_bit;
sbit LCD_D6_Direction at TRISB6_bit;
sbit LCD_D7_Direction at TRISB7_bit;
// End LCD module connections

#define PushButton PORTCbits.RC1


void main()
{
   unsigned char Count = 0;
   unsigned char Txt[4];

   TRISC = 2;                        // RC1 input
   AD1PCFG = 0xFFFF;                 // PORT B digital
```

Figure 7.40
(Continued on next page)

```
Lcd_Init();                        // Initialize LCD
Lcd_Cmd(_LCD_CURSOR_OFF);          // Turn OFF cursor
Lcd_Cmd(_LCD_CLEAR);               // Clear LCD

for(;;)                            // Endless loop
{
    while(PushButton == 1);        // Wait until push button is pressed
    Count++;                       // Increment Count
    ByteToStr(Count, Txt);         // Convert Count to string
    Lcd_Out(1, 1, Txt);            // Display Count
    while(PushButton == 0);        // Wait until push button is released
}
}
```

Figure 7.40
Program listing.

7.8 Project 7.8—Timer Interrupt-Based Counting

7.8.1 Project Description

This project shows how to use the timer interrupts of the PIC32MX460F512L 32-bit microcontroller. Here, Timer 1 is configured to generate interrupts every 100 ms. Inside the interrupt service routine, a variable is incremented and the result is displayed on the LCD.

The block diagram of the project is same as in Figure 7.16.

7.8.2 Project Hardware

The circuit diagram of the project is shown in Figure 7.17. The microcontroller is powered from a 3.3 V supply, and an 8 MHz crystal is used to provide the clock pulses. A voltage translator chip (74VCC3245) is used as in Project 7.2 to interface the LCD to the microcontroller.

In this project, the LV-32MX V6 development board is used. The following jumpers should be configured on the board to enable the on-board LCD:

```
DIP switch SW20 positions 1–6, set to ON
```

7.8.3 Project PDL

The operation of the project is described in the PDL shown in Figure 7.42. Inside the main program, the LCD is initialized and cursor is turned OFF. Variable *Count* is cleared to zero, and the LCD is cleared.

The main program then configures and enables Timer 1 to generate interrupts at every 100 ms. The rest of the program is handled by the interrupt service routine (called COUNTER) where variable *Count* is incremented and displayed on the LCD.

```
/*****************************************************************************
                        CONTACT DEBOUNCING
                        ===================

In this project an LCD is connected to PORT B of the microcontroller as follows:

    PORT B      LCD
    RB7         D7
    RB6         D6
    RB5         D5
    RB4         D4
    RB3         E
    RB2         RS

In addition, a push-button switch is connected to port pin 1 of PORT C (RC1). In this
version of the program the built-in "Button" statement is used to eliminate the switch
contact bouncing problem. When the switch is pressed now the count goes up by one.


Author:         Dogan Ibrahim
Date:           July 2012
File:           DEBOUNCE.C
*****************************************************************************/
// LCD module connections
sbit LCD_RS at LATB2_bit;
sbit LCD_EN at LATB3_bit;
sbit LCD_D4 at LATB4_bit;
sbit LCD_D5 at LATB5_bit;
sbit LCD_D6 at LATB6_bit;
sbit LCD_D7 at LATB7_bit;

sbit LCD_RS_Direction at TRISB2_bit;
sbit LCD_EN_Direction at TRISB3_bit;
sbit LCD_D4_Direction at TRISB4_bit;
sbit LCD_D5_Direction at TRISB5_bit;
sbit LCD_D6_Direction at TRISB6_bit;
sbit LCD_D7_Direction at TRISB7_bit;
// End LCD module connections

#define PushButton PORTCbits.RC1


void main()
{
    unsigned char Count = 0;
    unsigned char Txt[4];

    TRISC = 2;                          // RC1 input
    AD1PCFG = 0xFFFF;                   // PORT B digital

    Lcd_Init();                         // Initialize LCD
```

Figure 7.41
(Continued on next page)

```
        Lcd_Cmd(_LCD_CURSOR_OFF);              // Turn OFF cursor
        Lcd_Cmd(_LCD_CLEAR);                   // Clear LCD

        for(;;)                                // Endless loop
        {
            while(PushButton);                 // Wait for button to be pressed
            while(Button(&PORTC, 1, 2, 0));    // Eliminate contact bouncing
            Count++;                           // Increment Count
            ByteToStr(Count, Txt);             // Convert Count to string
            Lcd_Out(1, 1, Txt);                // Display Count
        }
    }
```

Figure 7.41

Program listing that eliminates switch contact bouncing.

7.8.4 Project Program

The program is called TMRTINT.C and the program listing is given in Figure 7.43. At the beginning of the program, the interface between the LCD and the microcontroller is defined as before. Then, variable *Count* is cleared to zero and a 4-byte text array called *Txt* is declared to store the value of *Count* as a string before it is sent to the LCD.

Inside the main program, PORT B pins are configured as digital I/O. The LCD is initialized, cursor turned OFF, and the LCD is cleared.

```
BEGIN
        Define LCD interface
        Clear Count to zero
        Configure PORT B as digital
        Initialize LCD
        Set cursor OFF
        Clear LCD
        Configure Timer 1 to generate interrupts at every 500 ms
        Enable interrupts
        DO FOREVER
                Wait for a Timer 1 external interrupt to occur
                Jump to Interrupt Service Routine (COUNTER)
        ENDDO
END


BEGIN/COUNTER
        Increment Count
        Convert Count to string
        Display Count on second row of LCD
        Clear Timer interrupt flag T1IF
END/COUNTER
```

Figure 7.42

PDL of the project.

```
/******************************************************************************
                    TIMER INTERRUPT BASED COUNTER
                    ==============================
```

In this project an LCD is connected to PORT B of the microcontroller as follows:

```
    PORT B    LCD
    RB7       D7
    RB6       D6
    RB5       D5
    RB4       D4
    RB3       E
    RB2       RS
```

The main program configures the I/O ports, initializes the LCD, turns OFF the cursor, and enables Timer 1 interrupts.

When a Timer 1 interrupt is detected the program jumps to the Interrupt Service Routine (called COUNTER). Here, a counter is incremented and the total count is displayed on the LCD. In addition, Timer 1 interrupt T1IF flag is cleared so that further interrupts can be detected by the microcontroller. In this project, the interrupt priority is set to 1.

The heading "COUNT" is displayed starting from row 1, column 5 of the LCD. A byte is used to store the event count. i.e. it is assumed that the event count is no more than 255.

In this project, Timer 1 generates an interrupt at every 100 ms. The timer prescaler is set to 256. The clock rate is 80 MHz (8 MHz crystal is used but the frequnecy is multiplied by 10 using the PLL logic). The value to be loaded into the period register PR1 is calculate dfrom:

 PR1 = Required Timer delay/(Clock period x Prescaler)

At 80 MHz, T = $0.0125 \times 10-6$

Thus, PR1 = 100 ms/($0.0125 \times 10-6 \times 256$) = 31250


```
    Author:        Dogan Ibrahim
    Date:          July 2012
    File:          TMRINT.C
    ******************************************************************************/
// LCD module connections
sbit LCD_RS at LATB2_bit;
sbit LCD_EN at LATB3_bit;
sbit LCD_D4 at LATB4_bit;
sbit LCD_D5 at LATB5_bit;
sbit LCD_D6 at LATB6_bit;
sbit LCD_D7 at LATB7_bit;

sbit LCD_RS_Direction at TRISB2_bit;
sbit LCD_EN_Direction at TRISB3_bit;
```

Figure 7.43

(Continued on next page)

```
sbit LCD_D4_Direction at TRISB4_bit;
sbit LCD_D5_Direction at TRISB5_bit;
sbit LCD_D6_Direction at TRISB6_bit;
sbit LCD_D7_Direction at TRISB7_bit;
// End LCD module connections

  unsigned char Count = 0;
  unsigned char Txt[4];

//
// This is the Timer 1 Interrupts Service Routine. Whenever a Timer 1 interrupt
// occurs, the program jumps here automatically. The priority level if the
// interrupt is set to 1. Notice that the interrupt flag of the interrupting device
// must be cleared before exiting from the interrupt service routine, so that further
// interrupts can be accepted from the same source.
//
void COUNTER() iv IVT_TIMER_1 ilevel 1 ics ICS_SOFT
{
  Count++;                            // Increment Count
  ByteToStr(Count, Txt);              // Convert Count to string
  Lcd_Out(1, 1, Txt);                 // Display Count
  IFS0bits.T1IF = 0;                  // Clear Timer 1 interrupt flag
}


//
// Start of main program
//
void main()
{
  AD1PCFG = 0xFFFF;                   // Configure PORT B as digital

  Lcd_Init();                         // Initialize LCD
  Lcd_Cmd(_LCD_CURSOR_OFF);           // Turn OFF cursor
  Lcd_Cmd(_LCD_CLEAR);                // Clear LCD
//
// Configure Timer 1 to generate interrupts at every 100 ms
//
  T1CONbits.ON = 0;                   // Disable Timer 1
  TMR1 = 0;                           // Clear TMR1 register
  T1CONbits.TCKPS = 3;                // Select prescaler = 256
  T1CONbits.TCS = 0;                  // Select internal clock
  PR1 = 31250;                        // Load period register
  IFS0bits.T1IF = 0;                  // Clear Timer 1 interrupt flag
  IPC1bits.T1IP = 1;                  // Set priority level to 1
  IEC0bits.T1IE = 1;                  // Enable Timer 1 interrupts
  T1CONbits.ON = 1;                   // Enable Timer 1
  EnableInterrupts();                 // Enable interrupts

  for(;;)                             // Wait for Timer 1 interrupts
  {

  }
}
```

Figure 7.43
Program listing.

The remaining code in the main program configures Timer 1 to generate interrupts at every 100 ms.

As was discussed in Chapter 2, the value to be loaded into the period register PR1 is calculated from the following equation:

```
PR1 = Required Timer delay/(Clock period × Prescaler value)
```

In all the projects in this book, the actual crystal frequency is 8 MHz, but the microcontroller operates at 80 MHz by using the PLL logic. In this project, the prescaler is chosen as 256.

At 80 MHz, the period is given by, $T = 0.0125 \times 10^{-6}$ s, or $T = 0.0125 \times 10^{-3}$ ms

Thus, $PR1 = 100$ ms$/(0.0125 \times 10^{-3} \times 256) = 31,250$.

The following bits need to be configured for the microcontroller to generate Timer 1 interrupts at every 100 ms:

```
T1CONbits.ON = 0;          // Disable Timer 1
TMR1 = 0;                  // Clear TMR1 register
T1CONbits.TCKPS = 3;       // Select prescaler = 256
T1CONbits.TCS = 0;         // Select internal clock
PR1 = 31250;               // Load period register
IFS0bits.T1IF = 0;         // Clear Timer 1 interrupt flag
IPC1bits.T1IP = 1;         // Set priority level to 1
IEC0bits.T1IE = 1;         // Enable Timer 1 interrupts
T1CONbits.ON = 1;          // Enable Timer 1
EnableInterrupts();        // Enable interrupts
```

The main program then waits in a **for** loop forever. When a Timer 1 interrupt is detected, the program jumps to the interrupt service routine, named COUNTER.

The interrupt service routine has been created using the Interrupt Assistant of the mikroC Pro for PIC32 compiler IDE, as described in Chapter 2. Here, the routine is given the name COUNTER, Timer_1 has been selected as the interrupt source, interrupt priority level set to 1, and context switching has been set to occur in software. The interrupt service routine code is as follows. Here, *Count* is incremented by 1, then converted into a string in character array *Txt* and is displayed in row 1 of the LCD. Notice that the Timer 1 interrupt flag T1IF must be cleared so that further interrupts can be accepted from Timer 1:

```
void COUNTER() iv IVT_TIMER_1 ilevel 1 ics ICS_SOFT
{
  Count++;                  // Increment Count
  ByteToStr(Count, Txt);    // Convert Count to string
  Lcd_Out(1, 1, Txt);       // Display Count
  IFS0bits.T1IF = 0;        // Clear Timer 1 interrupt flag
}
```

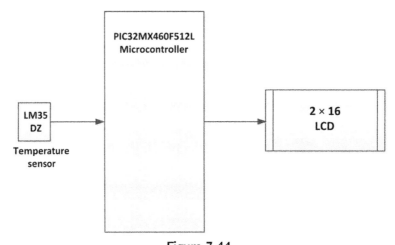

Figure 7.44
Block diagram of the project.

7.9 Project 7.9—Temperature Measurement and Display on LCD

7.9.1 Project Description

This project shows how the A/D converter of the PIC32MX460F512L microcontroller can be used to measure the analog temperature and display it on the LCD. The temperature is read every second and then displayed in floating point format on the LCD as follows:

```
T = nn.nnn
```

The block diagram of the project is same as in Figure 7.44. In this project, an LM35DZ-type temperature sensor is used with analog output. LM35DZ can measure temperature in the range of 2–100 °C. The output voltage is proportional to the measured voltage and is expressed as follows:

```
Vo = 10 mV/°C
```

Thus, for example, at 10 °C, the output voltage is 100 mv, and at 25 °C, the output voltage is 250 mV and so on.

7.9.2 Project Hardware

The circuit diagram of the project is shown in Figure 7.45. The output of the LM35DZ temperature sensor chip is connected to analog input AN0 (or RB0) of the PIC32MX460F512L microcontroller. The microcontroller is powered from a 3.3 V supply, and an external 8 MHz crystal is used to provide the clock pulses. With internal PLL clock circuitry, the actual microcontroller clock rate is 80 MHz. An LCD is connected to PORT B pins as before.

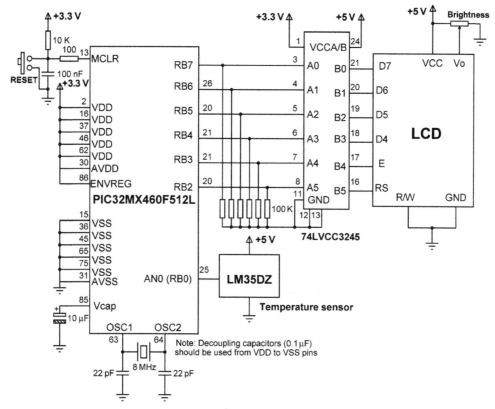

Figure 7.45
Circuit diagram of the project.

In this project, the A/D converter is operated from the supply voltage of +3.3 V. With a 10-bit A/D converter (1024 steps) and +3.3 V reference voltage, it is possible to detect a voltage change of 3.22 mV at the A/D converter inputs. Notice that the LM35DZ sensor requires a minimum of +4 V to operate.

A voltage translator chip (74VCC3245) is used as in Project 7.2 to interface the LCD to the microcontroller.

In this project, the LV-32MX V6 development board is used. The following jumpers should be configured on the board to enable the on-board LCD:

```
DIP switch SW20 positions 1–6, set to ON
In addition, provide +5 V for the LM35DZ sensor from the board
```

7.9.3 Project PDL

The operation of the project is described in the PDL shown in Figure 7.46. Here, AN0 (RB0) pin is configured as analog input and remaining PORT B pins are configured as digital.

```
BEGIN
        Define LCD interface
        Configure AN0 (RB0) as analog input
        Configure RB2–RB7 as digital
        Initialize LCD
        Set cursor OFF
        Clear LCD
        DO FOREVER
                Read channel 0 (AN0) data
                Convert into °C
                Display the temperature on LCD
        ENDDO
END
```

Figure 7.46
PDL of the project.

The LCD is initialized and cursor is turned OFF. Then an endless loop is formed where the temperature is read from AN0, converted into degree Celcius, and displayed on the LCD.

7.9.4 Project Program

The program is called TEMPERATURE.C and the program listing is given in Figure 7.47. At the beginning of the program, the interface between the LCD and the microcontroller is defined as before. Character array *Txt*, variables *ADC_Data*, and *mV* are declared. AN0 (RB0) is then configured as an analog channel. The remaining bits of PORT B are configured as digital since the LCD is connected to these pins. Setting TRISB = 1 configures AN0 as an input pin. The LCD is then initialized, cursor turned OFF, and the display is cleared. Then the A/D converter is initialized by calling to built-in library function *ADC1_Init*.

The remainder of the code is executed forever in a **for** loop. Inside the loop, the sensor data are read and converted into digital by calling built-in function *ADC1_Get_Sample (0)* where the channel number is specified as an argument. This value is then converted into actual physical voltage in millivolts. Since the A/D converter is 10-bits wide and the reference voltage is +3.3 V (3300 mV), it is necessary to multiply the read value by 3300/1023 in order to convert into real voltage. The actual temperature is then found by dividing the result by 10 (the sensor output is 10 mV/°C). The result is converted into a string held by array *Txt* and is then displayed on the LCD. The above process is repeated after a delay of 1 s.

Figure 7.48 shows a typical display of the temperature.

7.10 Project 7.10—Playing a Melody

7.10.1 Project Description

This project shows how to use the mikroC PRO for PIC32 Sound library to generate sound on a piezoelectric speaker (or buzzer). In this project, a program is written to play a simple melody, the Jingle Bells. Figure 7.49 shows the block diagram of the project.

```
/****************************************************************************
                        TEMPERATURE DISPLAY
                        ===================
```

In this project an LCD is connected to PORT B of the microcontroller as follows:

```
PORT B     LCD
RB7        D7
RB6        D6
RB5        D5
RB4        D4
RB3        E
RB2        RS
```

The main program configures the I/O ports, initializes the LCD, and turns OFF the cursor.

An LM35DZ type analog temperature sensor is connected to analog input AN0 of the microcontroller. The program reads the sensor data every second, converts into temperature, and then displays it on the LCD.

```
Author:        Dogan Ibrahim
Date:          July 2012
File:          TEMPERATURE.C
****************************************************************************/
// LCD module connections
sbit LCD_RS at LATB2_bit;
sbit LCD_EN at LATB3_bit;
sbit LCD_D4 at LATB4_bit;
sbit LCD_D5 at LATB5_bit;
sbit LCD_D6 at LATB6_bit;
sbit LCD_D7 at LATB7_bit;

sbit LCD_RS_Direction at TRISB2_bit;
sbit LCD_EN_Direction at TRISB3_bit;
sbit LCD_D4_Direction at TRISB4_bit;
sbit LCD_D5_Direction at TRISB5_bit;
sbit LCD_D6_Direction at TRISB6_bit;
sbit LCD_D7_Direction at TRISB7_bit;
// End LCD module connections


//
// Start of main program
//
void main()
{
    unsigned char Txt[] = "T=        ";
    unsigned ADC_Data;
    float mV;

    AD1PCFG = 0xFFFE;                          // Configure AN0 (RB0) analog, RB1–RB7 digital
```

Figure 7.47

(Continued on next page)

```
TRISB = 1;                          // Configure AN0 (RB0) as input
Lcd_Init();                         // Initialize LCD
Lcd_Cmd(_LCD_CURSOR_OFF);           // Turn OFF cursor
Lcd_Cmd(_LCD_CLEAR);                // Clear LCD
ADC1_Init();                        // Initialize A/D converter

for(;;)                             // DO FOREVER
{
  ADC_Data = ADC1_Get_Sample(0);    // Read analog input AN0 (Channel 0)
  mV = ADC_Data * 3300.0 / 1024.0;  // Convert to mV
  mV = mV / 10.0;                   // Convert to temperature in °C
  FloatToStr(mV, Txt+2);            // Convert to string
  Lcd_Out(1, 1, Txt);               // Display temperature
  Delay_Ms(1000);                   // Wait 1 second
}
}
```

Figure 7.47
Program listing.

Figure 7.48
A typical display of the temperature.

7.10.2 Project Hardware

The circuit diagram of the project is shown in Figure 7.50. The circuit is very simple: bit 3 of PORT C (RC3) is connected to a piezoelectric speaker through a transistor.

In this project, the LV-32MX V6 development board is used. There is no need to set any jumper on the board. The buzzer used in the project is the easyBUZZ card (see Figure 7.51), manufactured by mikroElektronika. easyBUZZ is a small board including a piezoelectric buzzer, a transistor, and resistors. The piezoelectric buzzer is designed to operate from +3.3 to +5 V, at the frequency range of 20 Hz—20 kHz. The board has a 10-way connector and it should be plugged in to the PORT C connector at the right-hand side of the LV-32MX V6

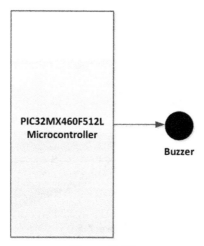

Figure 7.49
Block diagram of the project.

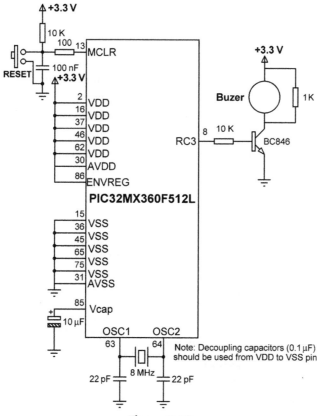

Figure 7.50
Circuit diagram of the project.

Figure 7.51
easyBUZZ board.

board. easyBUZZ has an eight-way DIP switch for selecting the connection to the microcontroller I/O port. For connecting to RC3, switch 3 should be set to ON position.

7.10.3 Project PDL

The operation of the project is described in the PDL shown in Figure 7.52. The musical notes are stored in an array called *Notes*. Similarly, the corresponding duration of each note is stored in another array called *Tim*. The program configures port pin RC3 as output, and initializes the sound library. A loop is then formed which sends each musical note and its duration to the buzzer. An internote gap is included between each output. At the end, there is silence for 5 s and the melody is repeated.

```
BEGIN
     Define musical notes
     Store notes for melody Jingle Bells in array Notes
     Store the corresponding duration of notes in array Tim
     Configure port pin RC3 as output
     Initialize the Sound library
     DO FOREVER
          DO FOR ALL NOTES
               Read a note from array Notes
               Read the duration of the note from array Tim
               Send the note and its duration to piezospeaker
               Include internote delay
          ENDDO
          Wait 5 seconds
     ENDDO
END
```

Figure 7.52
PDL of the project.

7.10.4 Project Program

The program is called MELODY.C and the program listing is given in Figure 7.53. At the beginning of the program, the frequencies of musical notes in the middle octave are defined as follows:

```
#define C 261
#define D 293
#define E 329
#define F 349
#define G 392
#define A 440
#define B 493
```

The musical notes of the popular melody Jingle Bells is as follows:

```
E E E E E E G C D E F F F F F F E E E E D D E D G E E E E E E G C D E X
```

Note that the letter 'X' at the end of the array denotes the end of the array. The corresponding duration of each note in the melody is as follows. Note that these are the weights of the notes, not the actual physical duration:

```
1 1 2 1 1 2 1 1 1 1 4 1 1 1 1 1 1 1 1 1 1 1 1 2 2 1 1 2 1 1 2 1 1 1 1 4
```

The program then configures port pin RC3 as output, and initializes the Sound library. The remaining part of the code is executed continuously inside a **for** loop. Here, the musical notes and their corresponding duration are read from the two arrays and sent to the piezospeaker using the built-in **Sound_Play** function. Here, variable **k** is incremented at each iteration to point to the next note. The durations are multiplied by 200 to convert them to actual physical durations in milliseconds:

```
Sound_Play(Notes[k], Tim[k]*200);          // Play a note
```

While playing musical notes, it is necessary to insert internote delays so that the notes are not mixed up. In this project 150 ms internote delay is used. After the end of the melody, there is 5 s of silence and the **for** loop is repeated.

7.11 Project 7.11—Playing a Melody Using Push-Button Switches

7.11.1 Project Description

In the previous project we have seen how to use the mikroC PRO for PIC32 Sound library to generate musical tones, and how a simple melody can be created. In this project we will use push-button switches to generate musical notes so that we can play simple melodies. Eight push-button switches connected to PORT B will be programmed to generate the musical notes. Only one octave is used (C4—C5) in this project for simplicity, but this can easily be extended if desired.

```
/******************************************************************************
                                MELODY PLAY
                                ===========

In this project a piezospeaker is connected to pin 3 of PORT C (RC3).

The program uses the Sound library of mikroC PRO for PIC32 to play a simple melody
(Jingle Bells).

The frequencies of the middle octave musical notes are:

C4 = 261, D4 = 293, E4 =329, F4 = 349, G4 = 392, A4 = 440, B4 = 493

The melody Jingle Bells has the following notes. The duration of each note is also given below.
The durations are multiplied by a fixed delay value of 200 ms. The internote gap (silence) is
taken as 150 ms:

Notes:      E, E, E, E, E, E, E, G, C, D, E, F, F, F, F, F, E, E, E, E, E, D, D, E, D, G
Duration:   1, 1, 2, 1, 1, 2, 1, 1, 1, 1, 4, 1, 1, 1, 1, 1, 1, 1, 1, 1, 1, 1, 1, 2, 2


Author:     Dogan Ibrahim
Date:       July 2012
File:       MELODY.C
******************************************************************************/
//
// Define the frequencies of musical notes
//
#define C 261
#define D 293
#define E 329
#define F 349
#define G 392
#define A 440
#define B 493


//
// Start of main program
//
void main()
{
    unsigned int Notes[] = {E, E, E, E, E, E, E, G, C, D, E, F, F, F, F, F, E,
                E, E, E, D, D, E, D, G, E, E, E, E, E, E, E, G, C, D, E, 'X'};
    unsigned char Tim[] = {1, 1, 2, 1, 1, 2, 1, 1, 1, 1, 4, 1, 1, 1, 1, 1, 1,
                1, 1, 1, 1, 1, 2, 2, 1, 1, 2, 1, 1, 2, 1, 1, 1, 1, 4};
    unsigned char k;

    TRISC3_bit = 0;                         // Configure RC3 as output
    Sound_Init(&PORTC, 3);                  // Initialize sound library
```

Figure 7.53

(Continued on next page)

```
   for(;;)                                  // DO FOREVER
   {
     k = 0;                                 // Start from the beginning
     while(Notes[k] != 'X')                 // Do while there are more notes
     {
       Sound_Play(Notes[k], Tim[k]*200);    // Play a note
       k++;                                 // Get next note
       Delay_Ms(150);                       // Internote gap in milliseconds
     }
     Delay_Ms(5000);                        // Wait 5 seconds
   }
}
```

Figure 7.53
Program listing.

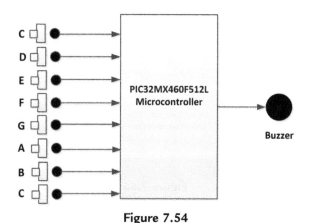

Figure 7.54
Block diagram of the project.

Figure 7.54 shows the block diagram of the project.

7.11.2 Project Hardware

The circuit diagram of the project is shown in Figure 7.55. The piezoelectric speaker is connected to port pin RC3 through a transistor as in the previous project. Eight push-button switches are connected to lower PORT B pins to play the musical notes.

As in the previous project, the hardware in this project is based on the LV-32MX V6 development board and the easyBUZZ piezoelectric buzzer board. The easyBUZZ board should be connected to PORT C connector at the edge of LV-32MX V6 board and DIP switch 2 on the easyBUZZ board should be set to ON position.

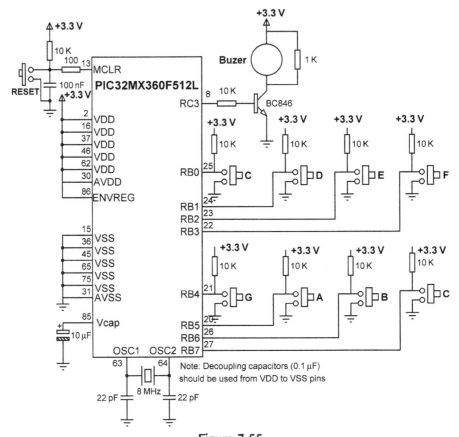

Figure 7.55
Circuit diagram of the project.

The following jumpers should be selected on the LV-32MX V6 board so that the push-button switch outputs go from logic 1 to logic 0 when they are pressed:

```
Jumper J15, set to Ground
DIP switch SW3 position 1–8, set to ON
Jumper J3, set to Pull-Up
```

7.11.3 Project PDL

The operation of the project is described in the PDL shown in Figure 7.56. At the beginning, we define the frequencies of musical notes, configure PORT B lower 8 bits as digital output, and also configure port pin RC3 as output (this is where the buzzer is connected to). The rest of the code is executed in an endless loop. Inside this, the program waits until a button is pressed, and the pressed button is identified. Then, the frequency corresponding to the pressed

BEGIN
    Define musical notes
    Configure PORT B as digital input
    Configure port pin RC3 as output
    Initialize the Sound library
    DO FOREVER
        Wait until a button is pressed
        Identify the pressed button
        Call **Sound_Play** and pass frequency and duration of the button
    ENDDO
END

Figure 7.56
PDL of the project.

button and the duration are passed to **Sound_Play** function to generate the required note on the piezoelectric speaker.

7.11.4 Project Program

The program is called PLAY.C and the program listing is given in Figure 7.57. At the beginning of the program, the frequencies of musical notes in the middle octave are defined as follows:

```
#define C1 261
#define D 293
#define E 329
#define F 349
#define G 392
#define A 440
#define B 493
#define C2 523
```

Then, PORT B where the push-button switches are connected to is configured as digital output. Port pin RC3 where the piezoelectric speaker is connected to is configured as an output pin.

The rest of the program is executed endlessly in a **for** loop. Inside this loop, the program waits until a button is pressed, using the following statement:

```
while((PORTB && 0xFF) == 0xFF);   // Wait until a button is pressed
```

When no buttons are pressed, the lower 8 bits of PORT B has the value 0xFF. Whenever a button is pressed, the program continues to the next statement where the pressed button is identified:

```
key = ~(PORTB & 0xFF);   // Get low 8 bits of PORT B
```

If, for example, button RB0 is pressed, then PORT B lower 8 bits will have the binary value "11111110". This is complemented to give "00000001" and stored in variable called *key*.

```
/**********************************************************************
                  MELODY PLAY WITH PUSH BUTTON SWITCHES
                  =====================================
```

In this project a piezoelectric speaker is connected to pin 3 of PORT C (RC3).

In addition, 8 push-button switches are connected to PORT B pins of the PIC32MX460F512L microcontroller.

The push-button switches are programmed to generate the musical tones in one octave. The button assignments are as follows:

RB0: C, RB1: D, RB2: E, RB3: F, RB4: G, RB5: A, RB6: B, RB7: C

The frequencies of the middle octave musical notes are:

C4 = 261, D4 = 293, E4 =329, F4 = 349, G4 = 392, A4 = 440, B4 = 493, C5 = 523

Thus, for example, pressing button RB5 will generate note "A" with frequency 440 Hz.

In this project we are not debouncing the switch contacts for simplicity. Also, only one key must be pressed at any time.


```
Author:       Dogan Ibrahim
Date:         July 2012
File:         PLAY.C
**********************************************************************/
//
// Define the frequencies of musical notes
//
#define C1 261
#define D  293
#define E  329
#define F  349
#define G  392
#define A  440
#define B  493
#define C2 523


//
// Start of main program
//
void main()
{
    unsigned char key;
    unsigned int Duration = 200;

    AD1PCFG = 0xFFFF;                    // Configure PORT B as digital
    TRISB = 0xFF;                        // Configure lower PORT B as inputs
```

Figure 7.57

(Continued on next page)

```
        TRISC3_bit = 0;                        // Configure RC3 as output
        Sound_Init(&PORTC, 3);                 // Initialize sound library
    //
    // Wait until a button is pressed in lower PORT B (0 to 7). Normally PORT B is at
    // logic 1 because all the switches are pulling up to logic 1. Thus PORT B is normally
    // at 0xFF. When a button is pressed, if for example RB0, then lower PORT B becomes
    // "11111110". Then, this data is complemented to give "00000001". A switch statement
    // is used to test which button is pressed (1 to 128), and then Sound_Play statement
    // is used to generate the required tone.
    //
        for(;;)                                // DO FOREVER
        {
            while((PORTB && 0xFF) == 0xFF);    // Wait until a button is pressed
            key = ~(PORTB & 0xFF);             // Get low 8 bits of PORT B
            switch(key)
            {
                case 1:                        // RB0 pressed ?
                    Sound_Play(C1, Duration);  // Play C1
                    break;
                case 2:                        // RB1 pressed ?
                    Sound_Play(D, Duration);   // Play D
                    break;
                case 4:                        // RB2 pressed ?
                    Sound_Play(E, Duration);   // Play E
                    break;
                case 8:                        // RB3 pressed ?
                    Sound_Play(F, Duration);   // Play F
                    break;
                case 16:                       // RB4 pressed ?
                    Sound_Play(G, Duration);   // Play G
                    break;
                case 32:                       // RB5 pressed ?
                    Sound_Play(A, Duration);   // Play A
                    break;
                case 64:                       // RB6 pressed ?
                    Sound_Play(B, Duration);   // Play B
                    break;
                case 128:                      // RB7 pressed ?
                    Sound_Play(C2, Duration);  // Play C2
                    break;
            }
        }
    }
```

Figure 7.57
Program listing.

This variable is then used in a **switch** statement to send the frequency and duration of the corresponding note to function **Sound_Play** as follows:

```
switch(key)
{
 case 1:                              // RB0 pressed ?
   Sound_Play(C1, Duration);         // Play C1
   break;
```

```
case 2:                          // RB1 pressed ?
  Sound_Play(D, Duration);       // Play D
  break;
case 4:                          // RB2 pressed ?
  Sound_Play(E, Duration);       // Play E
  break;
case 8:                          // RB3 pressed ?
  Sound_Play(F, Duration);       // Play F
  break;
case 16:                         // RB4 pressed ?
  Sound_Play(G, Duration);       // Play G
  break;
case 32:                         // RB5 pressed ?
  Sound_Play(A, Duration);       // Play A
  break;
case 64:                         // RB6 pressed ?
  Sound_Play(B, Duration);       // Play B
  break;
case 128:                        // RB7 pressed ?
  Sound_Play(C2, Duration);      // Play C2
  break;
}
```

Notice that the duration is set to 200 ms, but the note will sound as long as the key is pressed.

7.12 Project 7.12—Generating Sine Wave Using D/A Converter

7.12.1 Project Description

In this project, we will see how to generate a low-frequency sine wave using the built-in trigonometric **sin** function, and then send the output to a D/A converter. The generated sine wave has amplitude of 1 V, frequency of 100 Hz, and offset of 1.2 V.

Figure 7.58 shows the block diagram of the project.

7.12.2 Project Hardware

The circuit diagram of the project is shown in Figure 7.59. The project is based on the MCP4921 D/A converter. This is a 12-bit converter with a typical conversion time of 4.5 μs. The chip is controlled from the SPI bus of the microcontroller. PIC32MX460F512L microcontroller has two built-in SPI bus controllers. In this project SPI2 is used, which consists of I/O pins RG6 (Clock), RG8 (data out), and RG7 (Data in). Pin RG7 is not used in this project as there is no input from the D/A converter.

As shown in Figure 7.59, pins RG6 and RG8 of the microcontroller are connected to pins SCK and SDI of the D/A converter, respectively. The CS input of the D/A converter is

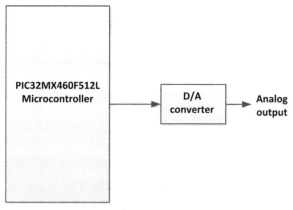

Figure 7.58
Block diagram of the project.

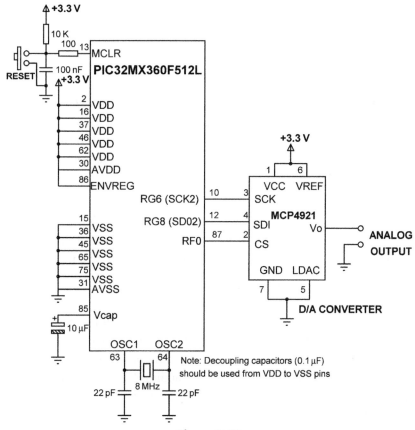

Figure 7.59
Circuit diagram of the project.

connected to pin RF0 of the microcontroller. The converter is operated from a +3.3 V supply voltage and the reference input is also connected to the same supply voltage. With +3.3 V reference voltage, one bit of the converter corresponds to 3.3 V/4096 = 0.0008 V or 0.8 mV. Similarly, +1 V corresponds to decimal value 4096/3.3 = 1241.

7.12.3 Project PDL

Before developing the PDL for this project, it is important to understand how the project works.

The frequency of the sine wave to be generated is 100 Hz. This wave has a period of 10 ms, or 10,000 μs. If we assume that the sine wave will consists of 100 samples, then each sample should be output at 10,000/100 = 100 μs intervals. Thus, we will configure Timer 1 to generate interrupts at every 100 μs, and inside the interrupt service routine we will output a new sample of the sine wave. The sample values will be calculated using the trigonometric **sin** function of the mikroC PRO for PIC32 compiler.

The sin function will have the format:

$$\sin\left(\frac{2\pi \times Count}{T}\right)$$

where, T is the period of the waveform and is equal to 100 samples. *Count* is a variable that ranges from 0 to 100 and is incremented by one inside the interrupt service routine every time a timer interrupt occurs. Thus, the sine wave is divided into 100 samples and each sample is output at 100 μs. The above formula can be rewritten as follows:

$$\sin(0.0628 \times Count)$$

It is required that the amplitude of the waveform should be 1 V. With a reference voltage of +3.3 V and a 12-bit D/A converter, 1 V is equal to decimal number 1241. Thus, we will multiply our sine function with the amplitude at each sample to give the following:

$$1241 * \sin(0.0628 \times Count)$$

The D/A converter used in this project is unipolar and cannot output negative values. Therefore, an offset is added to the sine wave to shift it so that it is always positive. The offset should be larger than the maximum negative value of the sine wave, which is −1241 when sine above is equal to 1. In this project we are adding 1.2 V offset which corresponds to a decimal value of 1500 at the D/A output. Thus, at each sample we will calculate and output the following value to the D/A converter:

$$1500 + 1241 * \sin(0.0628 \times Count)$$

```
BEGIN/MAIN
        Define D/A chip select connections
        Disable D/A converter
        Initialize SPI library
        Configure Timer 1 for 100 µs interrupts
        Enable interrupts
        DO FOREVER
                Wait for Timer 1 interrupts
        ENDDO
END/MAIN


BEGIN/TMR
        Calculate value of next sine sample using Count
        Multiply by amplitude
        Add offset
        Enable D/A converter
        Send high byte of data
        Send low byte of data
        Disable D/A converter
        Increment Count
        IF COUNT = 100 THEN
                Clear Count to zero
        ENDIF
        Clear Timer 1 interrupt flag
END/TMR
```

Figure 7.60
PDL of the project.

The D/A converter is 12-bits wide and data are normally sent to the converter with the 4-bit high byte sent first, followed by the 8-bit low byte. Data are sent using the built-in SPI function **SPI2_Write**. The SPI2 library must be initialized before the write function can be used. Also, the D/A converter CS input (RF0) must be lowered for it to be enabled and accept data.

We can now develop the PDL of the project. Figure 7.60 shows the PDL of the project. At the beginning, the D/A chip select connections to the microcontroller are defined. The D/A converter is disabled and the SPI bus library is initialized. The Timer 1 is configured to interrupt at every 100 µs. The main program then enables interrupts and waits in an endless loop for timer interrupts to occur.

The timer interrupt service routine calculates the value of the next sine wave sample, multiplies by the amplitude, and adds the offset, and sends the sample to the D/A converter. If the sample count is >100, it is set back to 0. Then, the timer interrupt flag is then cleared to 0, so that further interrupts can be accepted from this source.

7.12.4 Project Program

The program is called SINE.C and the program listing is given in Figure 7.61. At the beginning of the program, the chip select connection of the D/A converter chip is defined.

```
/*****************************************************************************
                        GENERATE SINE WAVE
                        ==================
```

In this project a Digital-to-Analog (D/A) converter chip is connected to the
microcontroller output and a 100 Hz sine wave with an amplitude of 1 V and an
offset of 1.2 V is generated in real-time at the output of the D/A converter.

The MCP4921 12-bit D/A converter is used in the project. This converter can operate
from +3.3 to +5 V, with a typical conversion time of 4.5 ms. Operation
of the D/A converter is based on th standard SPI interface.

mikroC PRO for PIC32 trigonometric "sin" function is used to calculate the sine
points in real time. 100 Hz waveform has the period T = 10 ms, or, T = 10,000 μs. If
we take 100 points to Sample the sine wave, then each Sample occurs at 100 μs.
Therefore, we need a timer interrupt service routine that will generate interrupts
at every 100 μs, and inside this routine we will calculate a new sine point and send it
to the D/A converter. The result is that we will get a 100 Hz sine wave. Because the
D/A converter is unipolar, we have to shift the output waveform to a level greater
than its maximum negative value so that the waveform is always positive and can be
output by the D/A converter.


```
Author:        Dogan Ibrahim
Date:          July 2012
File:          SINE.C
*****************************************************************************/
// DAC module connections
sbit Chip_Select at LATF0_bit;
sbit Chip_Select_Direction at TRISF0_bit;
// End DAC module connections

#define T 100                               // 100 samples
#define R 0.0628                            // 2*PI/T
#define Amplitude 1241                      // 1 V*4096/3.3 V

unsigned char temp, Count = 0;
float Sample;
unsigned int DAC;

//
// Timer 1 interrupt service routine. The program jumps here every 100 μs. In this
// routine, a new sine Sample is calculated, offset is added, and the Sample is
// sent to the D/A converter
//
void TMR() iv IVT_TIMER_1 ilevel 1 ics ICS_SOFT
{
    Sample = Amplitude*sin(R*Count);
    DAC = 1500+Sample;                      // Add offset = 1500 (1.2 V)
    Chip_Select = 0;                        // Select DAC chip
```

Figure 7.61

(Continued on next page)

```
        // Send High Byte
        temp = (DAC >> 8) & 0x0F;              // Store DAC[11..8] to temp[3..0]
        temp |= 0x30;                          // Define D/A setting
        SPI2_Write(temp);                      // Send high byte via SPI

        // Send Low Byte
        temp = DAC;                            // Store DAC[7..0] to temp[7..0]
        SPI2_Write(temp);                      // Send low byte via SPI

        Chip_Select = 1;                       // Deselect D/A converter chip

        Count++;
        if(Count == 100)Count = 0;
        IFS0bits.T1IF = 0;                     // Clear Timer 1 interrupt flag
}


//
// Start of man program
//
void main()
{
        Chip_Select_Direction = 0;            // Configure CS pin as output
        Chip_Select = 1;                      // Disable D/A converter
        SPI2_Init();                          // Initialize SPI2

//
// Configure Timer 1 for 100 µs interrupts
//
        T1CONbits.ON = 0;                     // Disable Timer 1
        TMR1 = 0;                             // Clear TMR1 register
        T1CONbits.TCKPS = 3;                  // Select prescaler = 256
        T1CONbits.TCS = 0;                    // Select internal clock
        PR1 = 31;                             // Load period register (for 100 µs)
        IFS0bits.T1IF = 0;                    // Clear Timer 1 interrupt flag
        IPC1bits.T1IP = 1;                    // Set priority level to 1
        IEC0bits.T1IE = 1;                    // Enable Timer 1 interrupts
        T1CONbits.ON = 1;                     // Enable Timer 1
        EnableInterrupts();                   // Enable interrupts

        for(;;)                               // Wait for Timer 1 interrupts
        {
        }
}
```

Figure 7.61

Program listing.

Then the sine wave amplitude is set to 1241, and variable R is defined as $2\pi/100$. The chip select direction is configured as output, D/A converter is disabled by setting its chip select input, and SPI2 is initialized.

The program then configures Timer 1, so that it generates interrupts at every 100 µs. With reference to Chapter 2, assuming a prescaler of 256, the value to be loaded into period register PR1 is calculated as follows:

```
PR1 = Required delay/(clock period × 256)
```

With 80 MHz clock, the clock period is 0.0125×10^{-3} ms. Thus,

```
PR1 = 0.1 ms/(0.0125 × 10⁻³ × 256) = 31
```

Interrupts are enabled and the program waits until interrupts occur.

The Timer 1 interrupt service routine is called TMR. Here, the value of the next sine sample is calculated, multiplied by the amplitude, and then the offset is assessed as described in the PDL section above. These operations are done using floating point arithmetic where the result is stored in variable *Sample*. The result is then copied to integer variable DAC. The D/A chip is enabled, and high 4 bits of the result in DAC is sent to the D/A converter with the following statements:

```
temp = (DAC≫8)&0x0F;     // Store DAC[11..8] to temp[3..0]
temp |= 0x30;            // Define D/A setting
SPI2_Write(temp);       // Send high byte via SPI
```

Then the lower 8 bits are sent to the D/A converter with the following statements:

```
temp = DAC;             // Store DAC[7..0] to temp[7..0]
SPI2_Write(temp);       // Send low byte via SPI
```

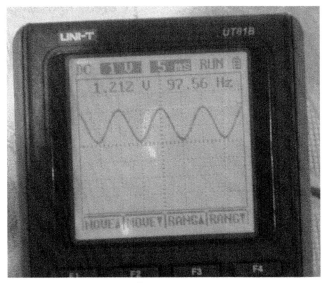

Figure 7.62
Output waveform generated.

The D/A is then disabled by setting its chip select input to 1. The value of variable *Count* is incremented and if its equal to 100, it is cleared back to 0.

Figure 7.62 shows the output sine waveform generated by the program on an oscilloscope.

Notice that a parallel D/A converter could have been used in this project to decrease the conversion time and hence to increase the overall system performance.

7.13 Project 7.13—Communicating with a PC Using the RS232 PORT

7.13.1 Project Description

In this project we will see how we can communicate with a PC over the serial link. Initially, a small program is given to receive a lower case character, convert it to upper case and then send it back and display on the PC screen.

The main project is about designing a calculator with a menu. Initially a menu will be displayed on the PC screen and the user will be requested to select an arithmetic operation and enter two numbers. The program will carry out the required operation and display the result on the PC screen.

There are two ways of communicating with a PC: either using an RS232 interface or using a USB-RS232-type interface. Early versions of the LV-32MX V6 board are equipped with 2 × RS232 ports, while the later versions are equipped with 2 × USB-RS232 ports. This project will explain the use of RS232-type interface with a nine-way D-type connector.

Figure 7.63 shows the block diagram of the project.

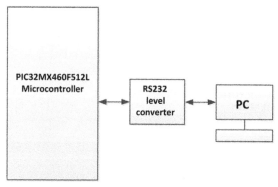

Figure 7.63
Block diagram of the project.

7.13.2 Project Hardware

The circuit diagram of the project is shown in Figure 4.8. The PIC32MX460F512L microcontroller has $2 \times$ UART ports for serial communication:

```
U1RX (RF2)
U1TX (RF8)

U2RX (RF4)
U2TX (RF5)
```

In this project, U1RX (RF2) is used to receive serial data from the PC, while U1TX (RF3) is used to transmit serial data to the PC. Connection to the PC serial port is by using a nine-way D-type connector.

Most PCs nowadays do not have serial communications ports. A small device called USB to RS232 converter (available from most computer shops) can be connected to the PC USB port to provide RS232 port capability to the PC.

The following jumpers should be selected on the LV-32MX V6 board so that the RS232 port A (UART1) is enabled on the board:

```
DIP switch SW5 position 1 and 2, set to ON
```

7.13.3 Project PDL (Simple Project)

In this simple project a lower case letter is received from the PC, converted to upper case, and is displayed on the PC. Figure 7.64 shows the PDL of this project.

7.13.4 Project Program

The program is called RS232-SIMPLE.C and the program listing is given in Figure 7.65. At the beginning of the program, UART1 is initialized to 4800 bps. Then the heading

BEGIN/MAIN
    Initialize UART1
    DO FOREVER
        Get a lower case character from UART1 (serial port)
        Convert the character to upper case
        Send the character to UART1 (serial port)
    ENDDO
END/MAIN

BEGIN/NEWLINE
    Send a carriage return character
    Send a line-feed character
END/NEWLINE

Figure 7.64
PDL of the simple project.

```
/******************************************************************************
                      SIMPLE SERIAL COMMUNICATION
                      ============================
```

In this project one of the microcontroller serial ports (UART1) is connected to a PC through a voltage translator chip. The project receives a lower case character from the PC, converts this character to upper case, and then sends it back to the PC to display it on the PC screen.

A typical run of the program is shown below:

```
CONVERT LOWER CASE TO UPPER CASE
================================
Enter a lower case character: a
Upper case is: A

Enter a lower case character:


Author:      Dogan Ibrahim
Date:        July 2012
File:        RS232-SIMPLE.C
******************************************************************************/
#define Enter 13

//
// This function creates Newline
//
void Newline(void)
{
  UART1_Write(13);
  UART1_Write(10);
}


//
// Start of main program
//
void main()
{
  unsigned char MyKey, c;

  UART1_Init(4800);                                          // Initialize UART1

  Newline();                                                 // Newline
  UART1_Write_Text("CONVERT LOWER CASE TO UPPER CASE");      // Write heading
  Newline();
  UART1_Write_text("================================");
  Newline();

  for(;;)                                                    // DO FOREVER
  {
```

Figure 7.65

(Continued on next page)

```
            Newline();                                           // Newline
            UART1_Write_Text("Enter a lower case character: ");  // Prompt for a character

            do                                                   // Loop to receive a character
            {
              if(UART1_Data_Ready())                             // If a character is ready
              {
                 MyKey = UART1_Read();                           // Get the character
                 if(MyKey == Enter)break;                        // If ENTER key
                 UART1_Write(MyKey);                             // Echo the character
                 c = MyKey;                                      // Save received character
              }
            }while(1);

            c = toupper(c);                                      // Convert to upper case
            Newline();                                           // Newline
            UART1_Write_Text("Upper case is: ");
            UART1_Write(c);                                      // Send character via UART
            Newline();                                           // Newline
        }
    }
```

Figure 7.65
Program listing.

"CONVERT LOWER CASE TO UPPER CASE" is displayed on the PC screen. The rest of the program is executed in an endless loop. Function *Newline* sends a carriage return and a line-feed character to the screen.

Inside the loop the user is prompted to enter a character. The program checks continuously until a character is available in the UART receive buffer. The received character is echoed on the PC screen. The program waits until the ENTER key is pressed after a character is entered, and the received character is stored in variable c as shown in the code below:

```
do                                   // Loop to receive a character
{
  if(UART1_Data_Ready())             // If a character is ready
  {
    MyKey = UART1_Read();            // Get the character
    if(MyKey == Enter)break;         // If ENTER key then exit loop
    UART1_Write(MyKey);              // Echo the character
    c = MyKey;                       // Save received character
  }
}while(1);
```

The received character is converted to upper case using built-in function **toupper**, and is then sent to the PC via the UART, as in the following code:

```
c = toupper(c);                      // Convert to upper case
Newline();                           // Newline
UART1_Write_Text("Upper case is: ");
UART1_Write(c);                      // Send character via UART
Newline();                           // Newline
```

A typical run of the program is given below which converts lower case 'a' to 'A':

CONVERT LOWER CASE TO UPPER CASE.

=====================================

Enter a lower case character: a

Upper case is: A

Enter a lower case character:

Testing the Program

The program can easily be tested using a terminal emulator software such as **HyperTerminal** or a similar one which is usually distributed free of charge with some Windows operating systems, or alternatively a copy can be obtained from the Internet. The steps to test the program are given below (these steps assume that the PC serial port COM2 is used):

- Connect the RS232 output from the microcontroller to the serial input port of your PC (e.g. COM2). If your PC does not have a serial port, you may like to use a USB to RS232 converter device.
- Start HyperTerminal terminal emulation software and give a name to the session.
- *Select File → New connection → Connect using* and select your serial port name (e.g. COM2).
- Select the Baud rate as 4800, data bits as 8, no parity bits, and one stop bit.
- Reset the microcontroller.

If you are not sure of the available serial communications ports on your PC, you can check this by going to *Control Panel → Device Manager → Ports*.

An example output of the program is shown in Figure 7.66 using the HyperTterminal software.

7.13.5 Project PDL (Calculator Project)

This is a more complex project. In this project an integer calculator program is developed which can perform the basic four mathematical operations of addition, subtraction, multiplication, and division on the PC screen. The user is given a menu to select the required operation. Then two numbers are accepted from the PC keyboard and the result of the operation is displayed on the PC screen.

```
CONVERT LOWER CASE TO UPPER CASE
================================

Enter a lower case character: b
Upper case is: B

Enter a lower case character: g
Upper case is: G

Enter a lower case character: t
Upper case is: T
```

Figure 7.66
HyperTerminal screen.

START/MAIN
      Configure UART1 to 4800 Baud
      DO FOREVER
            Display "CALCULATOR PROGRAM"
            Display "Enter First Number: "
            Read first number
            Display "Enter Second Number: "
            Read second number
            Display "Operation: "
            Read operation
            Perform operation
            Display "Result= "
            Display the result
      ENDDO
END/MAIN

START/NEWLINE
      Send carriage return to UART
      Send line-feed to UART
END/NEWLINE

Figure 7.67
PDL of the simple project.

A sample calculation is as follows:

```
CALCULATOR PROGRAM

   Enter First Number: 10
  Enter Second Number: 5
     Enter Operation: +
            Result = 15
```

The PDL of the calculator program is shown in Figure 7.67.

7.13.6 Project Program

The program listing of the project is shown in Figure 7.68 (CALCULATOR.C). The program consists of a main program and a function called *Newline*. The function

```
/******************************************************************************
                    CALCULATOR WITH PC INTERFACE
                    ============================
```

This is a simple integer calculator project. User enters the numbers through the PC
keyboard. Results are displayed on the PC screen.

The following operations can be performed:

```
    + – * /
```

This program uses the built in UART1 of the microcontroller. The UART1 is configured
to operate at 4800 Baud rate.


```
Author:        Dogan Ibrahim
Date:          August 2011
File:          CALCULATOR.C
******************************************************************************/

#define Enter 13
#define Plus '+'
#define Minus '–'
#define Multiply '*'
#define Divide '/'

//
// This function sends carriage return and line-feed to UART
//
void Newline()
{
   UART1_Write(0x0D);                           // Send carriage return
   UART1_Write(0x0A);                           // Send line-feed
}



//
// Start of MAIN program
//
void main()
{
    unsigned char MyKey, i,j,kbd[5],op[12];
    unsigned long Calc, Op1, Op2,Key;
    char *kb;
    unsigned char msg1[] = " CALCULATOR PROGRAM";
    unsigned char msg2[] = "      Enter First Number: ";
    unsigned char msg3[]= "Enter Second Nummber: ";
    unsigned char msg4[] = "      Enter Operation: ";
    unsigned char msg5[] = "            Result = ";
//
// Initialize UART1 to 4800 bps
```

Figure 7.68

(Continued on next page)

```
//
  UART1_Init(4800);                        // Baud rate = 4800

  for(;;)                                   // DO FOREVER
  {
    MyKey = 0;
    Op1 = 0;
    Op2 = 0;

    Newline();                              // Send newline
    Newline();                              // Send newline
    UART_Write_Text(msg1);                  // "CALCULATOR PROGRAM"
    Newline();                              // Send newline
    Newline();                              // Send newline
//
// Get the first number
//
    UART1_Write_Text(msg2);                 // "Enter First Number : "
    do                                      // Get first number
    {
      if(UART1_Data_Ready())                // If a character ready
      {
        MyKey = UART1_Read();               // Get a character
        if(MyKey == Enter)break;            // If ENTER key
        UART1_Write(MyKey);                 // Echo the character
        Key = MyKey – '0';
        Op1 = 10*Op1 + Key;                 // First number in Op1
      }
    }while(1);

    Newline();

//
// Get the second character
//
    UART1_Write_Text(msg3);                 // "Enter Second Number :"
    do                                      // Get second number
    {
      if(UART1_Data_Ready())
      {
        MyKey = UART1_Read();               // Get a character
        if(Mykey == Enter)break;            // If ENTER key
        UART1_Write(MyKey);                 // Echo the character
        Key = MyKey – '0';
        Op2 = 10*Op2 + Key;                 // Second number in Op2
      }
    }while(1);

    Newline();
//
```

Figure 7.68
(Continued on next page)

```
// Get the operation
//
    UART1_Write_Text(msg4);                    // "Enter Operation : "
    do
    {
      if(UART1_Data_Ready())
      {
        MyKey = UART1_Read();                  // Get a character
        if(MyKey == Enter)break;               // If ENTER key
        UART1_Write(MyKey);                    // Echo the character
        Key = MyKey;
      }
    }while(1);

//
// Perform the required operation
//
    Newline();
    switch(Key)                                // Calculate
    {
      case Plus:
          Calc = Op1 + Op2;                    // If ADD
          break;
      case Minus:
          Calc = Op1 - Op2;                    // If Subtract
          break;
      case Multiply:
          Calc = Op1 * Op2;                    // If Multiply
          break;
      case Divide:
          Calc = Op1 / Op2;                    // If Divide
          break;
    }

    LongToStr(Calc, op);                       // Convert to string in op
//
// Remove leading spaces and display the result
//
    kb = Ltrim(op);                            // Remove leading spaces
    UART1_Write_Text(msg5);                    // "Result = "
    UART1_Write_Text(kb);                      // Display result

  }
}
```

Figure 7.68
Program listing of the calculator.

sends a carriage return and line feed to the UART to move the cursor to the next line of the screen.

At the beginning of the program, various messages used in the program are defined as *msg1—msg5*. UART1 is then initialized to 4800 Baud using mikroC PRO for PIC library

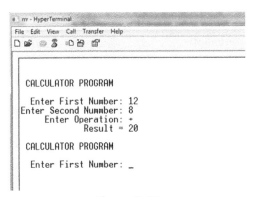

Figure 7.69
Typical display.

routine **UART1_Init**. Then the heading *CALCULATOR PROGRAM* is displayed on the PC monitor. The program reads the first number from the keyboard using the library function **UART1_Read**. Function **UART1_Data_Ready** checks when a new data byte is ready before reading it. Variable *Op1* stores the first number. Similarly, another loop is formed and the second number is read into variable *Op2*. The program then reads the operation to be performed $(+ - * /)$. The required operation is performed inside a Switch statement and the result is stored in variable *Calc*. The program then converts the result into string format by calling library function **LongToStr**. Leading blanks are removed from this string and the final result is stored in *kb* and sent to the UART to display on the PC screen.

Figure 7.69 shows a typical display from the program.

Using USB—RS232 Interface

Because some PCs and microcontrollers nowadays do not have serial communications ports, some manufacturers offer USB—RS232 interfaces which provide serial port capability to

```
START/MAIN
        Initialize LCD
        Turns cursor off
        Call function LCD_Scroll_Message
END/MAIN



START/LCD_Scroll_Message
        DO FOREVER
                Shift and store characters in buffer Txt
                Display contents of buffer Txt
        ENDDO
END/LCD_Scroll_Message
```

Figure 7.70
PDL of the project.

```
/******************************************************************************
                    LCD SCROLL TEXT
                    ===============

This project shows how text can be scrolled on the LCD. In this program a buffer is used to
store the text. The characters are shifted left inside the buffer and are then sent to the LCD display.

One second delay is used between each output.

Author:      Dogan Ibrahim
Date:        September 2012
File:        SCROLL.C
******************************************************************************/

// LCD module connections
sbit LCD_RS at LATB2_bit;
sbit LCD_EN at LATB3_bit;
sbit LCD_D4 at LATB4_bit;
sbit LCD_D5 at LATB5_bit;
sbit LCD_D6 at LATB6_bit;
sbit LCD_D7 at LATB7_bit;

sbit LCD_RS_Direction at TRISB2_bit;
sbit LCD_EN_Direction at TRISB3_bit;
sbit LCD_D4_Direction at TRISB4_bit;
sbit LCD_D5_Direction at TRISB5_bit;
sbit LCD_D6_Direction at TRISB6_bit;
sbit LCD_D7_Direction at TRISB7_bit;
// End LCD module connections


//
// This function shifts the characters left and then displays the buffer (Txt)
// on the LCD. One second delay is used between each output. Argument "Msg" is
// passed by the main program
//
void LCD_Scroll_Message(const char Msg[])
{
 char Txt[17];
 unsigned int i, Head,flag;

 for(;;)
 {
     flag = 0;
     Head = 0;

     while(!flag)
     {
         for(i = 0; i < 16; i++)
         {
          Txt[i] = Msg[Head + i];
```

Figure 7.71

(Continued on next page)

```
                      if(Msg[Head + i + 1] == 0x0) flag = 1;
                    }

                    Head++;
                    lcd_out(1, 1, Txt);
                    Delay_ms(1000);
              }
         }
    }
    //
    // Start of man program
    //
    void main()
    {
        LCD_Init();                                     // Initialize LCD
        LCD_Cmd(_LCD_CURSOR_OFF);                       // Turn cursor OFF

        LCD_Scroll_Message("THIS IS A SCROLLING DISPLAY          ");

    }
```

Figure 7.71
Program listing.

USB devices on your PC or the microcontroller. For example, the latest version of the LV-32MX V6 board has 2 × USB ports with RS232 capabilities. Most of these ports are designed using Future Technology Devices International Ltd (FTDI) chips which enable two USB devices to be connected together and then to communicate using the standard RS232 protocols. The actual serial port name provided by the FTDI chip can be found from the Ports section of the device Manager in your Control Panel.

7.14 Project 7.14—Scrolling LCD Display

7.14.1 Project Description

This project shows how text can be scrolled on the LCD. The block diagram of the project is same as in Figure 7.16.

7.14.2 Project Hardware

The circuit diagram of the project is as shown in Figure 7.17.

7.14.3 Project PDL

There are several ways of scrolling text on the LCD. One method loads new characters to the LCD from the right-hand side, and then shifts the existing characters left by one column position. Another method stores the characters in a buffer, shifts these characters left by one position within the buffer, and then displays the contents of the entire buffer. In this project, the second method is used for simplicity.

Figure 7.70 shows the PDL of the project. The scrolling is performed in a function called *LCD_Scroll_Message*, where main program calls the function by passing the message to be displayed as an argument.

7.14.4 Project Program

The program listing (called SCROLL.C) is shown in Figure 7.71. The main program initializes the LCD and turns off the cursor. Function *LCD_Scroll_Message* is then called to display and scroll the following message:

```
THIS IS A SCROLLING DISPLAY
```

Function *LCD_Scroll_Message* shifts and stores the characters to be displayed in a buffer called *Txt*. One second delay is inserted between each display. The function never returns to the main program.

Advanced PIC32 Projects

Chapter Outline

Designing Embedded Systems with 32-Bit PIC Microcontrollers and MikroC.
Copyright © 2014 Elsevier Ltd. All rights reserved.

8.9 Project 8.9—Calculating Timing in Digital Signal Processing 436

In Chapter 7, we have looked at simple PIC32 microcontroller projects. Most of these projects were very small, or required no high speed of processing, or did not use any complicated peripherals.

In this chapter, we will be looking at the design of more advanced projects where either more performance is required, or the project uses a complicated peripheral. The design of the projects follows the same steps given in Chapter 7, where for each project we will have a description, block diagram, circuit diagram, project development laboratory (PDL), full-program listing, and description of the operation of the program.

8.1 Project 8.1—Generating High-Frequency Sine Waveform

8.1.1 Project Description

In Project 7.14, we have seen how to create a sine waveform in real time and send it out through the digital to analog (D/A) converter chip. In that project, we calculated the sample points using the mikroC Pro for PIC32 compiler built-in **sin** function.

Because the **sin** function uses floating point, arithmetic it takes a lot of processing time and the method presented in Project 7.14 may not be suitable to generate high-frequency sine waves in real time. In this project, we will generate a sine wave at 2 kHz, amplitude 1 V, and offset 1.2 V, using two different methods to calculate the samples. It is possible to generate waveforms with very high frequencies using either of the two methods:

Method 1
This method is based on calculating the sine wave samples using an offline approach. The samples will be calculated and loaded into the appropriate array before we enter the real-time loop, which is handled by a timer interrupt. Inside the real-time loop, we will send the samples to the D/A converter.
Method 2
In this method, we will calculate the sine wave samples using a spreadsheet program such as Excel, and then we will load an array with the calculated values. Inside the real-time loop, we will send the samples to the D/A converter as in Method 1.

8.1.2 Method 1 in Detail

The block diagram of the project is same as given in Figure 7.58.

Project Hardware (Method 1)

The circuit diagram of the project is shown in Figure 7.59. The project is based on the MCP4921 D/A converter. This is a 12-bit converter with a typical conversion time of 4.5 μs. The chip is controlled from the SPI bus of the microcontroller. PIC32MX460F512L microcontroller has two built-in SPI bus controllers. In this project, SPI2 is used, which consists of input−output pins RG6 (Clock) and RG8 (data out).

Pins RG6 and RG8 of the microcontroller are connected to pins SCK and SDI of the D/A converter, respectively. The chip select (CS) input of the D/A converter is connected to pin RF0 of the microcontroller. The D/A converter is operated from a +3.3 V supply voltage and the reference input is also connected to the same supply voltage. With +3.3 V reference voltage, 1 bit of the converter corresponds to 3.3 V/4096 = 0.0008 V or 0.8 mV.

Project PDL (Method 1)

Before developing the PDL for this project, it is important to understand how the project works.

The frequency of the sine wave to be generated is 2 kHz. This waveform has a period of 500 μs. If we assume that the sine wave will consist of 100 samples, then each sample should have output at 500/100 = 5 μs intervals. Thus, we will configure Timer 1 to generate interrupts at every 5 μs (or 0.00 5ms), and inside the interrupt service routine (ISR), we will output a new sample of the sine wave. The sample values will be calculated once and loaded into an array before entering the real-time ISR, using the built-in trigonometric **sin** function of the mikroC PRO for PIC32 compiler.

The sin function will have the following format:

$$\sin\left(\frac{2\pi \times Count}{T}\right)$$

where, T is the period of the waveform and is equal to 100 samples. *Count* is a variable that ranges from 0 to 100 and is incremented by one inside the ISR every time a timer interrupt occurs. Thus, the sine wave is divided into 100 samples and each sample is output at 1 μs. The above formula can be rewritten as follows:

$$\sin(0.0628 \times Count)$$

It is required that the amplitude of the waveform should be 1 V. With a reference voltage of +3.3 V and a 12-bit D/A converter, 1 V is equal to decimal number 1241. Thus, we will multiply our sine function with the amplitude at each sample to give the following:

$$1241 * \sin(0.0628 \times Count)$$

The D/A converter used in this project is unipolar and cannot output negative values. Therefore, an offset is added to the sine wave to shift it, so that it is always positive. The offset should be larger than the maximum negative value of the sine wave, which is -1241 when sin above is equal to 1. In this project, we are adding 1.2 V offset which corresponds to a decimal value of 1500 at the D/A output. Thus, at each sample, we will calculate and output the following value to the D/A converter:

$$1500 + 1241 * \sin(0.0628 \times Count)$$

The 100 sine wave samples are calculated using the above formula and then be stored in an array that can be used inside the real-time loop.

The D/A converter is 12-bit wide and data are normally sent to the converter with the 4-bit high byte sent first, followed by the 8-bit low byte. Data are sent using the built-in SPI function **SPI2_Write**. The SPI2 library must be initialized before the write function can be used. Also, the D/A converter CS input (RF0) must be lowered for it to be enabled and accept data.

We can now develop the PDL of the project. Figure 8.1 shows the PDL of the project. At the beginning, the D/A CS connections to the microcontroller are defined. The D/A converter is disabled and the SPI bus library is initialized. Function *Calculate_Sine_Samples* is called to calculate and store the samples in an array. The Timer 1 is configured to interrupt at every 5 μs. The main program then enables interrupts and waits in an endless loop for timer interrupts to occur.

The timer ISR gets value of the next sine wave sample from the array, multiplies by the amplitude, adds the offset, and sends the sample to the D/A converter. If the sample count is equal to 100, it is set back to zero. Then, the timer interrupt flag is then cleared to zero, so that further interrupts can be accepted from this source.

Project Program (Method 1)

The program is called SINE2K.C and the program listing is given in Figure 8.2. At the beginning of the program, the CS connection of the D/A converter chip is defined. Then the sine wave amplitude is set to 1241, and variable R is defined as $2\pi/100$. The CS direction is configured as output, and D/A converter is disabled by setting its CS input.

The D/A converter used in this project is controlled with the SPI bus. Because the speed of the SPI bus is limited, the data transfer rate is also limited on the bus and we cannot have very high data rates. Some D/A converters are connected to a microcontroller in parallel form using 8 bits and very high speeds can be obtained when using such converters. The default initialization of the SPI bus sets the bus clock rate to System_Clock/64, which is too low to

BEGIN/MAIN
    Define D/A chip select connections
    Disable D/A converter
    Initialize SPI library
    Call Calculate_Sine_Samples
    Configure Timer 1 for 5 μs interrupts
    Enable interrupts
    DO FOREVER
        Wait for Timer 1 interrupts
    ENDDO
END/MAIN

BEGIN/CALCULATE_SINE_SAMPLES
    Use sin function to calculate 100 equally spaced sine samples
    Multiply the samples by the amplitude
    Add offset to each sample
    Store samples in an array DAC
END/CALCULATE_SINE_SAMPLES

BEGIN/TMR
    Get value of next sine sample from array DAC indexed by Count
    Enable D/A converter
    Send high byte of data
    Send low byte of data
    Disable D/A converter
    Increment Count
    IF COUNT = 100 **THEN**
        Clear Count to zero
    ENDIF
    Clear Timer 1 interrupt flag
END/TMR

Figure 8.1
PDL of the project.

operate our D/A converter for a signal of 2 kHz. In this project, the advanced version of the SPI bus initialization routine is used and the SPI bus clock rate is set to System_Clock/4 as shown below:

```
//
//Set SPI2 to the Master Mode, data length is 8-bit, clock = Fcy/4,
//          data sampled in the middle of interval, clock IDLE state high,
//          data transmitted at low to high clock edge
//
SPI2_Init_Advanced(_SPI_MASTER, _SPI_8_BIT, 4, _SPI_SS_DISABLE,
                   _SPI_DATA_SAMPLE_MIDDLE, _SPI_CLK_IDLE_LOW,
                   _SPI_IDLE_2_ACTIVE);
```

The main program then calls the function *Calculate_Sine_Samples* to calculate equally spaced 100 samples of the sine wave. The samples are multiplied by the amplitude and offset is then added. The result is stored in an array called *DAC*, so that it can be used by the real-time code of the timer ISR.

```
/****************************************************************************
                    GENERATE HIGH FREQUENCY SINE WAVE
                    =================================
```

In this project a Digital-to-Analog (D/A) converter chip is connected to the
microcontroller output and a 2 kHz sine wave with an amplitude of 1 V and an
offset of 1.2 V is generated in real-time at the output of the D/A converter.

The MCP4921 12-bit D/A converter is used in the project. This converter can operate
from +3.3 V to +5 V, with a typical conversion time of 4.5 μs. Operation
of the D/A converter is based on the standard SPI interface.

mikroC PRO for PIC32 trigonometric "sin" function is used to calculate the sine
points offline, i.e. before entering the real-time loop. 2 kHz waveform has the
period T = 500 μs. If we take 100points to Sample the sine wave, then each Sample
occurs at 5 μs. Therefore, we need a timer interrupt service routine that will generate
interrupts at every 5us, and inside this routine we will output a new sine wave
Sample to the D/A converter. Because the D/A converter is unipolar, we have to
shift the output waveform to a level greater than its maximum negative value so
that the waveform is always positive and can be output by the D/A converter.

```
Author:      Dogan Ibrahim
Date:        August 2012
File:        SINE2K.C
****************************************************************************/
// DAC module connections
sbit Chip_Select at LATF0_bit;
sbit Chip_Select_Direction at TRISF0_bit;
// End DAC module connections

#define T 100                               // 100 samples
#define R 0.0628                            // 2*PI/T
#define Amplitude 1241                      // 1 V*4096/3.3 V

unsigned char temp,Count = 0;
float Samples;
unsigned int DAC[100];

//
// This function calculates 100 sine wave samples for the frequency of 2 kHz and
// stores the Sample values in array Samples
//
void Calculate_Sine_Samples(void)
{
  unsigned char i;

  for(i = 0; i < 100; i++)
  {
    Samples = Amplitude*sin(R*i);           // Calculate sine values
    DAC[i] = 1500 + Samples;                // Add offset
```

Figure 8.2
(Continued on next page)

```
   }
}


//
// Timer 1 interrupt service routine. The program jumps here every 0.005 ms (5 µs).
// In this routine, a new sine Sample is calculated, offset is added, and the Sample
// is sent to the D/A converter
//
void TMR() iv IVT_TIMER_1 ilevel 1 ics ICS_SOFT
{
    Chip_Select = 0;                                   // Select DAC chip

  // Send High Byte
    temp = (DAC[Count] >> 8) & 0x0F;                   // Store [11..8] to temp[3..0]
    temp |= 0x30;                                      // Define D/A setting
    SPI2_Write(temp);                                  // Send high byte via SPI

    // Send Low Byte
    temp = DAC[Count];                                 // Store DAC[7..0] to temp[7..0]
    SPI2_Write(temp);                                  // Send low byte via SPI

    Chip_Select = 1;                                   // Disable D/A converter chip

    Count++;                                           // Increment sample count
    if(Count == 100)Count = 0;                         // Reset if end of samples
    IFS0bits.T1IF = 0;                                 // Clear Timer 1 interrupt flag
}



//
// Start of man program
//
void main()
{
    Chip_Select_Direction = 0;                         // Configure CS pin as output
    Chip_Select = 1;                                   // Disable D/A converter
//
//Set SPI2 to the Master Mode, data length is 16-bit, clock = Fcy (no clock scaling),
//             data sampled in the middle of interval, clock IDLE state high,
//             data transmitted at low to high clock edge
//
    SPI2_Init_Advanced(_SPI_MASTER, _SPI_8_BIT, 4, _SPI_SS_DISABLE,
                    _SPI_DATA_SAMPLE_MIDDLE, _SPI_CLK_IDLE_LOW,
                    _SPI_IDLE_2_ACTIVE);

    Calculate_Sine_Samples();                          // Calculate the samples

//
```

Figure 8.2

(Continued on next page)

```
// Configure Timer 1 for 0.5ms interrupts
//
    T1CONbits.ON = 0;                    // Disable Timer 1
    TMR1 = 0;                            // Clear TMR1 register
    T1CONbits.TCKPS = 0;                 // Select prescaler = 1
    T1CONbits.TCS = 0;                   // Select internal clock
    PR1 = 400;                           // Load period register (for 5us)
    IFS0bits.T1IF = 0;                   // Clear Timer 1 interrupt flag
    IPC1bits.T1IP = 1;                   // Set priority level to 1
    IEC0bits.T1IE = 1;                   // Enable Timer 1 interrupts
    T1CONbits.ON = 1;                    // Enable Timer 1
    EnableInterrupts();                  // Enable interrupts

    for(;;)                              // Wait for Timer 1 interrupts
    {
    }
}
```

Figure 8.2
Program listing.

The program then configures Timer 1 so that it generates interrupts at every 5 μs. With reference to Chapter 2, assuming a prescaler of 1, the value to be loaded into period register PR1 is calculated as (notice that choosing a lower prescaler rate gives a higher PR1 value and more accurate timing) follows:

```
PR1 = Required delay/(clock period × 1)
```

With 80 MHz clock, the clock period is 0.0125×10^{-3} ms. Thus,

```
PR1 = 0.005 ms/(0.0125 × 10⁻³ × 1) = 400
```

Interrupts are enabled and the program waits until interrupts occur.

The Timer 1 ISR is called TMR. Here, CS is enabled, and high 4 bits of the sine wave sample data are sent to the D/A converter. Then the low 8 bits of the sample data are sent to the D/A converter:

```
    Chip_Select = 0;                 // Select DAC chip
// Send High Byte
    temp = (DAC[Count]>>8)&0x0F;      // Store [11..8] to temp[3..0]
    temp |= 0x30;                     // Define D/A setting
    SPI2_Write(temp);                 // Send high byte via SPI

// Send Low Byte
    temp = DAC[Count];                // Store DAC[7..0] to temp[7..0]
    SPI2_Write(temp);                 // Send low byte via SPI
```

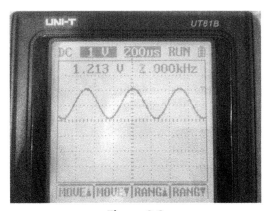

Figure 8.3
Output waveform generated.

The D/A is then disabled as there are no more data to send to it. The sample count is incremented and if it is equal to 100, it is reset back to zero ready for the next sample. The timer interrupt flag is cleared just before exiting the ISR, so that further interrupts can be accepted from the timer:

```
Chip_Select = 1;                 // Disable D/A converter chip
Count++;                         // Increment sample count
if(Count == 100)Count = 0;       // Reset if end of samples
IFS0bits.T1IF = 0;               // Clear Timer 1 interrupt flag
```

Figure 8.3 shows the output sine waveform generated by the program on an oscilloscope.

8.1.3 Method 2 in Detail

The block diagram of the project is same as shown in Figure 7.58.

Project Hardware (Method 2)

The circuit diagram of the project is as shown in Figure 7.59.

Project PDL (Method 2)

Before developing the PDL for this project, it is important to understand how this second method works.

The frequency of the sine wave to be generated is 2 kHz. This waveform has a period of 500 μs. If we assume that the sine wave will consist of 100 samples, then each sample should have output at $500/100 = 5$ μs intervals. Thus, we will configure Timer 1 to generate interrupts at every 5 μs (or 0.005 ms), and inside the ISR, we will output a new sample of the sine wave. The sample values will be calculated once using a spreadsheet program such as Excel and then loaded into an array before entering the real-time ISR.

As in method 1, the sine wave samples will be multiplied by the amplitude and offset will be added. The final result will be stored in an array. The following formula will be used to calculate the required 100 samples:

$$1500 + 1241 * \sin(0.0628 \times Count)$$

We can now develop the PDL of the project. Figure 8.4 shows the PDL of the project. At the beginning, the D/A CS connections to the microcontroller are defined. The D/A converter is disabled and the SPI bus library is initialized. The precalculated sine wave sample values are loaded into array DAC.

The Timer 1 is configured to interrupt at every 5 μs. The main program then enables interrupts and waits in an endless loop for timer interrupts to occur.

The timer ISR gets value of the next sine wave sample from the array, multiplies by the amplitude, adds the offset, and sends the sample to the D/A converter. If the sample count is equal to 100, it is set back to zero. Then, the timer interrupt flag is then cleared to zero so that further interrupts can be accepted from this source.

```
BEGIN/MAIN
        Define D/A chip select connections
        Initialize array DAC with sine wave samples
        Disable D/A converter
        Initialize SPI library
        Configure Timer 1 for 5 μs interrupts
        Enable interrupts
        DO FOREVER
                Wait for Timer 1 interrupts
        ENDDO
END/MAIN


BEGIN/TMR
        Get value of next sine sample from array DAC indexed by Count
        Enable D/A converter
        Send high byte of data
        Send low byte of data
        Disable D/A converter
        Increment Count
        IF COUNT = 100 THEN
            Clear Count to zero
        ENDIF
            Clear Timer 1 interrupt flag
END/TMR
```

Figure 8.4
PDL of the project.

Project Program (Method 2)

In this method, the sine wave samples are calculated using the Excel spreadsheet program. The sample values are obtained by using the following table function (notice that the samples are converted into integer numbers as this is how they are used in the program):

```
= Offset + INT(Amplitude * Sin(Counter*2*Pi/SampleCount))
```

In our example, we have the following:

```
= 1500 + INT(1241*Sin(0.0628*Counter))
```

Figure 8.5 shows the spreadsheet program where column A is the Counter, filled with values from 0 to 100 using the "Auto fill option". The above formula is copied to 100 locations in column B.

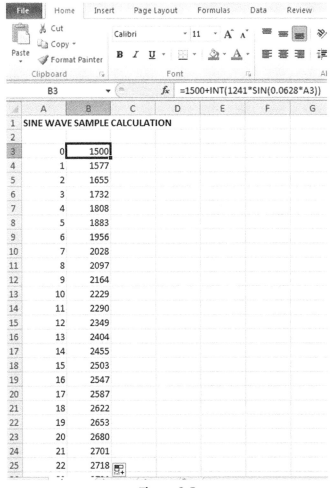

Figure 8.5
Sine wave samples.

The sample values in column B of the table are copied into the array DAC in the program.

The program is called SINE2K2.C and the program listing is given in Figure 8.6. At the beginning of the program, the CS connection of the D/A converter chip is defined. Then the sine wave samples are copied into the integer array DAC. The CS direction is configured as output, and D/A converter is disabled by setting its CS input.

As in method 1, the advanced version of the SPI bus library is used and the SPI bus clock rate is set to System_Clock/4. The Timer 1 is then configured to interrupt at every 5 μs. The remainder of the main program waits for timer interrupts to occur.

The ISR code is same as the one in method 1, where the high 4 bit and the low byte of the sample data are sent to the D/A converter.

8.2 Project 8.2—Generating Pulse-Width Modulation Waveform

8.2.1 Project Description

Pulse-width modulation (PWM) is a powerful technique for controlling analog circuits with a microcontroller's digital outputs. PWM is used in many applications, ranging from communications to power control and conversion. For example, the PWM is commonly used to control the speed of electric motors, the brightness of lights, in ultrasonic cleaning applications, and many more.

A PWM is basically a digital unipolar square wave signal where the duration of the ON time can be adjusted (or modulated) as desired. This way the power delivered to the load can be controlled from a microcontroller.

Figure 8.7 shows a typical PWM signal. The ON and the OFF times are sometimes referred to as the MARK (or M) and SPACE (or S) times of the signal, respectively. Here, we are interested in three parameters: signal amplitude, signal frequency (or period), and the signal duty cycle.

The amplitude is usually fixed by the logic 1 level of the microcontroller output which depends on the power supply voltage. In some applications, it may be necessary to use external circuitry to increase the amplitude. In this project, we will be using the output logic 1 level of the microcontroller which is +3.3 V.

The frequency depends on the application. In this project, we will generate a PWM signal with a frequency of 40 kHz. This is the frequency commonly used in most ultrasonic applications, such as distance measurement, ultrasonic cleaning, and so on.

The duty cycle, denoted by D, is the ratio of the ON time to the period of the signal, i.e. $D = M/T$. D can range from 0 to 1 and is sometimes expressed as a percentage, i.e. from 0% to

```
/**********************************************************************************
                        GENERATE HIGH FREQUENCY SINE WAVE
                        ==================================
```

In this project a Digital-to-Analog (D/A) converter chip is connected to the
microcontroller output and a 2 kHz sine wave with an amplitude of 1 V and an
offset of 1.2 V is generated in real-time at the output of the D/A converter.

The MCP4921 12-bit D/A converter is used in the project. This converter can operate
from +3.3 V to +5 V, with a typical conversion time of 4.5 μs. Operation
of the D/A converter is based on the standard SPI interface.

2 kHz waveform has the period T = 500 μs. If we take 100 points to Sample the sine
wave, then each Sample occurs at 5 μs. Therefore, we need a timer interrupt service
routine that will generate interrupts at every 5 μs, and inside this routine we will
output a new sine wave.

Sample to the D/A converter. Because the D/A converter is unipolar, we have to
shift the output waveform to a level greater than its maximum negative value so
that the waveform is always positive and can be output by the D/A converter.

In this version of the program (Method 2), an Excel spread-sheet program is used
to calculate the sine wave sample values, and these values are loaded into integer
array DAC from the Excel column.

```
     Author:       Dogan Ibrahim
     Date:         August 2012
     File:         SINE2K2.C
**********************************************************************************/
// DAC module connections
sbit Chip_Select at LATF0_bit;
sbit Chip_Select_Direction at TRISF0_bit;
// End DAC module connections

unsigned char temp,Count = 0;

unsigned int DAC[]={1500,1577,1655,1732,1808,1883,1956,2028,2097,2164,2229,2290,
2349,2404,2455,2503,2547,2587,2622,2653,2680,2701,2718,2731,2738,2740,2738,2731,2719,
2702,2680,2654,2623,2588,2548,2504,2457,2405,2350,2292,2230,2166,2099,2029,1958,1885,
1810,1734,1657,1579,1501,1424,1346,1269,1193,1118,1045,973,904,837,772,710,652,597,545,
497,453,413,378,347,320,298,281,269,261,259,261,268,280,297,318,344,375,410,450,494,541,
592,647,706,767,832,898,968,1039,1112,1187,1263,1340,1418};

//
// Timer 1 interrupt service routine. The program jumps here every 0.005 ms (5 μs).
// In this routine, a new sine Sample is calculated, offset is added, and the Sample
// is sent to the D/A converter
//
void TMR() iv IVT_TIMER_1 ilevel 1 ics ICS_SOFT
{
```

Figure 8.6

(Continued on next page)

```
        Chip_Select = 0;                                // Select DAC chip

    // Send High Byte
        temp = (DAC[Count] >> 8) & 0x0F;                // Store [11..8] to temp[3..0]
        temp |= 0x30;                                   // Define D/A setting
        SPI2_Write(temp);                               // Send high byte via SPI

        // Send Low Byte
        temp = DAC[Count];                              // Store DAC[7..0] to temp[7..0]
        SPI2_Write(temp);                               // Send low byte via SPI

        Chip_Select = 1;                                // Disable D/A converter chip

        Count++;                                        // Increment sample count
        if(Count == 100)Count = 0;                      // Reset if end of samples
        IFS0bits.T1IF = 0;                              // Clear Timer 1 interrupt flag
    }


    //
    // Start of man program
    //
    void main()
    {
        Chip_Select_Direction = 0;                      // Configure CS pin as output
        Chip_Select = 1;                                // Disable D/A converter
    //
    //Set SPI2 to the Master Mode, data length is 16-bit, clock = Fcy (no clock scaling),
    //        data sampled in the middle of interval, clock IDLE state high,
    //        data transmitted at low to high clock edge
    //
        SPI2_Init_Advanced(_SPI_MASTER, _SPI_8_BIT, 4, _SPI_SS_DISABLE,
                _SPI_DATA_SAMPLE_MIDDLE, _SPI_CLK_IDLE_LOW,
                _SPI_IDLE_2_ACTIVE);

    //
    // Configure Timer 1 for 0.5ms interrupts
    //
        T1CONbits.ON = 0;                               // Disable Timer 1
        TMR1 = 0;                                        // Clear TMR1 register
        T1CONbits.TCKPS = 0;                             // Select prescaler = 1
        T1CONbits.TCS = 0;                               // Select internal clock
        PR1 = 400;                                       // Load period register (for 0.005ms)
        IFS0bits.T1IF = 0;                               // Clear Timer 1 interrupt flag
        IPC1bits.T1IP = 1;                               // Set priority level to 1
        IEC0bits.T1IE = 1;                               // Enable Timer 1 interrupts
        T1CONbits.ON = 1;                                // Enable Timer 1
        EnableInterrupts();                              // Enable interrupts

        for(;;)                                          // Wait for Timer 1 interrupts
        {

        }
    }
```

Figure 8.6
Program listing.

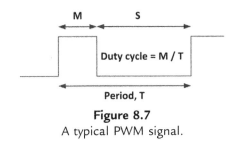

Figure 8.7
A typical PWM signal.

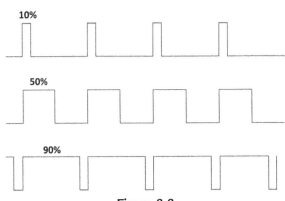

Figure 8.8
PWM signals with different duty cycles.

100%. The power supplied to the load is controlled by varying the duty cycle. Figure 8.8 shows signals with different duty cycles.

Many electronic circuits such as electric motors, solenoids, LEDs and so on average the applied ON−OFF signal in their operation. The average voltage used by these circuits can be expressed as follows:

```
Vavg = D*Von
```

where Von is the logic voltage level during the ON time. With a 3.3 V supply voltage, the average voltage becomes

```
Vavg = 3.3 × D
```

Thus, the average voltage supplied to the load is directly proportional to the duty cycle of the PWM signal output by the microcontroller.

In this project, the duty cycle of the PWM waveform is configured to be 50%.

The block diagram of the project is shown in Figure 8.9.

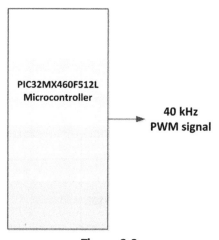

Figure 8.9
Block diagram of the project.

8.2.2 Project Hardware

The circuit diagram of the project is shown in Figure 8.10. The PIC32MX460F512L microcontroller has five pins named OC1, OC2, OC3, OC4 and OC5 that can be used to generate PWM signal in hardware and independent of the CPU operations. Thus the CPU can perform other tasks while the PWM signal has output continuously from the microcontroller.

In this project, the 40 kHz PW1 signal is output from pin OC1 (or RD0, pin 72) of the microcontroller.

If you are using the LV-32MX V6 development board, then the following jumper must be set as follows:

```
DIP switch SW20, jumper 8, set to ON
```

8.2.3 Project PDL

Before developing the PDL for this project, it is important to understand how the project works.

The PWM module inside the PIC32 microcontroller uses a timer to control the signal frequency and duty cycle. The period of the generated PWM signal is given by the following:

```
PWM Period = [(PR + 1) × TPB × (Timer prescaler value)]
```

where PR is the value loaded into the period register, and T_{PB} is the clock period of the peripheral clock. Assuming a peripheral clock frequency of 80 MHz, the peripheral clock period is given by the following:

```
TPB = 1/(80 × 10⁻⁶) = 0.0125 × 10-6 s
```

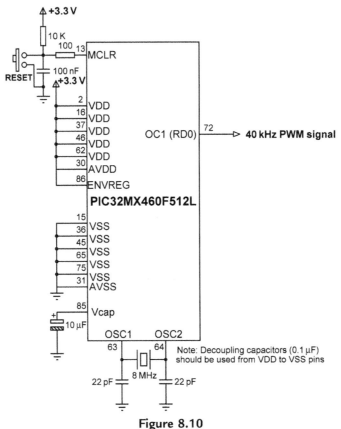

Figure 8.10
Circuit diagram of the project.

We are normally interested in finding the value to be loaded into the period register. Assuming that we are using Timer 2 (Timer 1 cannot be used in PWM mode), the value loaded into this register is calculated as follows:

$$PR2 = \frac{PWM\ Period}{0.0125 \times 10^{-6} \times (Timer\ 2\ prescaler\ value)} - 1$$

Also,

```
PWM frequency = 1/[PWM Period]
```

The PWM duty cycle is specified by writing to the OCxRS register. The maximum PWM duty cycle resolution is calculated using the following formula:

$$Maximum\ PWM\ resolution = \frac{Log_{10}\left(\frac{F_{PB}}{FPWM \times (Timer\ 1\ prescaler\ value)}\right)}{Log_{10}2} bits$$

where, F_{PB} is the peripheral clock frequency, and FPWM is the clock frequency of the PWM signal to be generated.

The specifications of this project are summarized below:

- Frequency of the PWM signal $= 40$ MHz (or, period $= 0.025 \times 10^{-3}$ s)
- Peripheral bus clock frequency $= 80$ MHz (or 0.0125×10^{-6} s)
- Timer to be used $=$ Timer 2
- Timer 2 prescaler value $= 1$
- Duty cycle $= 50\%$.

The value to be loaded into period register PR2 is calculated as follows:

$$PR2 = \frac{0.025 \times 10^{-3}}{0.0125 \times 10^{-6} \times 1} - 1 = 1999$$

The maximum PWM resolution is calculated as follows:

$$Maximum\ PWM\ resolution = \frac{Log_{10}\left(\frac{80 \times 10^{6}}{40 \times 10^{3} \times 1}\right)}{Log_{10}2} = 11\ bits$$

The steps for configuring the PWM module for the above specifications are given below (notice that since we are using output pin OC1, the register OCxCON is the register OC1CON):

- Calculate the PWM period (0.025×10^{-3})
- Calculate the PWM duty cycle (50%)
- Use Timer 2 in 16-bit mode
- Clear register OC1CON, bit 5 (OC32) for 16-bit operation
- Load PR2 with decimal 1999
- Load OC1RS low 16-bits with duty cycle (50% duty cycle corresponds to 1999/2 $=$ decimal 1000)
- No interrupts required
- Set OCM bits of OC1CON to six to enable PWM mode of operation
- Select Timer 2 as the timer source
- Clear TCKPS bits of T2CON to set Timer 2 prescaler to 1
- Set bit ON of T2CON to enable Timer 2
- Set bit ON of OC1CON to enable OC1CON.

Figure 8.11 shows the PDL of this project.

BEGIN/MAIN
Configure pin OC1 (RD0) as output
Call to Configure_PWM_Module
DO FOREVER
Wait here as the PWM module does all the work
ENDDO
END/MAIN

BEGIN/CONFIGURE_PWM_MODULE
Configure register OC1CON as in the text
Configure Timer 2 as in the text
END/CONFIGURE_PWM_MODULE

Figure 8.11
PDL of the project.

8.2.4 Project Program

Figure 8.12 shows the program listing (called PWM.C). At the beginning of the program pin OC1 (RD0) of the microcontroller is configured as an output pin. Then, function CONFIGURE_PWM_MODULE is called to configure the OC1CON and the TIMER 2 registers to generate the 40 kHz PWM wave with the 50% duty cycle. The following code shows how the configuration is done:

```
void Configure_PWM_Module(void)
{
  T2CONbits.T32 = 0;           // Timer 2 in 16 bit mode
  OC1CONbits.OC32 = 0;         // 16 bit operation
  PR2 = 1999;                  // Load PR2
  OC1RS = 1000;                // Load duty cycle
  OC1CONbits.OCM = 6;          // Enable PWM module
  OC1CONbits.OCTSEL = 0;       // TIMER 2 is the source
  T2CONbits.TCKPS = 0;         // Set Timer 2 prescaler = 1
  T2CONbits.ON = 1;            // Enable Timer 2
  OC1CONbits.ON = 1;           // Enable OC1CON
}
```

The rest of the main program consists of an endless loop where the PWM module runs in the background.

Figure 8.13 shows the PWM waveform generated in this project.

8.3 Project 8.3—Changing the Brightness of an LED

8.3.1 Project Description

This project is very similar to Project 8.2. Here, the generated PWM signal is used to drive an LED. The duty cycle of the signal is changed continuously so that the average voltage given to the LED, and hence the brightness of the LED, changes continuously.

```
/********************************************************************
                    GENERATE PWM OUTPUT
                    ===================

In this project a Pulse Width Modulated (PWM) waveform is generated from the
microcontroller. Once configured and running, the PWM module is independent of the
CPU and runs in the background so that the CPU is free to carry out other tasks.

In this project the OC1 output (RD0) of the PIC32MX460F512L microcontroller is
configured to generate a 40 kHz PWM wave with a duty cycle of 50%.




Author:        Dogan Ibrahim
Date:          August 2012
File:          PWM.C
*********************************************************************/
//
// This funtion configures the PWM module to generate a 40 kHz signal with a duty cycle of
// 50% on pin OC1 of the microcontroller. Timer 2 is used for the timing of the PWM signal.
//
void Configure_PWM_Module(void)
{
  T2CONbits.T32 = 0;                        // Timer 2 in 16 bit mode
  OC1CONbits.OC32 = 0;                      // 16 bit operation
  PR2 = 1999;                               // Load PR2
  OC1RS = 1000;                             // Load duty cycle
  OC1CONbits.OCM = 6;                       // Enable PWM module
  OC1CONbits.OCTSEL = 0;                    // TIMER 2 is the source
  T2CONbits.TCKPS = 0;                      // Set Timer 2 prescaler = 1
  T2CONbits.ON = 1;                         // Enable Timer 2
  OC1CONbits.ON = 1;                        // Enable OC1CON
}



//
// Start of main program
//
void main(void)
{
  TRISD0_bit = 0;                           // Configure OC1 as output

  Configure_PWM_Module();                   // Configure the PWM module

  for(;;)                                   // Wait here, PWM module does all the work
  {
  }
}
```

Figure 8.12
Program listing of the project.

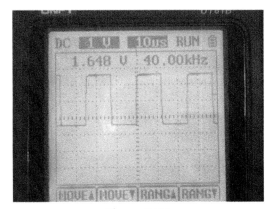

Figure 8.13
The generated PWM waveform.

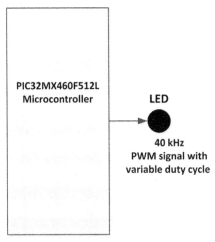

Figure 8.14
Block diagram of the project.

The block diagram of the project is shown in Figure 8.14.

8.3.2 Project Hardware

The circuit diagram of this project is very similar to the one in Project 8.2. Here, an LED is connected to port pin OC1 through a current-limiting resistor. Figure 8.15 shows the circuit diagram of the project.

8.3.3 Project PDL

In this project, the same PWM frequency given in Project 8.2 is used. The duty cycle is changed continuously so that the brightness of the LED changes. The PDL of the project is shown in Figure 8.16.

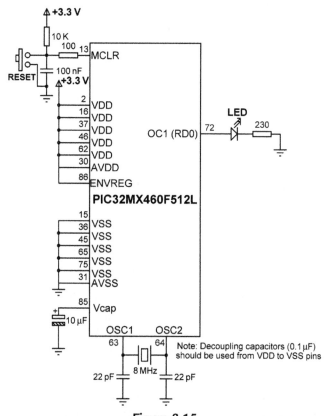

Figure 8.15
Circuit diagram of the project.

BEGIN/MAIN
    Configure pin OC1 (RD0) as output
    Call to Configure_PWM_Module
    DO FOREVER
        Change duty cycle from 0% to 100% in 4 seconds
    ENDDO
END/MAIN

BEGIN/CONFIGURE_PWM_MODULE
    Configure register OC1CON as in the text
    Configure Timer 2 as in the text
END/CONFIGURE_PWM_MODULE

Figure 8.16
PDL of the project.

8.3.4 Project Program

The program listing is shown in Figure 8.17 (LED-BRIGHTNESS.C). Most of the code is given in Figure 8.12. Here, in addition, the duty cycle is changed continuously inside the main program. As a result of this, the average voltage supplied to the LED changed from 0 V (0% duty cycle) to +3.3 V (100% duty cycle). A variable called *Dim* is used to change the duty cycle in steps of 50, from 0 to the maximum allowable (1999). A delay of 100 ms is used between each output.

8.4 Project 8.4—Using a Thin Film Transistor Display

8.4.1 Project Description

This project shows how to use a thin film transistor (TFT) display in 32-bit microcontroller applications. In this project, various shapes, characters, and text are displayed on the TFT.

The display used in this project is the MI0283QT, which is 320×240 pixel color TFT display with touch-panel controller (the touch-panel feature is not used in this project). The display is controlled with the HX8347D controller. The LV-32MX V6 development board includes this display on-board, and readers who own this board will find it very easy to complete the project.

Figure 8.18 shows the TFT display X and Y coordinates. The origin (0, 0) is at the top-left hand of the display. X coordinate runs from left to right from 0 to 239, and the Y coordinate runs from top to bottom from 0 to 319.

The TFT display is controlled using the TFT library of the mikroC PRO for PIC32 language. This is a large library which contains functions to write characters and text any required coordinate of the display, to draw shapes such as circles and rectangles with required sizes, to display bitmap images and so on. Some of the important TFT functions will be described in this project. Interested readers can get further information from either the mikroC Pro for PIC32 User Manual or the help menu of the compiler IDE. The program given in this chapter is large and requires the full version of the compiler.

8.4.2 Some mikroC PRO for PIC32 TFT Functions

Some useful TFT functions are summarized in this section.

TFT_Init

This function initializes the TFT library in 8-bit data transfer mode for use with the HX8347D-type controller (there are other initialization routines in the library for other types of controllers). The initialization is based on using the Parallel Master Port (PMP) of the

```
/******************************************************************************
                    CHANGING BRIGHTNESS OF AN LED
                    ==============================
```

In this project a Pulse Width Modulated (PWM) waveform is generated from the
microcontroller. Once configured and running, the PWM module is independent of the
CPU and runs in the background so that the CPU is free to carry out other tasks.

In this project the OC1 output (RD0) of the PIC32MX460F512L microcontroller is
configured to generate a 40 kHz PWM wave. An LED is connected to pin OC1. The duty
cycle of the PWM signal is changed continuously so that the brightness of the LED
changes from full OFF to fully ON in about 4 s (the duty cycle is changed
from 0 to 1999 in steps of 50, with 100 ms between each change)

```
    Author:      Dogan Ibrahim
    Date:        August 2012
    File:        LED_BRIGHTNESS.C
*******************************************************************************/
//
// This funtion configures the PWM module to generate a 40 kHz PWM signal
//
void Configure_PWM_Module(void)
{
  T2CONbits.T32 = 0;              // Timer 2 in 16 bit mode
  OC1CONbits.OC32 = 0;            // 16 bit operation
  PR2 = 1999;                     // Load PR2
  OC1RS = 0;                      // Load duty cycle (0 to start with)
  OC1CONbits.OCM = 6;             // Enable PWM module
  OC1CONbits.OCTSEL = 0;          // TIMER 2 is the source
  T2CONbits.TCKPS = 0;            // Set Timer 2 prescaler = 1
  T2CONbits.ON = 1;               // Enable Timer 2
  OC1CONbits.ON = 1;              // Enable OC1CON
}


//
// Start of main program
//
void main(void)
{
  unsigned int Dim = 0;

  TRISD0_bit = 0;                 // Configure OC1 as output

  Configure_PWM_Module();         // Configure the PWM module

//
// Change LED brightness continuously
//
```

Figure 8.17
(Continued on next page)

```
for(;;)                          // DO FOREVER
{
  OC1RS = Dim;                   // Change duty cycle
  Delay_Ms(100);                 // Wait 100 ms
  Dim = Dim + 50;                // Duty cycle
  if(Dim > 1999)Dim = 0;         // Reset duty cycle
}
}
```

Figure 8.17
Program listing of the project.

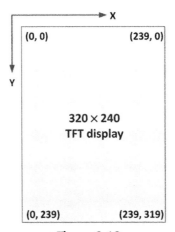

Figure 8.18
Coordinates of the TFT display.

microcontroller. The display width and height must be entered in pixels as arguments to the function. An example is given below:

```
TFT_Init(240, 320);
```

TFT_Set_Font

This function is used to change the font color and font orientation. The valid values are the following:

Font Color: CL_AQUA, CL_BLACK, CL_BLUE, CL_FUSCHIA, CL_GRAY, CL_GREEN, CL_LIME, CL_MAROON, CL_NAVY, CL_OLIVE, CL_PURPLE, CL_RED, CL_SILVER, CL_TEAL, CL_WHITE, CL_YELLOW.

Font Orientation: FO_HORIZONTAL, FO_VERTICAL.

In the following example, the font color is set to blue with the orientation horizontal:

```
TFT_Set_Font(TFT_defaultFont, CL_BLUE, FO_HORIZONTAL);
```

TFT_Write_Char

This function displays a single character at the specified display coordinates. In the following example, character "X" is displayed to coordinate x = 20 and y = 50.

```
TFT_Write_Char('X', 20, 50);
```

TFT_Write_Text

This function displays the specified text at the given display coordinates. In the following example, text "TFT DISPLAY" is displayed starting from coordinate x = 10 and y = 12.

```
TFT_Write_Text("TFT DISPLAY", 10, 12);
```

TFT_Fill_Screen

This function changes the background of the display. Valid colors are as in TFT_Set_Font. In the following example, the display background is set to white:

```
TFT_Fill_Screen(CL_WHITE);
```

TFT_Dot

This function displays a dot at the specified display coordinates. The color of the dot can be any of the colors specified in function TFT_Set_Font. In the following example, a red dot is placed at coordinate x = 10 and y = 20.

```
TFT_Dot(10, 20, CL_RED);
```

TFT_Set_Pen

This function sets the color and thickness of a shape such as a circle and rectangle. In the following example, the color is set to red, and the thickness is set to 5:

```
TFT_Set_Pen(CL_RED, 5);
```

TFT_Line

This function draws a straight line from a given coordinate to another coordinate. In the example below, a line is drawn from (0, 0) to (10, 10).

```
TFT_Line(0, 0, 10, 10);
```

TFT_Rectangle

This function draws a rectangle at the specified coordinates. The parameters are the following:

- x_upper_left: x coordinate of the upper left rectangle corner.
- y_upper_left: y coordinate of the upper left rectangle corner.

- x_bottom_right: x coordinate of the lower right rectangle corner.
- y_bottom_right: y coordinate of the lower right rectangle corner.

In the example given below, a rectangle is drawn where the top-left coordinates are (20, 20) and the bottom-right coordinates are (50, 50).

```
TFT_Rectangle(20, 20, 50, 50);
```

TFT_Circle

This function draws a circle with the center at the given coordinates, and with the specified radius. In the following example, a circle is drawn at center (50, 50) and radius 10.

```
TFT_Circle(50, 50, 10);
```

TFT_Set_Brush

This function is used to set color and the gradient which will be used to fill in circles or rectangles. The parameters are as follows:

- Brush fill mode (0 = disable and 1 = enable)
- Brush fill color (any of the colors specified in TFT_Set_Font)
- Gradient (0 = disable, 1 = enable)
- Gradient orientation (LEFT_TO_RIGHT or TOP_TO_BOTTOM)
- Starting gradient color (any of the colors specified in TFT_Set_Font)
- Ending gradient color (any of the colors specified in TFT_Set_Font).

In the following example, the gradient is enabled from black to white color, left–right orientation:

```
TFT_Set_Brush(0, 0, 1, LEFT_TO_RIGHT, CL_BLACK, CL_WHITE);
```

The block diagram of the project is shown in Figure 8.19.

8.4.3 Project Hardware

Figure 8.20 shows the circuit diagram of the project. The TFT display is controlled from the PMP port pins of the PIC32MX460F512L microcontroller. The display is controlled in 16-bit mode where 16 PMP port pins (PMD0 to OMD15) of the microcontroller are connected to the TFT data pins.

The configuration pins IM0, IM1, IM2 and IM3 of the display control the communications mode as shown in Table 8.1. In this project, since 16-bit communication is used, IM1 is connected to VCC (+3 V) while the other pins are connected to ground.

The LED backlight pins LED-A1, LED-A2, LED-A3 and LED-A4 are connected to +5 V. LED-K is connected to ground.

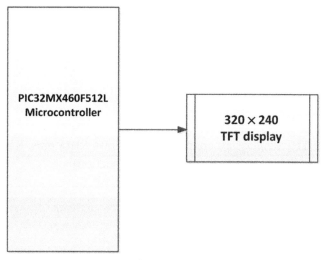

Figure 8.19
Block diagram of the project.

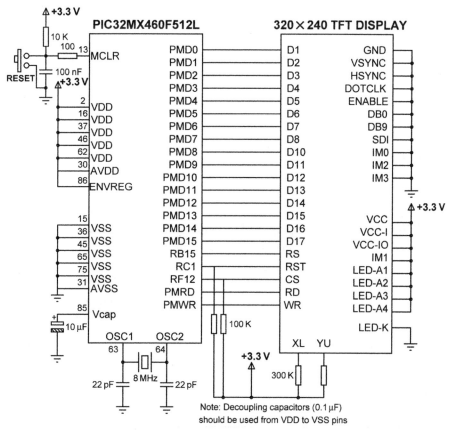

Figure 8.20
Circuit diagram of the project.

Table 8.1: Selecting the TFT Display Communications Mode

IM3	IM2	IM1	IM0	Mode	Display Pin Number
0	0	1	0	16-bit interface	DB[17:10], DB[8:1]
0	0	1	1	8-bit interface	DB[17:10]
0	1	0	X	SPI interface	SDI, SDO
1	0	1	0	18-bit interface	DB[17:0]
1	0	1	1	9-bit interface	DB[17:9]

The other pins of the TFT display are connected as follows (data pins, RS, RST, and CS are configured in software as we shall see in the program section):

GND, VSYNC, HSYNC, DOTCLK, ENABLE, DB9, DB0, SDI	Ground
VCC, VCC-IO, VCC-I	+3.3 V
RS	RB15
RST	RC1
CS	RF12
SDO, FMARK	No connection

Notice that in 16-bit interface, PMD0—PMD15 pins of the microcontroller are connected to the DB1—DB8 and DB10—DB17 pins of the display. Pins DB0 and DB9 of the display are not used.

If you are using the LV-32MX V6 development board, the following jumpers should be configured:

```
DIP switch SW17, all switched set to ON
DIP switch SW18, all switches set to ON
DIP switch SW19, switches 2—8 set to ON
```

8.4.4 Project PDL

The PDL of the project is shown in Figure 8.21. Here, the TFT is initialized and then configured for operation in 16-bit mode by calling function *Setup_PMP* and built-in TFT function **TFT_Set_Active**. The display background is set to white and then, the following characters, text and shapes are displayed on the TFT in black color:

Coordinate	Display
(10, 10)	Character "R"
(75, 120)	Text "TFT DISPLAY"
(90, 50)	Circle with radius 20
(15, 160) to (80, 160)	Line
(20, 150)	Dot
(22, 150)	Dot
(24, 150)	Dot
(20, 200) to (50, 250)	Rectangle
(40, 100)	Filled circle with radius 10

BEGIN/MAIN
    Configure the interface between the display and the microcontroller
    Configure PMP port for 16-bit operation
    Initialize TFT library
    Draw white background
    Display character "R" at coordinate (10, 10)
    Write text "TFT DISPLAY" at coordinate (75, 120)
    Draw circle with center at (90, 50), radius 20
    Draw line from (15, 160) to (80, 160)
    Draw a dot at coordinate (20, 150)
    Draw a dot at coordinate (22, 150)
    Draw a dot at coordinate (24, 150)
    Draw a rectangle at (20, 100, 50, 250)
    Set the brush to fill mode with left-to-right gradient
    Draw a filled circle with center at (40, 100), radius 10
END/MAIN

BEGIN/SETUP_PMP
    Configure PMP port for 16-bit operation
END/SETUP_PMP

Figure 8.21
PDL of the project.

8.4.5 Project Program

The program listing of the project is given in Figure 8.22 (program TFT-SAMPLE.C). At the beginning of the main program, the interface between the TFT display and the microcontroller is configured using **sbit** statements. Then, PMP port is configured for 16-bit operation. Then the TFT library is initialized and the TFT display is filled with white background.

The program then calls the TFT library functions to display the characters, texts, and shapes as shown in Figure 8.21.

Figure 8.23 shows the display for this project.

8.5 Project 8.5—Plotting a Graph on the TFT Display

8.5.1 Project Description

This project shows how we can plot a graph on the TFT display. The graph of function $y = x^2 - 1$ is plotted on the display as x varies between -2 and $+2$. The block diagram of the project is as shown in Figure 8.19.

8.5.2 Project Hardware

The circuit diagram of the project is as shown in Figure 8.20.

```
/***************************************************************************
                     USING THE TFT DISPLAY
                     ====================

This project shows how to use the 320 × 240 colour TFT display by displaying various
characters, texts, and shapes on the display.

The LV-32MX V6 development board contains a TFT display and readers who own such a board
will find it easy to complete this project.

The mikroC PRO for PIC32 TFT library is used in the project. This library includes many
functions for writing text, drawing lines, circles, rectangles, dots and so on, on the TFT display.

The Parallel Master Port (PMP) is used in the library in 16-bit communication mode. Details of
the hardware interface are given in the text.


     Author:        Dogan Ibrahim
     Date:          September 2012
     File:          TFT-SAMPLE.C
***************************************************************************/
//
// TFT display connections
//
unsigned short TFT_DataPort at LATE;
sbit TFT_WR at LATD4_bit;
sbit TFT_RD at LATD5_bit;
sbit TFT_CS at LATF12_bit;
sbit TFT_RS at LATB15_bit;
sbit TFT_RST at LATC1_bit;
//
// TFT display port directions
//
unsigned short TFT_DataPort_Direction at TRISE;
sbit TFT_WR_Direction at TRISD4_bit;
sbit TFT_RD_Direction at TRISD5_bit;
sbit TFT_CS_Direction at TRISF12_bit;
sbit TFT_RS_Direction at TRISB15_bit;
sbit TFT_RST_Direction at TRISC1_bit;
//
// End of TFT display connections
//

//
// The following functions are used by the TFT library while using the PMP port
// communications.
//
// Wait if port is busy
//
void PMPWaitBusy()
{
```

Figure 8.22

(Continued on next page)

```
    while(PMMODEbits.BUSY);
}


//
// Function Set Index handler
//
void SetIndex(unsigned short index)
{
  TFT_RS = 0;
  PMDIN = index;
  PMPWaitBusy();
}


//
// Function write Command handler
//
void WriteCommand(unsigned short cmd )
{
  TFT_RS = 1;
  PMDIN = cmd;
  PMPWaitBusy();
}


//
// Function Write Data handler
//
void WriteData(unsigned int _data)
{
  TFT_RS = 1;
  PMDIN = _data;
  PMPWaitBusy();
}



//
// This function configures the Parallel Master Port (PMP) for 16 bit communication
//
void SETUP_PMP(void)
{
  PMMODE = 0;                              // Clear all bits (default)
  PMAEN  = 0;                              // Configure PMA0–15 as Port I/O
  PMCON  = 0;                              // Clear all bits (default)
  PMMODEbits.MODE = 2;                     // Master 2 with 16-bit PMD, PMRD,PMWR
  PMMODEbits.WAITB = 0;                    // Default Data setup wait
  PMMODEbits.WAITM = 1;                    // Wait 2TPB
  PMMODEbits.WAITE = 0;                    // Default Data hold wait
  PMMODEbits.MODE16 = 1;                   // 16 bit mode (single 16 bit transfers)
  PMCONbits.CSF = 0;                       // Configure PMCS1,PMCS2 addr 14 and 15
  PMCONbits.PTRDEN = 1;                    // PMRD,PMWR port enabled
  PMCONbits.PTWREN = 1;                    // PMWR/PMENB port enabled
  PMCONbits.PMPEN = 1;                     // Enable PMP
```

Figure 8.22

(Continued on next page)

```
    }


    //
    // Start of main program
    //
    void main()
    {
     Setup_PMP();                                          // Configure PMP port
     TFT_Set_Active(SetIndex,WriteCommand,WriteData);      // Set pointers to user-defined routines

     AD1PCFG = 0xFFFF;
     TFT_Init(240,320);                                    // Initialize TFT library
    //
    // Draw White background
    //
     TFT_Fill_Screen(CL_WHITE);
    //
    // Write character 'R' at coordinate (10, 10)
    //
     TFT_Write_Char('R', 10, 10);
    //
    // Write text 'TFT DISPLAY' at coordinate (75, 120)
    //
     TFT_Write_Text('TFT DISPLAY', 75, 120);
    //
    // Draw circle at center (90, 50), radius 20
    //
     TFT_Circle(90, 50, 20);
    //
    // Draw line from (15, 160) to (80, 160)
    //
     TFT_Line(15, 160, 80, 160);
    //
    // Draw 3 horizontal dots starting at coordinate (20, 150)
    //
    TFT_Dot(20, 150, CL_BLACK);
    TFT_Dot(22, 150, CL_BLACK);
    TFT_Dot(24, 150, CL_BLACK);
    //
    // Draw rectangle at coordinates (20, 200) to (50, 250)
    //
     TFT_Rectangle(20, 200, 50, 250);
    //
    // Fill a circle
    //
     TFT_Set_Brush(1, CL_BLACK, 1, LEFT_TO_RIGHT, CL_WHITE, CL_BLACK);
     TFT_Circle(40, 100, 10);
    }
```

Figure 8.22
Program listing of the project.

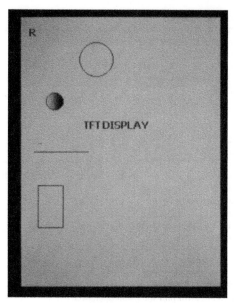

Figure 8.23
Display for the project.

8.5.3 Project PDL

Figure 8.24 shows the PDL of the project. After configuring the interface between the TFT display and the microcontroller, function *SETUP_PMP* is called in the main program to configure the PMP for 16 bits of operation. Then, the X and Y values of the function to be plotted are stored in floating point arrays. The main program then calls the function *PlotXY* to plot the graph of the function.

Inside the function *PlotXY*, the maximum and minimum X and Y values are calculated for scaling the graph. Then the X and Y axes are drawn and the X, Y points are plotted as dots. Finally, the axis ticks and labels are drawn and the function type is displayed at the bottom of the graph.

8.5.4 Project Program

The program listing of the project is given in Figure 8.26 (program GRAPH.C). At the beginning of the program, the interface between the TFT display and the microcontroller is defined using **sbit** statements. The main program calls function *SETUP_PMP* and built-in function *TFT_Set_Active* to configure the PMP port for 16-bit communication.

The TFT library is then initialized. The function to be plotted is divided into 100 segments and the X and Y values of each segment are calculated and stored in variables

BEGIN/MAIN
> Configure the interface between the display and the microcontroller
> Configure PMP port for 16-bit operation
> Initialize TFT library
> Calculate X and Y values of the function to be plotted
> Call function PlotXY to plot the graph

END/MAIN

BEGIN/PLOTXY
> Draw white background
> Draw X and Y axes
> Calculate minimum and maximum values of X and Y values
> Plot the graph using dots
> Draw axes ticks
> Draw axes ticks
> Write function type at the bottom of the display

END/PLOTXY

BEGIN/SETUP_PMP
> Configure PMP port for 16-bit operation

END/SETUP_PMP

Figure 8.24
PDL of the project.

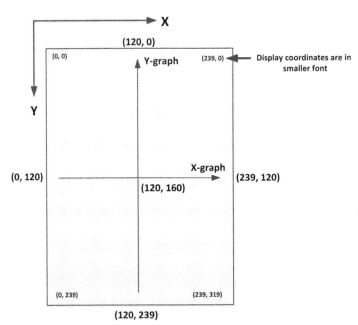

Figure 8.25
Graph coordinates.

```
/*****************************************************************************
              PLOTTING A GRAPH ON THE TFT DISPLAY
              ===================================
```

This project shows how to use the 320 × 240 colour TFT display to draw the graph of a simple function.

The LV-32MX V6 development board contains a TFT display and readers who own such a board will find it easy to complete this project.

In this project the function y = x*x−1 is plotted on the TFT display in the region x = −2 to x = +2 by taking 100 sample points.

The mikroC PRO for PIC32 TFT library is used in the project. This library includes many functions for writing text, drawing lines, circles, rectangles, dots and so on on the TFT display.

The Parallel Master Port (PMP) is used in the library in 16-bit communication mode. Details of the hardware interface are given in the text.

```
Author:        Dogan Ibrahim
Date:          September 2012
File:          GRAPH.C
*****************************************************************************/
//
// TFT display connections
//
unsigned short TFT_DataPort at LATE;
sbit TFT_WR at LATD4_bit;
sbit TFT_RD at LATD5_bit;
sbit TFT_CS at LATF12_bit;
sbit TFT_RS at LATB15_bit;
sbit TFT_RST at LATC1_bit;
//
// TFT display port directions
//
unsigned short TFT_DataPort_Direction at TRISE;
sbit TFT_WR_Direction at TRISD4_bit;
sbit TFT_RD_Direction at TRISD5_bit;
sbit TFT_CS_Direction at TRISF12_bit;
sbit TFT_RS_Direction at TRISB15_bit;
sbit TFT_RST_Direction at TRISC1_bit;
//
// End of TFT display connections
//

float XValues[100], YValues[100];
float Stp, Starting_X, Ending_X;

//
```

Figure 8.26

(Continued on next page)

```
// The following functions are used by the TFT library while using the PMP port
// communications.
//
// Wait if port is busy
//
void PMPWaitBusy()
{
  while(PMMODEbits.BUSY);
}


//
// Function Set Index handler
//
void SetIndex(unsigned short index)
{
  TFT_RS = 0;
  PMDIN = index;
  PMPWaitBusy();
}


//
// Function write Command handler
//
void WriteCommand(unsigned short cmd )
{
  TFT_RS = 1;
  PMDIN = cmd;
  PMPWaitBusy();
}


//
// Function Write Data handler
//
void WriteData(unsigned int _data)
{
  TFT_RS = 1;
  PMDIN = _data;
  PMPWaitBusy();
}



//
// This function configures the Parallel Master Port (PMP) for 16 bit communication
//
void SETUP_PMP(void)
{
  PMMODE = 0;                       // Clear all bits (default)
  PMAEN = 0;                        // Configure PMA0–15 as Port I/O
  PMCON  = 0;                       // Clear all bits (default)
  PMMODEbits.MODE = 2;              // Master 2 with 16-bit PMD, PMRD,PMWR
  PMMODEbits.WAITB = 0;             // Default Data setup wait
```

Figure 8.26

(Continued on next page)

```
    PMMODEbits.WAITM = 1;                          // Wait 2TPB
    PMMODEbits.WAITE = 0;                          // Default Data hold wait
    PMMODEbits.MODE16 = 1;                         // 16 bit mode (single 16 bit transfers)
    PMCONbits.CSF = 0;                             // Configure PMCS1,PMCS2 addr 14 and 15
    PMCONbits.PTRDEN = 1;                          // PMRD,PMWR port enabled
    PMCONbits.PTWREN = 1;                          // PMWR/PMENB port enabled
    PMCONbits.PMPEN = 1;                           // Enable PMP
}


//
// This function plots the axes and data points for the required function
//
void PlotXY()
{
    int i;
    float XMAX, YMAX, XMIN, YMIN, Xpos, Ypos;
    TFT_Fill_Screen(CL_WHITE);
//
// Draw X and Y axes
//
    TFT_Line(0,160,239,160);                       // Draw X axis
    TFT_Line(120,0,120,319);                       // Draw Y axis
//
// Find MAX and MIN values so we can scale the graph
//
    XMAX = 0;
    XMIN = 0;
    YMAX = 0;
    YMIN = 0;

    for(i = 0; i < 100; i++)
    {
     if(XValues[i] > XMAX) XMAX = XValues[i];
     if(XValues[i] < XMIN) XMIN = XValues[i];
     if(YValues[i] > YMAX) YMAX = YValues[i];
     if(YValues[i] < YMIN) YMIN = YValues[i];
    }
    XMAX = fabs(XMAX);
    XMIN = fabs(XMIN);
    YMAX = fabs(YMAX);
    YMIN = fabs(YMIN);
    if(XMAX < XMIN)XMAX = XMIN;
    if(YMAX < YMIN)YMAX = YMIN;
//
// Now plot the graph. 100 points are used for the graph
//
    for(i = 0; i < 100; i++)
    {
    Xpos = 120.0 + XValues[i]*120.0 / XMAX;
    Ypos = 160.0 - YValues[i]*160.0 / YMAX;
```

Figure 8.26

(Continued on next page)

```
        TFT_Dot(Xpos, Ypos, CL_BLACK);
    }
//
// Draw axes ticks and labels (only −1, and +1 are labelled)
//
    for(i = −1; i <= 1; i++)
    {
      Xpos = 120.0 + i*120.0 / XMAX;
      TFT_Dot(Xpos, 161, CL_BLACK);                    // Insert an axis tick
      Ypos = 160.0 − i*160.0/ YMAX;
      TFT_Dot(121, Ypos, CL_BLACK);                    // Insert an axis tick
      if(i == −1)
      {
        TFT_Write_Text("−1", Xpos, 160);               // Insert label "−1"
        TFT_Write_Text("−1", 122, Ypos);               // Insert label "−1"
      }
      else if(i == 1)
      {
        TFT_Write_Text("+1", Xpos, 160);               // Insert label "+1"
        TFT_Write_Text("+1", 122, Ypos);               // Insert label "+1"
      }
    }
//
// Write the function below the graph
//
    TFT_Write_Text("Y = X*X − 1", 10, 300);            // Write the function at the bottom

}


//
// Start of main program
//
void main()
{
  unsigned char N, i;
  float x;

  Setup_PMP();                                         // Configure PMP port
  TFT_Set_Active(SetIndex,WriteCommand,WriteData);     // Set pointers to user-defined routines

  AD1PCFG = 0xFFFF;
  TFT_Init(240,320);                                   // Initialize TFT library
//
// Find X and Y values of the function Y = x*x − 2. Take 100 samples.
//
  N = 100;                                             // Number of points
  Starting_X = −2.0;                                   // Starting X values
  Ending_X = 2.0;                                      // Ending X value
  Stp = fabs(Starting_X − Ending_X) / N;               // step in X
```

Figure 8.26

(Continued on next page)

```
x = Starting_X;
i = 0;

do
{
  XValues[i] = x;                            // X values of the function
  YValues[i] = x*x – 1;                      // Y values of the function
  x = x + Stp;                               // Increment the step
  i++;
}while(i != N);

  PlotXY();                                  // Plot the graph
}
```

Figure 8.26
Program listing.

XValues and *YValues*, respectively. The main program then calls function *PlotXY* to plot the graph.

Function *PlotXY* draws a white display background, and calculates the maximum and minimum values of the X and Y values. The display origin is then moved to the middle of the display so that negative values can also be plotted, and the X and Y axes drawn. Figure 8.25 shows the new graph coordinates.

Data values are then converted to graph coordinates using the following two operations:

```
Xpos = 120.0 + XValues[i]*120.0/XMAX;
Ypos = 160.0 – YValues[i]*160.0/YMAX;
```

where, variable i runs from zero to the sample size. The data points are plotted in black color by placing dots at the calculated X and Y positions:

```
TFT_Dot(Xpos, Ypos, CL_BLACK);
```

The program then inserts axis labels at the graph coordinates $(1, 0), (-1, 0), (0, 1)$, and $(0, -1)$ by the following code:

```
for(i = –1; i <= 1; i++)
{
Xpos = 120.0 + i*120.0/XMAX;
TFT_Dot(Xpos, 161, CL_BLACK);          // Insert an axis tick
Ypos = 160.0–i*160.0/YMAX;
TFT_Dot(121, Ypos, CL_BLACK);          // Insert an axis tick
if(i == –1)
{
 TFT_Write_Text("–1", Xpos, 160);      // Insert axis label "–1"
 TFT_Write_Text("–1", 122, Ypos);      // Insert axis label "–1"
}
```

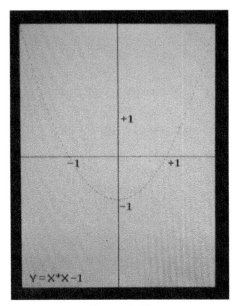

Figure 8.27
Output displayed by the program.

```
else if(i == 1)
{
  TFT_Write_Text("+1", Xpos, 160);      // Insert axis label "+1"
  TFT_Write_Text("+1", 122, Ypos);      // Insert axis label "+1"
}
}
```

Finally, the program displays the function type at the bottom of the display:

```
TFT_Write_Text("Y = X*X - 1", 10, 300);   // Write the function type at the bottom
```

Figure 8.27 shows the output displayed by the program.

8.6 Project 8.6—Using Secure Digital Cards

This project shows how to use secure digital (SD) cards in 32-bit microcontroller-based projects to store data. In this project, we create a file on the SD card and write the following text inside this file:

```
This project shows how to create a file on the SD card
```

Before going into details of the project, it is worthwhile to look at the characteristics of SD cards and their use in microcontroller-based applications.

The SD card is a flash memory storage device designed to provide high capacity, nonvolatile, and rewritable storage in small size. These devices are commonly used in most electronic

Table 8.2: Standard and Mini SD Cards

	Standard SD	miniSD
Dimensions	32 × 24 × 2.1 mm	21.5 × 20 × 1.4 mm
Card weight	2.0 g	1.0 g
Operating voltage	2.7–3.6 V	2.7–3.6 V
Write protect	Yes	No
Pins	9	11
Interface	SD or SPI	SD or SPI
Current consumption	<75 mA (write)	<40 mA (write)

consumer goods such as cameras, computers, GPS systems, mobile phones, PDAs and so on. The memory capacity of SD cards is increasing all the time. Currently, they are available at capacities from 256 MB to up to 32 GB or more. SD cards are in three sizes: *standard SD card*, *miniSD card*, and the *microSD card*. Table 8.2 lists the main specifications of the most commonly used standard and mini SD cards.

SD card specifications are maintained by the *SD Card Association* which has over 600 members. MiniSD and microSD cards are electrically compatible with the standard SD cards and they can be inserted in special adapters and used as standard SD cards in standard card slots.

SD card speeds are measured in three different ways: in kB/s (kilobytes per second), in MB/s (megabytes per second), or in an "x" rating similar to that of CD-ROMS where "x" is the speed corresponding to 150 kB/s. Thus, the various "x"-based speeds are the following:

- 4x: 600 kB/s
- 16x: 2.4 MB/s
- 40x: 6.0 MB/s
- 66x: 10 MB/s.

In this chapter, we shall be using the standard SD card only. The specifications of the smaller size SD cards are the same and are not described in this chapter any further.

SD cards can be interfaced to microcontrollers using two different protocols: SD card protocol and the Serial Peripheral Interface (SPI) protocol. The SPI protocol is the most commonly used protocol and is the one used in this chapter. The standard SD card has nine pins with the pin layout shown in Figure 8.28. Depending on the interface protocol, the pins have different functions. Table 8.3 gives the function of each pin both in SD mode and in SPI mode of operation.

Since the SD card projects described in this chapter are based on the SPI bus protocol, it is worthwhile to look at the specifications of this bus before looking at the design of the projects in greater detail.

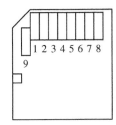

Figure 8.28
Standard SD card pin layout.

Table 8.3: Standard SD Card Pin Definitions

Pin	Name	SD Description	SPI Description
1	CD/DAT3/CS	Data line 3	Chip select
2	CMD/Data in	Command/response	Host to card command and data
3	VSS	Supply ground	Supply ground
4	VDD	Supply voltage	Supply voltage
5	CLK	Clock	Clock
6	VSS2	Supply voltage ground	Supply voltage ground
7	DAT0	Data line 0	Card to host data and status
8	DAT1	Data line 1	Reserved
9	DAT2	Data line 2	Reserved

8.6.1 The SPI Bus

SPI (serial peripheral interface) is a synchronous serial bus standard named by Motorola that operates in full duplex mode. Devices on an SPI bus operate in master–slave mode, where the master device initiates the data transfer, selects a slave, and provides clock to the slaves. The selected slave responds and sends its data to the master at each clock pulse. The SPI bus can operate with a single master device and with one or more slave devices. SPI is a simple interface and is also called a "four wire" interface.

The signals in SPI bus are named as follows:

- MOSI—master output, slave input
- MISO—master input, slave output
- SCLK—serial clock
- SS—slave select.

These signals are also named as follows:

- DO—data out
- DI—data in
- CLK—clock
- CD—chip select.

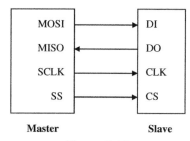

Figure 8.29
SPI master—slave connection.

Figure 8.29 shows the basic connection of a master and a slave device in SPI bus. Master sends out data on line MOSI, and receives data on line MISO. The slave is selected before data transfer can take place.

Figure 8.30 shows the case where more than one slave device is connected to the SPI bus. Here, each slave is selected individually by the master and although all the slaves receive clock pulses, only the selected slave device responds. If an SPI device is not selected, its data output goes into a high-impedance state, so that it does not interfere with the currently selected device on the bus.

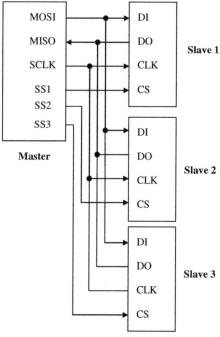

Figure 8.30
Multiple slave SPI bus.

Data transmission normally occurs in and out of the master and slave devices as the clock pulses are applied by the master. To begin a communication, the master first pulls the slave select line low for the desired slave device. Then the master issues clocks, and during each SPI clock cycle, a full duplex data transmission occurs. When there are no more data to be transmitted, the master stops toggling its clock output.

SPI bus is currently used by microcontroller interface circuits to talk to a variety of devices such as follows:

•   Memory devices (SD cards)
•   Sensors
•   Real-time clocks
•   Communications devices
•   Displays.

The advantages of the SPI bus are as follows:

•   Simple communication protocol
•   Full duplex communication
•   Very simple hardware interface.

In addition, the disadvantages of the SPI bus are as follows:

•   Requires four pins
•   No hardware flow control
•   No slave acknowledgment.

It is important to realize that there are no SPI standards governed by any international committee. As a result of this, there are several versions of the SPI bus implementation. In some applications, the MOSI and MISO lines are combined into a single data line, thus reducing the line requirements into three. Some implementations have two clocks, one to capture (or display) data, and another to clock it into the device. Also, in some implementations, the CS line may be active high rather than active low.

8.6.2 Operation of the SD Card in SPI Mode

When the SD card is operated in SPI mode only seven pins are used:

•   Two power supply ground (pins 3 and 6)
•   Power supply (pin 4)
•   Chip select (pin 1)
•   Data out (pin 7)
•   Data in (pin 2)
•   CLK (pin 5).

Three pins are used for the power supply, leaving four pins for the SPI mode of operation:

- Chip select (pin 1)
- Data out (pin 7)
- Data in (pin 2)
- CLK (pin 5).

At power-up, the SD card defaults to the SD bus protocol. The card is switched to the SPI mode if the CS signal is asserted during the reception of the reset command. When the card is in SPI mode, it only responds to SPI commands. The host may reset a card by switching the power supply off and on again.

mikroC PRO for PIC32 compiler provides a library of commands for initializing, reading, and writing to SD cards. In general, it is not necessary to know the internal structure of an SD card before it can be used since the available library functions can easily be used. It is however important to have some knowledge about the internal structure of an SD card so that it can be used efficiently. In this section, we shall be looking briefly at the internal architecture and the operation of SD cards.

An SD card has a set of registers that provide information about the status of the card. When the card is operated in SPI mode, these registers are as follows:

- Card Identification Register (CID)
- Card-Specific Data Register (CSD)
- SD Configuration Register (SCR)
- Operation Control Register (OCR).

The CID register consists of 16 bytes and it contains the manufacturer ID, product name, product revision, card serial number, manufacturer date code, and a checksum byte.

CSD register consists of 16 bytes and it contains card-specific data such as the card data transfer rate, read/write block lengths, read/write currents, erase sector size, file format, write protection flags, checksum, and so on.

SCR register is 8 bytes long and it contains information about the SD card's special features capabilities such as the security support, data bus widths supported, and so on.

OCR register is only 4 bytes long and it stores the VDD voltage profile of the card. The OCR shows the voltage range, in which the card data can be accessed.

All SD card SPI commands are 6 bytes long with the most significant bit (MSB) transmitted first. The first byte is known as the *command* byte, and the remaining 5 bytes are *command arguments*. Bit 6 of the command byte is set to "1" and the MSB is always "0". With the remaining 6 bits, we have 64 possible commands, named CMD0–CMD63.

In response to a command, the card sends a status byte known as R1. The MSB of this byte is always "0" and the other bits indicate the following error conditions:

- Card in idle state
- Erase reset
- Illegal command
- Communication cyclic redundancy check (CRC) error
- Erase sequence error
- Address error
- Parameter error.

Reading Data

SD card in SPI mode supports single block and multiple block read operations. The host should set the block length, and after a valid read command, the card responds with a response token, followed by a data block, and a CRC check. The block length can be between 1 and 512 bytes. The starting address can be any valid address range of the card.

In multiple block read operations, the card sends data blocks with each block having its own CRC check attached to the end of the block.

Writing Data

SD card in SPI mode supports single or multiple block write operations. After receiving a valid write command from the host, the card will respond with a response token, and will wait to receive a data block. A 1-byte "start block" token is added to the beginning of every data block. After receiving the data block, the card responds with a "data response" token and the card will be programmed as long as the data block has been received with no errors.

In multiple write operations, the host sends the data blocks one after the other one, each preceded with a "start block" token. The card sends a response byte after receiving each data block.

8.6.3 Card Size Parameters

In addition to the normal storage area on the card, there is also a protected area that relates to the secured copyright management. This area can be used by applications to save security-related data and can be accessed by the host using secured read/write commands. The card write protection mechanism does not affect this area.

Data can be written to or read from any sector of the card using raw sector access methods. In general, SD card data is structured as a file system and two DOS formatted partitions are placed on the card: *User Area*, and *Security Protected Area*. The size of each area depends upon the overall capacity of the card used.

A card can be inserted and removed from the bus without any damage. This is because all data transfer operations are protected by CRC codes and any bit changes as a result of inserting or removing a card can easily be detected. SD cards operate with a typical supply voltage of 2.7 V. The maximum allowed power supply voltage is 3.6 V. If the card is to be operated from a standard 5.0 V supply, a voltage regulator should be used to drop the voltage to 2.7 V.

Using an SD card requires the card to be inserted into a special card holder with external contacts (Figure 8.31). Connections can then be made easily to the required card pins.

8.6.4 mikroC PRO for PIC32 Language SD Card Library Functions

mikroC PRO for PIC32 language provides an extensive set of library functions to read and write data to SD cards (and also MultiMedia Cards (MMC)). The library is called the MMC library and it supports both the standard SD and high-capacity SDHC cards. Standard

Figure 8.31
SD card holder.

SD cards are available in capacities up to 2 GB, while the SDHC cards are available in 32 GB or even higher.

Using the library functions, cards can be formatted, data can be written or read from a given sector of the card, or the file system on the card can be used for more sophisticated applications. In this section, we will briefly look at the commonly used functions for creating and using files on the card. These functions are also known as FAT functions and they are based on the FAT16 filing system. Interested readers can get more detailed information about all the SD card functions by referring to the mikroC PRO for PIC32 Users' Manual, or from the compiler IDE help section.

SPIx_Init_Advanced and Mmc_FAT_Init

The SPI library must be initialized before the MMC FAT library is initialized. The following parameters should be set during the initialization of the SPI library:

- SPI Master
- Eight-bit mode
- Secondary prescaler 1
- Primary prescaler 64
- Slave Select disabled
- Data sampled in the middle of data output time
- Clock idle high
- Serial output data changes on transition from active clock state to idle clock state.

The MMC module connections (CS pin) should be defined before initializing the library. An example is shown below which assigns microcontroller pin RF0 to CS pin of the card:

```
// MMC module connections
sbit Mmc_Chip_Select at LATF0_bit;
sbit Mmc_Chip_Select_Direction at TRISF0_bit;
// MMC module connections
```

The following example shows how the SPI library is initialized:

```
SPI1_Init_Advanced(_SPI_MASTER, _SPI_8_BIT, 64, _SPI_SS_DISABLE,
                   _SPI_DATA_SAMPLE_MIDDLE, _SPI_CLK_IDLE_HIGH,
                   _SPI_ACTIVE_2_IDLE);
```

The MMC FAT library should then be initialized as shown below:

```
Mmc_Fat_Init;
```

The function returns the following:

```
0:   initialization successful
1:   FAT16 boot sector not found on the card
255: The card has not been detected
```

After the MMC library is initialized, it is possible to increase the SPI bus speed by reinitializing the SPI library as shown below:

```
// Reinitialize the SPI module at higher speed (change primary prescaler).
SPI1_Init_Advanced(_SPI_MASTER, _SPI_8_BIT, 8, _SPI_SS_DISABLE,
                    _SPI_DATA_SAMPLE_MIDDLE, _SPI_CLK_IDLE_HIGH,
                    _SPI_ACTIVE_2_IDLE);
```

Mmc_Fat_QuickFormat

This function formats the SD card and sets the specified label to the card. An example is given below which formats the card and sets its label to "MySDCard":

```
Mmc_QuickFormat("MySDCard");
```

Mmc_Fat_Assign

This function assigns a file for read, write, and delete operations. All subsequent FAT operations will be assigned on the assigned file. The *filename* and *file creation attributes* are arguments of this function. The filename must consist of a name and extension. The file creation attributes are listed in Table 8.4. Notice that these bits can be OR'ed to create multiple attributes.

The function returns the following:

```
0:  The file does not exist and a new file is created
1:  The fie already exists but a new blank file is created
2:  There are no more file handlers (currently opened file is closed)
```

In the following example, a new file called "TEST.TXT" is created:

```
Mmc_Fat_Assign("TEST.TXT", 0x80);
```

Mmc_Fat_Reset

This function resets the file pointer of the assigned file so that it can be read. An example is given below:

```
Mmc_Fat_Reset();
```

Table 8.4: File Creation Attributes

Mask	Description
0×01	File is read only
0×02	Hidden file
0×04	System file
0×08	Volume label
0×10	Subdirectory
0×20	Archive
0×40	Internal use only
0×80	File creation flag. If file does not exist and this flag is set, a new file with the specified name will be created

Mmc_Fat_Read

This function reads a byte from the currently assigned file opened for reading. Upon function execution, file pointers will be set to the next character in the file. Before using this function, the file must have been assigned and opened for reading. An example is given below which reads a character from a file:

```
Mmc_Fat_Read(&ch);
```

Mmc_Fat_ReadN

This function reads the specified number of characters from an opened file and stores in an array. An example is given below, which reads 250 characters from a file:

```
Mmc_Fat_ReadN(&Buffer, 250);
```

The number of bytes read is returned by the function.

Mmc_Fat_Rewrite

This function opens the currently assigned file for writing. If the file is not empty, its contents will be erased. An example is given below:

```
Mmc_Fat_Rewrite();
```

Mmc_Fat_Append

This function opens the currently assigned file for appending data to it. Upon execution, the file pointer will move to the end of the file. An example is given below:

```
Mmc_Fat_Append();
```

Mmc_Fat_Write

This function writes the specified number of characters to the opened file. The file must have been assigned and opened for writing. An example is given below which writes 20 characters stored in the array *MyData* to the file:

```
Mmc_Fat_Write(MyData, 20);
```

Mmc_Fat_Delete

This function deletes the currently assigned file. An example is given below:

```
Mmc_Fat_Delete();
```

Mmc_Fat_Open

This function combines several other functions and is used to open a file. The filename, mode, and file attributes of the file to be opened must be specified.

The mode can be one of the following:

```
FILE_WRITE, FILE_READ, FILE_APPEND
```

The file attributes are given in Table 8.4.

The function returns the following:

```
<0:  An error occurred
>=0:  File handle of the opened file
```

An example is given below which opens a new file called TEST.TXT for writing:

```
Mmc_Fat_Open("TEST.TXT", "FILE_WRITE", 0x80);
```

Mmc_Fat_Close

This function closes an open file. An example is given below:

```
Mmc_Fat_Close();
```

The function returns the following:

```
0:  Closing has been successful
1:  There was no assigned file
```

Mmc_Fat_EOF

This function is used in file read operations and it checks if the end-of-file is reached.

The function returns the following:

```
0:  End of file was reaches
1:  End of file was not reached
<0:  There was no assigned file
```

A typical use of this function is shown below where the code following the **if** statement is executed if end of the file is reached:

```
if(Mmc_Fat_EOF() == 0)
{
    .................
}
```

The block diagram of this project is shown in Figure 8.32.

8.6.5 Project Hardware

The circuit diagram of the project is shown in Figure 8.33. The connection between the SD card and the microcontroller are as follows:

SD Card Pin	Microcontroller Pin
CS	RG9
Din	SDO2 (RG8)
Dout	SDI2 (RG7)
SCK	SCK2 (RG6)

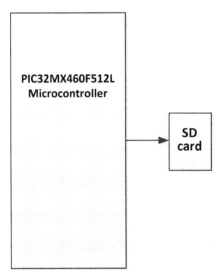

Figure 8.32
Block diagram of the project.

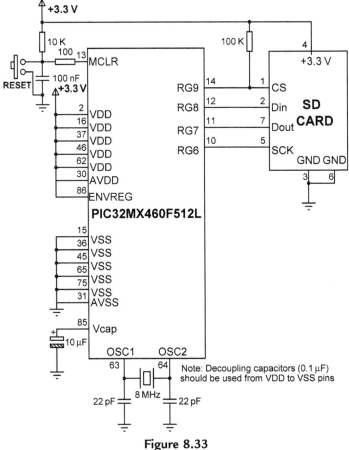

Figure 8.33
Circuit diagram of the project.

Notice that if the microcontroller is operated from a +5 V supply, then the voltage at its output pins are high and can damage the SD card inputs. Therefore, in such cases, the CS, Din and SCK inputs of the SD card should be connected to the microcontroller using three potential divide resistors to lower the voltage to around +3 V. Suitable resistors for the potential divider are 2.2 K from microcontroller output pins to SD card input pins, and 3.3 K from the SD card input pins to ground.

If you are using the LV-32MX V6 development board, then the following jumpers must be set on the board:

```
DIP switch SW13, switch positions 4—8, set to ON
```

8.6.6 Project PDL

The project is very simple and PDL is shown in Figure 8.34. At the beginning, the CS connection to the SD card is defined, and SPI and the MMC libraries are initialized. Then the file TEXT.TXT is created on the card and the required text is written inside this file.

8.6.7 Program Listing

The program listing (SD-TEXT.C) is shown in Figure 8.35. At the beginning of the project, the CS connection of the SD card is defined. Then the text to be written to the SD card and its length are stored in variables *MyText* and *TextLen*, respectively. The SPI bus is initialized at the standard speed and the program waits until the FAT library is initialized. After this, the speed of the SPI bus can be increased if desired.

A new file called TEXT.TXT is then created on the SD card, the file pointer is set to the beginning of the file and the contents of the file have been erased using functions:

```
Mmc_FAT_Assign("TEST.TXT", 0x80);
Mmc_FAT_Rewrite();
```

The required text is then written to the file using function:

```
Mmc_FAT_Write(MyText, TextLen);
```

BEGIN
Configure the chip select pin
Initialize the SPI library
Initialize the MMC library
Create file TEXT.TXT on the card
Write the required text inside the file
END

Figure 8.34
PDL of the project.

```
/*******************************************************************************
                    CREATING FILE ON SD CARD AND WRITING TEXT
                    =========================================

In this project an SD card is connected to the microcontroller. A file called TEXT.TXT is created
on the SD card and the following text is written inside this file.

                This project shows how to create a file on the SD card


Author:        Dogan Ibrahim
Date:          September 2012
File:          SD-TEXT.C
*******************************************************************************/
//
// Define MMC module connections
sbit Mmc_Chip_Select         at LATG9_bit;
sbit Mmc_Chip_Select_Direction at TRISG9_bit;
//
// End of MMC module connections
//


//
// Start of main program
//
 void main()
 {
    unsigned char MyText[] = "This project shows how to create a file on the SD card";
    unsigned char TextLen;

//
// Initialize the SPI library at standard speed
//
  SPI2_Init_Advanced(_SPI_MASTER, _SPI_8_BIT, 64, _SPI_SS_DISABLE,
                      _SPI_DATA_SAMPLE_MIDDLE, _SPI_CLK_IDLE_HIGH,
                      _SPI_ACTIVE_2_IDLE);
  Delay_ms(10);
  TextLen = strlen(MyText);                      // Length of the text
//
// Initialize the FAT library
//
  while(Mmc_Fat_Init());                         // Wait until initialized
//
// Reinitialize the SPI library at higher speed
//
   SPI2_Init_Advanced(_SPI_MASTER, _SPI_8_BIT, 8, _SPI_SS_DISABLE,
                       _SPI_DATA_SAMPLE_MIDDLE, _SPI_CLK_IDLE_HIGH,
                       _SPI_ACTIVE_2_IDLE);
//
// Create new file TEXT.TXT on the card
```

Figure 8.35

(Continued on next page)

```
//
    Mmc_FAT_Assign("TEXT.TXT", 0x80);              // Create the new file
    Mmc_FAT_Rewrite();
    Mmc_Fat_Write(MyText, TextLen);                // Write the required text
}
```

Figure 8.35
Program listing.

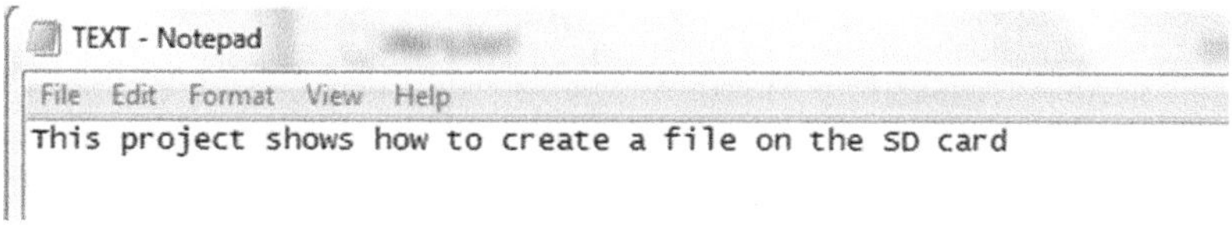

TEXT - Notepad

File   Edit   Format   View   Help

This project shows how to create a file on the SD card

Figure 8.36
Displaying contents of the file on the PC.

Testing the project is easy as the contents of the SD card can easily be viewed on a PC using a suitable SD card reader hardware (most laptops are nowadays equipped with SD card readers).

In the above program, the user has no idea when writing to the card is complete. It is good idea for example to turn on an LED to indicate when the writing is complete and the card can be removed safely.

Notice in this project that the MMC library function **Mmc_FAT_Open** could also be used to create the file as follows:

```
Mmc_FAT_Open("TEST.TXT", "FILE_WRITE", 0x80);
```

Figure 8.36 shows the contents of the file displayed on the PC.

8.7 Project 8.7—Storing Temperature Readings in a File on the SD Card

8.7.1 Project Description

In this project, the design of a temperature data logger system is described. The ambient temperature is read every 10 s using an analog temperature sensor and is stored in a file called TEMPS.TXT on an SD card. An LED is connected to port pin RB0. In addition, a push-button switch is connected to port pin RC1. Data collection starts when the button is pressed. During data collection, the LED is turned ON to indicate that the process is continuing and the SD card must not be removed from its holder. Data are collected for 1 min where the LED is turned OFF at the end of this period. Figure 8.37 shows the block diagram of the project.

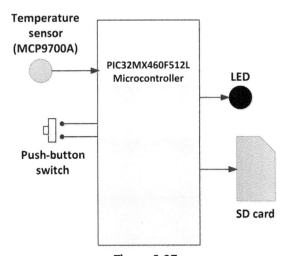

Figure 8.37
Block diagram of the project.

8.7.2 Project Hardware

The project is based on the MCP9700A-type low-power linear active analog thermistor temperature sensor device. This is small three-pin device which can measure the temperature in the range $-40\,°C$ to $+125\,°C$ with a typical accuracy of $\pm 2\,°C$. Two of the pins are the power supply and ground and the third pin is the analog output voltage. This sensor does not require an additional signal-conditioning circuit and can be connected directly to one of the A/D converter channels of the microcontroller. The operating voltage of the sensor is within the range of $+2.3$ to $+5.5$ V.

The sensor output voltage is given by the following:

$$V_{out} = T_C \times T_A + V_{oC}$$

where,

V_{out} is the output voltage
T_C is the temperature coefficient of the sensor (typically 10 mV/°C)
T_A is the ambient temperature (in °C)
V_{oC} is the sensor output voltage at 0 °C (typically 500 mV).

Rearranging the above formula we get,

```
TA = (Vout − VoC)/TC
```

or,

```
TA = (Vout − 0.5)/0.01
```

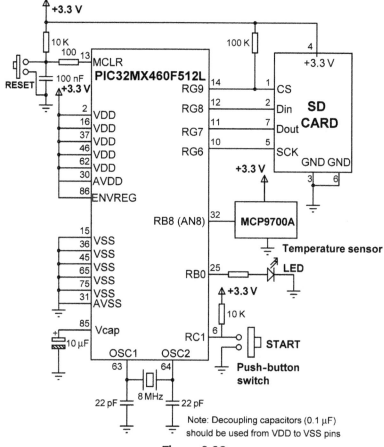

Figure 8.38
Circuit diagram of the project.

Thus, by measuring the output voltage, we can find the temperature of the sensor. For example, if the measured voltage is 0.750 V, then the temperature is

```
TA = (0.750 - 0.5)/0.01 = 25 °C
```

The circuit diagram of the project is shown in Figure 8.38. The temperature sensor is connected to analog channel 8 (RB8) of the microcontroller. The push-button switch and the LED are connected to the port pins RC1 and RB0, respectively.

If you are using the PIC32MX V6 development board, then the following switch must be set to connect output of the on-board MCP9700A temperature sensor to analog channel 8 (RB8) of the microcontroller, to enable the LED and the push-button switch connected to RC1:

```
DIP switch SW12, switch 8 set to ON
```

Also, enable the SD card interface by setting as follows:

```
DIP switch SW13, swich positions 4—8, set to ON
```

Also, enable the push-button switch RC1 by the following:

```
DIP switch SW5, switch 1 set to ON
Jumper SW5 set to pull-up
Jumper J15 connected to ground
```

Enable the LED by setting the following:

```
DIP switch SW12, switch PORT B set to ON
```

Insert the SD card into its holder.

8.7.3 Project PDL

Figure 8.39 shows the PDL of the project. In this project, after configuring the SD card CS connection, port RB8 is configured as analog input, port pin RC1 as digital input, and RB0 as digital output. The SPI library and MMC libraries are initialized and a new file is created on the SD card with the name TEMPS.TXT. The A/D converter module is also initialized. The program then waits for the push-button switch to be pressed. After this, the temperature is read every 10 s, converted into °C and stored on the SD card. This process repeats every 10 s until it terminates after 1 min.

8.7.4 Project Program

Figure 8.40 shows the program listing (TEMPS.C) of the project. At the beginning, the CS connection between the SD card and the microcontroller is defined. In addition, symbols *LED* and *SWITCH* are assigned to port pins RB0 and RC1, respectively. Inside the main program, port pin RB8 (AN8) is configured as analog, while other bits of PORT B are configured as digital, by setting the AD1PCFG register AD1PCFG to 0xFEFF. TRISB is set to

```
BEGIN
    Configure the chip select pin
    Initialize the SPI library
    Initialize the MMC library
    Initialize A/D converter module
    Wait until the switch is pressed
    DO FOR 1 minute
        Read temperature from channel 8
        Convert to °C
        Store in the file on SD card
        Wait 10 seconds
    ENDDO
END
```

Figure 8.39
Project PDL.

```
/****************************************************************************
                  STORE TEMPERATURE IN A FILE ON THE SD CARD
                  ==========================================
```

In this project the temperature is read from an analog temperature sensor every
10 s and is stored in a file called TEMPS.TXT on the SD card.

The temperature sensor used is the MCP9700A which can measure the temperature in
the range 0 to 70 °C with an accuracy of ±2°.

The output of the sensor is connected to analog channel 8 (RB8) of the microcontroller.
In addition, a switch is connected to port pin RC1 and an LED is connected to port
pin RB0.

The program starts when the switch is pressed. During the measurement the LED is
turned ON to indicate that the process is continuing. After 1 min the data
collection process stops and the LED turns OFF.

The ambient temperature is found using the following formula (see manufacturer's
data sheet on MCP9700A sensor):

  $T = (Vout - 0.5)/0.01$


```
Author:        Dogan Ibrahim
Date:          September 2012
File:          TEMPS.C
****************************************************************************/
//
// Define MMC module connections
sbit Mmc_Chip_Select        at LATG9_bit;
sbit Mmc_Chip_Select_Direction at TRISG9_bit;
//
// End of MMC module connections
//

#define LED PORTBbits.RB0
#define SWITCH PORTCbits.RC1


//
// Start of main program
//
 void main()
 {
   unsigned ADC;
   float V;
   unsigned char Duration, Txt[14];
   unsigned char NewLine[2] = {0x0D, 0x0A};
   unsigned char TextLen;
```

Figure 8.40

(Continued on next page)

```
        AD1PCFG = 0xFEFF;                        // Configure RB8 analog, others digital
        TRISB = 0x100;                           // RB0 is output, RB8 is input (analog)
        TRISC = 2;                               // RC1 is input
        PORTB = 0;                               // Turn OFF PORT B LEDs to start with
//
// Initialize the SPI library at standard speed
//
    SPI2_Init_Advanced(_SPI_MASTER, _SPI_8_BIT, 64, _SPI_SS_DISABLE,
                        _SPI_DATA_SAMPLE_MIDDLE, _SPI_CLK_IDLE_HIGH,
                        _SPI_ACTIVE_2_IDLE);
    Delay_ms(10);

//
// Initialize the FAT library
//
    while(Mmc_Fat_Init());                       // Wait until initialized
//
// Reinitialize the SPI library at higher speed
//
     SPI2_Init_Advanced(_SPI_MASTER, _SPI_8_BIT, 8, _SPI_SS_DISABLE,
                        _SPI_DATA_SAMPLE_MIDDLE, _SPI_CLK_IDLE_HIGH,
                        _SPI_ACTIVE_2_IDLE);
//
// Create new file TEXT.TXT on the card
//
    Mmc_FAT_Assign("TEMPS.TXT", 0x80);           // Create the new file
    Mmc_FAT_Rewrite();
//
// Initialize the A/D converter module
//
    ADC1_Init();
//
// Wait until the swtich is pressed
//
    while(SWITCH == 1);
//
// Beginning of the loop where the temperature is received, converted into C and
// stored in the file on the SD card
//
    Duration = 0;
    PORTBbits.RB0 = 1;                           // Turn ON RB0 LED

    for(;;)
    {
        ADC = ADC1_Get_Sample(8);                // Get a sample from channel 8
        V = ADC*3.3/1024.0;                      // Convert to Volts
        V = (V – 0.5)/0.01;                      // Convert to °C
        FloatToStr(V, Txt);                      // Convert to string
        TextLen = strlen(Txt);
        Mmc_FAT_Write(Txt, TextLen);             // Store in the file on SD card
        Mmc_FAT_Write(NewLine, 2);               // Insert newline characters
```

Figure 8.40

(Continued on next page)

```
            Delay_Ms(10000);                    // Wait 10 seconds
            Duration++;                          // Increment Duration (every 10 sec)
            if(Duration == 6)                    // If 1 min
            {
              PORTBbits.RB0 = 0;                 // Turn OFF RB0 LED
      Stp:    goto Stp;
            }
          }
        }
```

Figure 8.40
Program listing.

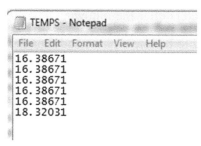

Figure 8.41
Contents of the file.

0x100 so that RB8 is input and other port pins are outputs. TRISC is set to 2 to configure RC1 as input, and all LEDs connected to PORT B are turned OFF.

The SPI library and the MMC libraries are then initialized, a new file called TEMPS.TXT is created on the SD card, and the A/D module is initialized. The program then waits until the SWITCH is pressed and turns ON the LED connected to port pin RB0 when the switch is pressed. The remainder of the program is executed in a **for** loop. Inside this loop, a new sample is obtained from the sensor and the temperature is calculated and stored in floating point variable *V*, which is then converted into a string using the **FloatToStr** statement and stored in character array *Txt*. This is then stored in the created file on the SD card. A newline character is inserted at the end of each data. The program waits for 10 s and the above process is repeated for 1 min. The LED connected to port pin RB0 is turned OFF at the end of the data collection.

Figure 8.41 shows the contents of the file displayed on the PC.

8.8 Project 8.8—Designing a Finite Impulse Response Filter

8.8.1 Project Description

This project shows how to design a finite impulse response (FIR)-type digital filter using the PIC32MX460F512L-type 32-bit microcontroller. The designed filter is low-pass and has a sampling frequency of 1 kHz, and the cutoff frequency is 100 Hz.

Before going into details of the design, it is worthwhile to look at the theory of FIR digital filters briefly.

8.8.2 FIR Filters

There are two types of digital filters used in digital signal processing (DSP) applications: FIR and infinite impulse response (IIR). FIR filters have no feedback in their structures and thus they are always stable and are used much more than IIRs. Each type of filter has its advantages and disadvantages.

Compared to IIR filters, FIR filters offer the following advantages:

- FIR filters can be designed to be linear phase and thus they do not distort the phase of the input signal.
- FIR filters are very simple to implement.
- FIR filters are inherently stable.
- FIR filters are suited to multirate applications.
- FIR filters can be implemented using fractional arithmetic where the filter coefficients have magnitudes <1.0.

Compared to IIR filters, FIR filters have the following disadvantages:

- FIR filters require more memory.
- FIR filters require more mathematical operations to achieve a given filter response.
- Certain filter responses cannot be implemented with FIR filters.

In an FIR filter, the output depends only on the inputs (and not the previous outputs). The output of such a filter is a weighted sum of the current and a finite number of previous values of the input. The output of an FIR filter at time is given in terms of its inputs by the following equation:

$$y[n] = h_0 x[n] + h_1 x[n-1] + h_2 x[n-2] + \cdots\cdots\cdots + h_N x[n-N]$$

or,

$$y[n] = \sum_{i=0}^{N} h_i x[n-i]$$

where

$x[n]$ is the input signal;
$y[n]$ is the output signal;
h_i is the filter coefficient that makes up the impulse response;
N is the filter order.

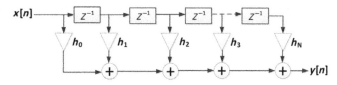

Figure 8.42
Nth order FIR filter structure.

In general, an Nth order filter has $(N + 1)$ terms on the right-hand side.

There are different methods to find the filter coefficients. Some of the commonly used methods are the following:

- Frequency sampling method
- Windowed design method
- Parks–McClellan method
- Equiripple method.

In the windowed design method, the filter is initially designed and then a window function is applied (e.g. Hanning window, Hamming window, Rectangular window, Kaiser window, and so on) and the filter coefficients are multiplied by the window function. This results in the frequency response of the filter being convolved with the frequency response of the window function.

Although the coefficients of an FIR filter can easily be calculated from the required impulse response, there are many PC-based software packages that help to design FIR filters. Most programs give the filter coefficients as well as they plot the frequency response of the filter to be designed. Some commonly used FIR filter design programs are MATLAB, SciPy, dsPICFD, Scope FIR and so on. In this section, we will be using the Scope FIR filter design program to design our filter and obtain the filter coefficients.

Figure 8.42 shows a typical implementation of an FIR filter of order N. Note that the top part is an N-stage delay line (where Z^{-1} denotes the unit delay operator) with $N + 1$ taps.

8.8.3 Using the Scope FIR Filter Design Program

In this section, we will calculate the filter coefficients of our FIR filter using the Scope FIR software. Scope FIR is developed by the Iowegian International Corporation (www.iowegan .com). A 30-day trial version of the software is available and can be downloaded from the

Iowegian website. In this project, we will design our low-pass FIR filter to have the following specifications:

Filter Type	Simple Parks–McClellan
Sampling frequency	1000 Hz
Passband upper frequency	100 Hz
Stopband lower frequency	200 Hz
Passband ripple	1 dB
Stopband attenuation	80 db
Filter order	18 (19 taps)

The definition of the various parameters used in the design is shown in Figure 8.43.

Start the program and create a new project by selecting the filter type as Simple Parks–McClellan, as shown in Figure 8.44.

Enter the filter specifications as shown in Figure 8.45 and click Design to design the filter. The actual ripple and the attenuation are calculated as 4.791 and 66.567 dB, respectively.

The frequency response, filter coefficients, and the impulse response of the filter to be designed will be displayed by the program. Figure 8.46 shows the frequency response, Figure 8.47 the filter coefficients, and Figure 8.48 the filter impulse response.

The block diagram of the project is shown in Figure 8.49. Here, a D/A converter is used to convert the digital signal at the output of the microcontroller to analog. The Velleman PCSGU250-type sine wave sweep frequency generator and Bode frequency plotter device are used to plot the frequency response of the designed filter.

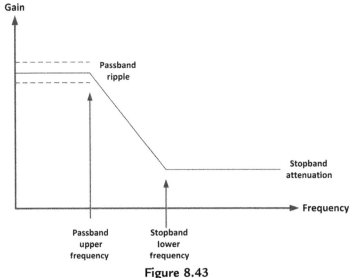

Figure 8.43
Definition of the parameters used in the design.

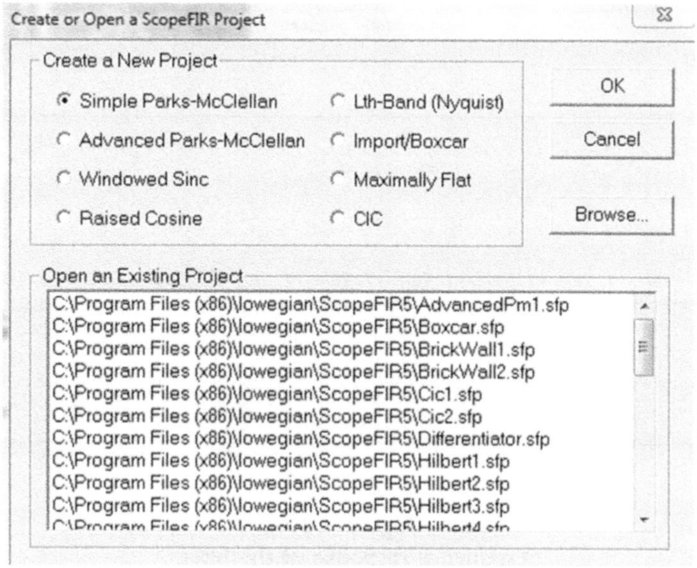

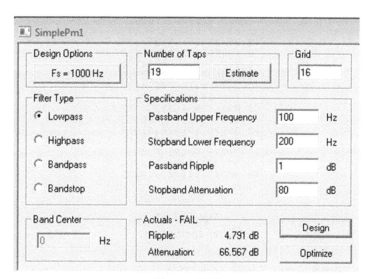

Figure 8.44
Select the filter type as Simple Parks–McClellan.

Figure 8.45
Filter specifications.

8.8.4 Project Hardware

Figure 8.50 shows the circuit diagram of the project. Input waveform is applied to analog input AN0 of the microcontroller. A serial D/A converter is used in the project and this

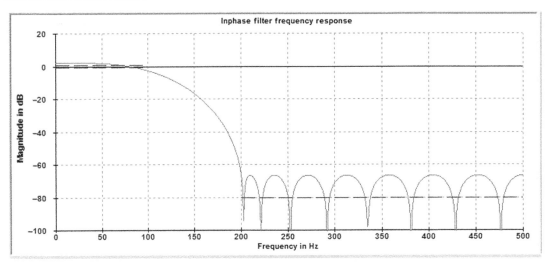

Figure 8.46
Frequency response of the filter.

```
-0.002471041247149621
-0.008783154222768062
-0.017791904818824537
-0.022224420307720615
-0.009757106876401019
 0.030571717070766433
 0.099168591591250937
 0.180514082103892610
 0.247238418679934180
 0.273085321838523910
 0.247238418679934180
 0.180514082103892610
 0.099168591591250937
 0.030571717070766433
-0.009757106876401019
-0.022224420307720615
-0.017791904818824537
-0.008783154222768062
-0.002471041247149621
```

Figure 8.47
Filter coefficients.

is controlled by the SPI bus. The output of the D/A converter is connected to the PCSGU250 device for plotting the frequency response.

8.8.5 Implementing the FIR Filter

There are several methods of implementing the FIR digital filter. Different PDLs and different programs are given to describe the operation of each method. The structure of the filter to be designed is shown in Figure 8.51 with $N = 18$, having 19 taps.

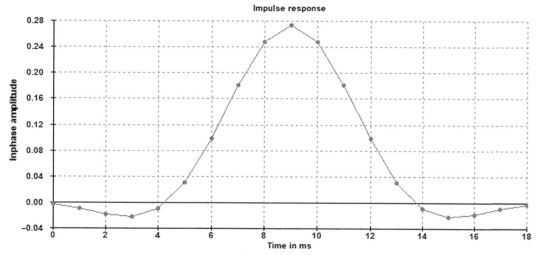

Figure 8.48
Filter impulse response.

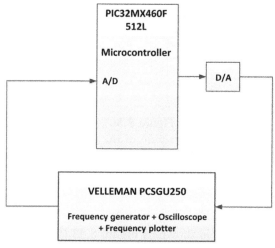

Figure 8.49
Block diagram of the project.

Method 1

This is probably the most straightforward method of implementing an FIR digital filter. As shown in the PDL in Figure 8.52, here the processing takes place inside a timer ISR where the timer is configured to interrupt at every sampling time. Inside the ISR, the input signal samples are received, converted into analog, the output samples are calculated, converted into analog and then output through the D/A converter. At the same time, the input samples are delayed by one sample time.

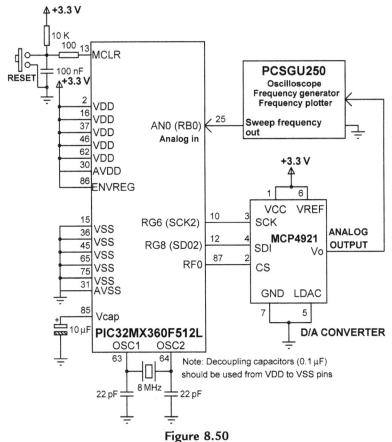

Figure 8.50
Circuit diagram of the project.

8.8.6 Program Listing

The program listing for Method 1 is shown in Figure 8.53 (FIR1.C). At the beginning of the program, the D/A converter CS connection is defined. The 19 filter coefficients are then stored in a floating point array called *h*:

```
float h[N] = {−0.00247, −0.00878, −0.01779, −0.02222, −0.009757,
              0.03057, 0.099168, 0.180514, 0.247238, 0.273085,
              0.247238, 0.180514, 0.099168, 0.03057, −0.009757,
              −0.02222, −0.01779, −0.00878, −0.00247};
```

Inside the main program, analog input AN0 (RB0) is configured as an analog input, D/A converter is disabled, and the SPI library and A/D converter module are initialized. Timer 1 is then configured to interrupt at every 1000 µs.

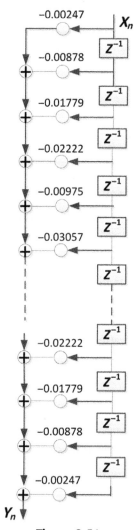

Figure 8.51
Structure of the FIR filter.

With a clock frequency of 80 MHz, the clock period is 0.0125×10^{-3} ms. The timer prescaler is set to 64, giving a value of 1250 for the period register:

```
PR1 = Delay/(Clock period × prescaler value)
```

or,

```
PR1 = 1 ms/(0.0125 × 10⁻³ × 64) = 1250
```

The main program then enters a loop and waits for timer interrupts to occur. The digital filtering operation is performed inside the timer ISR called *TMR*. Here, a new input

BEGIN/MAIN
    Configure the chip select pin
    Store filter coefficients in an array
    Configure AN0 (RB0) as analog input
    Initialize the SPI library
    Initialize A/D converter module
    Initialize Timer 1 for 1ms interrupts
    Wait for Timer 1 interrupts (TMR)
END/MAIN

BEGIN/TMR
    Get a new signal sample
    Calculate the output sample
    Send the output sample to the D/A converter
    Delay the input signals by one sample time
    Clear Timer 1 interrupt flag
END/TMR

Figure 8.52
The PDL for method 1.

sample is obtained from analog channel AN0, and the output sample is calculated using the multiply and accumulate (MAC) operations:

```
for(i = 0; i <= N; i++)
{
  yn = yn + h[i]*x[i];
}
```

The program then sends the output sample to the D/A converter. The input samples are shifted (delayed) by one sampling time using the following code:

```
for(i = 0; i < N; i++)
{
  x[N - i] = x[N - i - 1];
}
```

Just before exiting from the ISR, the timer interrupt flag is cleared so that further timer interrupts can be accepted by the processor.

Figure 8.54 shows the frequency response obtained from the designed filter.

Method 2

Shifting of the input samples in the program given in Figure 8.53 can result in large number of bytes to be moved in memory and as a result, excessive time can be wasted by this operation. This second method shows how we can use a circular buffer to store the input sample values and then simply change a pointer instead of moving the data in memory. This method is explained in most DSP books and the basic operation is summarized below by considering a 3rd order filter.

```
/******************************************************************************
                    FINITE IMPULSE RESPONSE FILTER DESIGN
                    ====================================

This project shows how a FIR type digital filter can be designed.

Analog sine wave signal is fed to the AN0 (RB0) analog input of the microcontroller.
A D/A converter is connected to the microcontroller through the SPI bus so that
the filtered signal is in analog form and can be plotted using a frequency
plotter (Bode plotter).

In this example a low-pass filter is designed with the following specifications:

        Filter Type:                    Simple Parks–McClellan
        Sampling frequency:             1000 Hz
        Passband upper frequency:       100 Hz
        Stopband lower frequency:       200 Hz
        Passband ripple:                1 dB
        Stopband attenuation:           80 db
        Filter order:                   18 (19 taps)

The FIR filter coefficients are obtained using the SCOPE FIR software package.

The filter response is plotted using a Velleman PCSGU250 type oscilloscope +
frequency generator + frequency plotter device.

The FIR filter is implemented using different methods. This is Method 1.



        Author:         Dogan Ibrahim
        Date:           September 2012
        File:           FIR1.C
******************************************************************************/

// DAC module connections
sbit Chip_Select at LATF0_bit;
sbit Chip_Select_Direction at TRISF0_bit;
// End DAC module connections

#define N 18                                    // Filter order = 18, having 19 taps

float Sample,xn, yn, x[N];
unsigned ADC;
unsigned int DAC;
float h[N+1] = {−0.00247, −0.00878, −0.01779, −0.02222, −0.009757,
        0.03057,  0.099168, 0.180514, 0.247238, 0.273085,
        0.247238, 0.180514, 0.099168, 0.03057, −0.009757,
        −0.02222, −0.01779, −0.00878, −0.00247};

//
// Timer 1 interrupt service routine. The program jumps here every 1000 µs (the sampling
```

Figure 8.53

(Continued on next page)

```c
// frequency is 1 kHz, i.e. Period = 1 ms = 1000 µs). Here, a new output is calculated and
// sent to the D/A converter
//
void TMR() iv IVT_TIMER_1 ilevel 1 ics ICS_SOFT
{
    unsigned char i;
    unsigned int temp;

    ADC = ADC1_Get_Sample(0);                    // Get a new input Sample from AN0
    x[0]=ADC;
    yn = 0.0;
//
// Calculate a new output yn
//
    for(i = 0; i <= N; i++)
    {
      yn = yn + h[i]*x[i];
    }
//
// Output the new Sample
//
    DAC = yn;
    Chip_Select = 0;                             // Select DAC chip

    // Send High Byte
    temp = (DAC >> 8) & 0x0F;                     // Store DAC[11..8] to temp[3..0]
    temp |= 0x30;                                 // Define D/A setting
    SPI2_Write(temp);                             // Send high byte via SPI

    // Send Low Byte
    temp = DAC;                                   // Store DAC[7..0] to temp[7..0]
    SPI2_Write(temp);                             // Send low byte via SPI

    Chip_Select = 1;                             // Deselect D/A converter chip
//
// Shift the input samples for the delay action
//
    for(i = 0; i < N; i++)
    {
      x[N–i] = x[N–i–1];
    }

    IFS0bits.T1IF = 0;                           // Clear Timer 1 interrupt flag
}


//
// Start of man program
//
void main()
{
```

Figure 8.53

(Continued on next page)

```
AD1PCFG = 0xFFFE;                              // AN0 (RB0) is analog
TRISB = 1;                                     // AN0 (RB0) is input
Chip_Select_Direction = 0;                     // Configure CS pin as output
Chip_Select = 1;                               // Disable D/A converter
SPI2_Init();                                   // Initialize SPI2
ADC1_Init();                                   // Initialize A/D converter

//
// Configure Timer 1 for 1000 μs (1 ms) interrupts
//
T1CONbits.ON = 0;                              // Disable Timer 1
TMR1 = 0;                                       // Clear TMR1 register
T1CONbits.TCKPS = 2;                           // Select prescaler = 64
T1CONbits.TCS = 0;                             // Select internal clock
PR1 = 1250;                                     // Load period register (for 1000 μs)
IFS0bits.T1IF = 0;                             // Clear Timer 1 interrupt flag
IPC1bits.T1IP = 1;                             // Set priority level to 1
IEC0bits.T1IE = 1;                             // Enable Timer 1 interrupts
T1CONbits.ON = 1;                              // Enable Timer 1
EnableInterrupts();                            // Enable interrupts

for(;;)                                         // Wait for Timer 1 interrupts
{
}
}
```

Figure 8.53
Program listing for method 1.

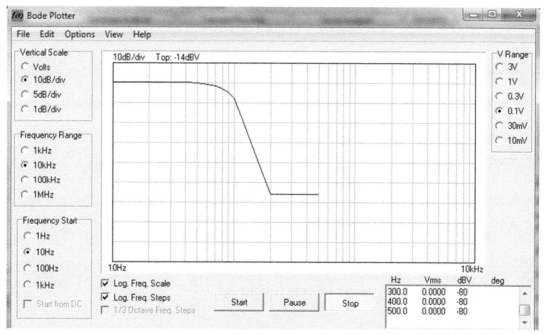

Figure 8.54
Frequency response of the designed filter.

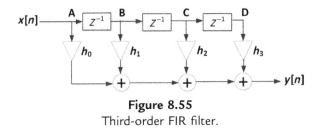

Figure 8.55
Third-order FIR filter.

Figure 8.55 shows a third-order FIR filter where A, B, C and D are the tapping points. The operations performed are summarized below:

Stage 1:

$$D = x[3], \quad C = x[2], \quad B = x[1], \quad A = x[0]$$

$$y = h_0 x[0] + h_1 x[1] + h_2 x[2] + h_3 x[3]$$

where $x[0]$ is where the last input sample is stored.

After the output, shift the input values so that

$$D = x[2], \quad C = x[1], \quad B = x[0], \quad A = new1$$

where, *new*1 is the new sample stored in tapping point A.

Stage 2:

$$y = h_0 new1 + h_1 x[0] + h_2 x[1] + h_3 x[2]$$

After the output, shift the input values so that

$$D = x[1], \quad C = x[0], \quad B = new1, \quad A = new2$$

where, *new*2 is the new sample stored in tapping point A.

Stage 3:

$$y = h_0 new2 + h_1 new1 + h_2 x[0] + h_3 x[1]$$

After the output, shift the input values so that

$$D = x[0], \quad C = new1, \quad B = new2, \quad A = new3$$

where, *new*3 is the new sample stored in tapping point A.

Stage 4:

$$y = h_0 new3 + h_1 new2 + h_2 new1 + h_3 x[0]$$

After the output, shift the input values so that

$$D = new1, \quad C = new2, \quad B = new3, \quad A = new4$$

where, $new4$ is the new sample stored in tapping point A.

Stage 5:

$$y = h_0 new4 + h_1 new3 + h_2 new2 + h_3 new1$$

and the above process repeats forever.

The important point to notice here is that the input sample is always multiplied by h_0 and the input values are shifted right by one sample time.

Figure 8.56 shows the filter operation where the variables multiplied with each other are shown with arrows. Examining this figure, it is clear that the filtering operation can be carried out by making the input array x to be a circular array. The last input samples are always loaded into the array location which is not needed for the next calculation and the input array

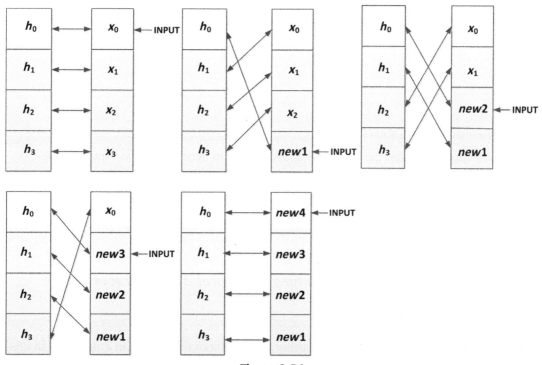

Figure 8.56

Schematic representation of the operations.

pointer moves up as new samples are received. At the same time, the array pointer moves down to carry out the multiplication with the filter coefficients.

The required operations inside the ISR are described by the following PDL. Notice here that round bracket means the contents of the array pointed to by the corresponding pointer. The input pointer (*i_pointer*) and the x-array pointer (*x_pointer*) are initialized outside the ISR:

```
y = 0
Input to x[i_pointer]
DO for all taps
      y = y + h[i] * (x_pointer)
      Increment x_pointer and wrap around if > N
ENDDO
Output y to D/A converter
Decrement i_pointer and wrap around if < 0
Clear Timer 1 interrupt flag
```

8.8.7 Program Listing

The program listing is shown in Figure 8.57. Most of the code is similar to the one in Figure 8.53, but here a circular buffer is used for the input samples. The filtering operation is implemented by the following code:

```
for(i = 0; i <= N; i++)
{
    yn = yn + h[i]*x[x_pointer];
    if(++x_pointer > N)x_pointer = 0;
}
i_pointer--;
if(i_pointer < 0)i_pointer = N;
```

Here, as mentioned earlier, the *x_pointer* moves down to access the next input sample, while new input samples are loaded into array location pointed to by *i_pointer*.

Notice that the overall system performance could be improved by using a D/A converter with faster conversion time (e.g. parallel D/A converter).

8.9 Project 8.9—Calculating Timing in Digital Signal Processing

8.9.1 Project Description

DSP algorithms require MAC operations using fixed-point or floating point arithmetic. In this project, we will calculate the time required to carry out floating point MAC operations using the mikroC Pro for PIC32 compiler, and the PIC32MX460F512L microcontroller

```
/***************************************************************************
                FINITE IMPULSE RESPONSE FILTER DESIGN
                ====================================
```

This project shows how a FIR type digital filter can be designed.

Analog sine wave signal is fed to the AN0 (RB0) analog input of the microcontroller.
A D/A converter is connected to the microcontroller through the SPI bus so that the filtered
signal is in analog form and can be plotted using a frequency plotter (Bode plotter).

In this example a low-pass filter is designed with the following specifications:

Filter Type:	Simple Parks–McClellan
Sampling frequency:	1000 Hz
Passband upper frequency:	100 Hz
Stopband lower frequency:	200 Hz
Passband ripple:	1 dB
Stopband attenuation:	80 db
Filter order:	18 (19 taps)

The FIR filter coefficients are obtained using the SCOPE FIR software package.

The filter response is plotted using a Velleman PCSGU250 type oscilloscope +
frequency generator + frequency plotter device.

The FIR filter is implemented using different methods. This is Method 2. In this method
a circular buffer is used to store the input samples. Input samples are delayed by changing
the pointer to this array instead of moving data around the memory. Two pointers are used:
i_pointer (or input pointer), and x_pointer (or x-array pointer).

This method results in faster execution time and thus higher order filters can be
implemented

```
Author:        Dogan Ibrahim
Date:          September 2012
File:          FIR2.C
***************************************************************************/
```

```
// DAC module connections
sbit Chip_Select at LATF0_bit;
sbit Chip_Select_Direction at TRISF0_bit;
// End DAC module connections

#define N 18                            // Filter order = 18 (19 tapping points)

float Sample,xn, yn, x[N];
unsigned ADC;
signed char i_pointer = 0, x_pointer = 0;
unsigned int DAC;
```

Figure 8.57

(Continued on next page)

```
float h[N+1] = {−0.00247, −0.00878, −0.01779, −0.02222, −0.009757,
                0.03057, 0.099168, 0.180514, 0.247238, 0.273085,
                0.247238, 0.180514, 0.099168, 0.03057, −0.009757,
               −0.02222, −0.01779, −0.00878, −0.00247};

//
// Timer 1 interrupt service routine. The program jumps here every 1000 μs (the sampling
// frequency is 1 kHz, i.e. Period = 1 ms = 1000 μs). Here, a new output is calculated and
// sent to the D/A converter
//
void TMR() iv IVT_TIMER_1 ilevel 1 ics ICS_SOFT
{
    unsigned char i;
    unsigned int temp;

    ADC = ADC1_Get_Sample(0);                     // Get a new input Sample from AN0
    x[i_pointer] = ADC;
    yn = 0.0;
//
// Calculate a new output yn. This process will also shift the input samples as
// required so that there is no need to move the samples around in memory. Index
// x_pointer is used as the pointer and the pointer is initialized outside the loop
//
    x_pointer = i_pointer;

    for(i = 0; i <= N; i++)
    {
      yn = yn + h[i]*x[x_pointer];
      if(++x_pointer > N)x_pointer = 0;
    }
    i_pointer−;
    if(i_pointer < 0)i_pointer = N;
//
// Output the new Sample
//
    DAC = yn;
    Chip_Select = 0;                              // Select DAC chip

    // Send High Byte
    temp = (DAC >> 8) & 0x0F;                      // Store DAC[11..8] to temp[3..0]
    temp |= 0x30;                                  // Define D/A setting
    SPI2_Write(temp);                              // Send high byte via SPI

    // Send Low Byte
    temp = DAC;                                    // Store DAC[7..0] to temp[7..0]
    SPI2_Write(temp);                              // Send low byte via SPI

    Chip_Select = 1;                              // Deselect D/A converter chip

    IFS0bits.T1IF = 0;                            // Clear Timer 1 interrupt flag
}
```

Figure 8.57

(Continued on next page)

```
//
// Start of man program
//
void main()
{
    AD1PCFG = 0xFFFE;                          // AN0 (RB0) is analog
    TRISB = 1;                                 // AN0 (RB0) is input
    Chip_Select_Direction = 0;                 // Configure CS pin as output
    Chip_Select = 1;                           // Disable D/A converter
    SPI2_Init();                               // Initialize SPI2
    ADC1_Init();                               // Initialize A/D converter

//
// Configure Timer 1 for 1000 µs (1 ms) interrupts
//
    T1CONbits.ON = 0;                          // Disable Timer 1
    TMR1 = 0;                                  // Clear TMR1 register
    T1CONbits.TCKPS = 2;                       // Select prescaler = 64
    T1CONbits.TCS = 0;                         // Select internal clock
    PR1 = 1250;                                // Load period register (for 1000 µs)
    IFS0bits.T1IF = 0;                         // Clear Timer 1 interrupt flag
    IPC1bits.T1IP = 1;                         // Set priority level to 1
    IEC0bits.T1IE = 1;                         // Enable Timer 1 interrupts
    T1CONbits.ON = 1;                          // Enable Timer 1
    EnableInterrupts();                        // Enable interrupts

    for(;;)                                    // Wait for Timer 1 interrupts
    {
    }
}
```

Figure 8.57

Program listing for method 2.

with 80 MHz clock. Various MAC operations are performed and the time it takes to carry out these operations is displayed on an LCD.

8.9.2 Project Hardware

The project hardware is same as shown in Figure 7.17. An LCD is connected to PORT B of the microcontroller and the microcontroller is configured to operate at 80 MHz clock rate using the built-in PLL module.

```
BEGIN
      Configure LCD connections to the microcontroller
      Disable Timer 1
      Configure Timer 1 to count at a rate of 0.0125 µs
      Start Timer 1
      Perform MAC operations
      Stop Timer 1
      Calculate elapsed time
      Display elapsed time on the LCD
END
```

Figure 8.58

Project PDL.

```
/******************************************************************************
                    DIGITAL SIGNAL PROCESSING MAC TIMING
                    ====================================

This project shows how to calculate the time it takes to carry out different number
of MAC operations.

An LCD is connected to PORT B of the PIC32MX460F512L microcontroller, operating at the
clock rate of 80 MHz (using the internal PLL). The program uses Timer 1 to calculate
the timing and then displayes the result on the LCD.


Author:        Dogan Ibrahim
Date:          September 2012
File:          TIMING.C
******************************************************************************/

// LCD module connections
sbit LCD_RS at LATB2_bit;
sbit LCD_EN at LATB3_bit;
sbit LCD_D4 at LATB4_bit;
sbit LCD_D5 at LATB5_bit;
sbit LCD_D6 at LATB6_bit;
sbit LCD_D7 at LATB7_bit;

sbit LCD_RS_Direction at TRISB2_bit;
sbit LCD_EN_Direction at TRISB3_bit;
sbit LCD_D4_Direction at TRISB4_bit;
sbit LCD_D5_Direction at TRISB5_bit;
sbit LCD_D6_Direction at TRISB6_bit;
sbit LCD_D7_Direction at TRISB7_bit;
// End LCD module connections


//
// Start of man program
//
void main()
{
    float y, Tim, a[500], b[500];
    unsigned char Txt[14];
    unsigned int i;
//
// Load some floating point numbers to a and b
//
    for(i = 0; i < 500; i++)
    {
      a[i] = 2.85*i;
      b[i] = 3.57*i;
    }
//
```

Figure 8.59

(Continued on next page)

```
// Configure Timer 1
//
    T1CONbits.ON = 0;              // Disable Timer 1
    TMR1 = 0;                      // Clear TMR1 register
    T1CONbits.TCKPS = 0;           // Select prescaler = 1
    T1CONbits.TCS = 0;             // Select internal clock
    PR1 = 0xFFFF;                  // Load period register

    LCD_Init();                    // Initialize LCD
    LCD_Out(1,1, "TIME (us)");
    y = 0.0;

    T1CONbits.ON = 1;              // Enable Timer 1
    for(i = 0; i < 10; i++)        // Start of MAC loop ( 10 operations here)
    {
      y = y + a[i] * b[i];
    }                              // End of MAC loop
    T1CONbits.ON = 0;              // Disable Timer 1

    Tim = 0.0125*TMR1;             // Calculate elapsed time in us
    FloatToStr(Tim, Txt);          // Convert to string
    Lcd_Out(2,1, Txt);             // Display elapsed time
}
```

Figure 8.59
Program listing for 10 MAC operations.

In this project, the LV-32MX V6 development board is used. The following jumpers should be configured on the board to enable the on-board LCD:

```
DIP switch SW20 positions 1–6, set to ON
```

8.9.3 Project PDL

The PDL of the project is shown in Figure 8.58. Timer 1 is configured to count up with a prescaler of 1 (i.e. with a clock of 80 MHz, the count rate is 0.0125 μs). The counter is enabled before the MAC operations. At the end of the MAC operations, the timer count is multiplied by 0.0125 μs in order to calculate the elapsed time.

8.9.4 Project Program

The program listing is shown in Figure 8.59 (TIMING.C). At the beginning of the program, the LCD connection to the microcontroller is defined. Floating point variables y, a and b are then declared where a and b are loaded with some floating point numbers. Up to 500 locations are reserved for each variable.

Timer 1 is then disabled and configured to count with a prescaler of 1, i.e. at a rate of 0.0125 μs. Timer 1 is enabled, a **for** loop is used to carry out a number of different MAC

Table 8.5: Timing Calculations

No of MAC Operations	Time (μs)
10	1.05
20	1.93
30	2.8
40	3.67
50	4.55
100	8.93
200	17.68
400	35.18
500	43.93

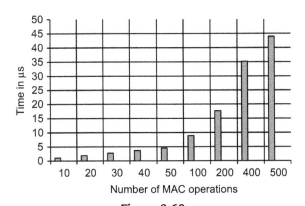

Figure 8.60
MAC timing. (For color version of this figure, the reader
is referred to the online version of this book.)

operations, and at the end, the elapsed time is calculated by multiplying the Timer 1 count
with the clock rate. The time is then displayed on the LCD.

Table 8.5 shows results of the tests with various number of MAC operations.

A graph of the time it takes to perform various number of MAC operations is shown in
Figure 8.60.

MPLAB and PIC32 Compiler

The MPLAB C compiler for PIC32 is an ANSI compliant C compiler developed by Microchip Inc. for PIC32 family of microcontrollers. The compiler supports fast floating point arithmetic operations, and provides libraries for a large number of peripherals, including Ethernet, CAN, USB, DSP libraries, and so on.

This Appendix describes the basic features of the MPLAB C compiler and gives an example of using this compiler with the PIC32 Starter Kit, the 32-bit microcontroller development board manufactured by Microchip Inc. A free evaluation version of the compiler is available from Microchip Inc. and can be downloaded from the Microchip website. The evaluation version has no code size limit; all libraries are available and are functional for 60 days. After 60 days, the compiler is still usable but certain optimization levels are disabled. The compiler is fully compatible with the MPLAB IDE.

MPLAB PIC32 compiler is integrated into the MPLAB IDE, and this IDE must be installed before the compiler is installed. To install both the MPLAB IDE and the MPLAB PIC32 compiler, download the installation packages from the Microchip Inc. website and install the products as you would install other packages.

Although both the MPLAB PIC32 and mikroC PRO for PIC32 are C languages, there are many differences between the two. Interested readers should refer to the MPLAB PIC32 Users' Guide for full details and use of the MPLAB PIC32 compiler. Since MPLAB PIC32 is one of the popular and commonly used C languages for the 32-bit PIC microcontrollers, this Appendix shows the steps required to write and compile a simple program using this compiler with the PIC32 Starter Kit development board.

PIC32 Starter Kit Development Board

PIC32 Starter Kit has been developed by Microchip Inc. for experimenting with the 32-bit PIC family of microcontrollers. The kit has the following basic features:

- PIC32MX360F512L 32-bit microcontroller
- Regulated +3.3 V power supply
- Crystal of 8 MHz for microcontroller clocking
- On-board debugging

- Three user-defined LEDs
- Three user-defined push-button switches
- Connector for expansion boards.

The push-button switches do not have hardware debounce circuitry and are connected as follows:

- SW1: Active-low switch connected to RD6
- SW2: Active-low switch connected to RD7
- SW3: Active-low switch connected to RD13

The normal state of the switches are logic 1 (+3.3 V), and when pressed, they change to logic 0.

The LEDs are connected to PORT D pins RD0 through RD2, where sending a logic 1 to an LED turns it ON.

The development board takes its power from the PC and communicates with the PC via a mini USB cable.

Figure A.1 shows the block diagram of the PIC32 Starter Kit. A picture of the kit is shown in Figure A.2.

PIC32 Starter Kit Project

In this section, a very simple project is given to show how to use the PIC32 Starter Kit and the MPLAB PIC32 compiler. The LED connected to port pin RD0 of the microcontroller (top

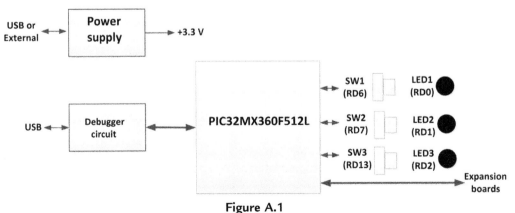

Figure A.1
PIC32 Starter Kit block diagram.

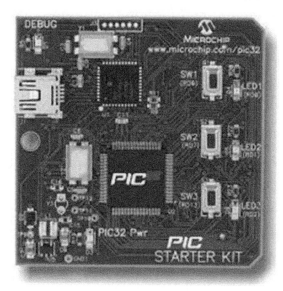

Figure A.2
Picture of the PIC32 Starter Kit.

right-hand side LED on the board) is turned on when the push-button switch SW1 (next to the LED) connected to pin RD6 of the microcontroller is pressed.

The required program (named LED-SWITCH.C) is shown in Figure A.3. Save this program in the folder that you will be using later for your project files (in this example, this is folder called ALEVS. i.e. C:\ALEVS). In this program, the MPLAB PIC32 peripheral library is used for simplicity, and this library is included at the beginning of the program (**plib.h**). Also, the configuration bits are set by the IDE as we shall see later.

Using the MPLAB PIC32 Compiler

The steps to compile and test the program given in Figure A.3 on the PIC32 Starter Kit is shown below.

> *Step 1*: Start the MPLAB IDE as shown in Figure A.4.
> *Step 2*: Click *Project → Project Wizard* to create a new project. Click *Next* and select the device type as PIC32MX360F512L as shown in Figure A.5.
> *Step 3*: Click *Next* and make sure the following are selected (Figure A.6):

```
Active Toolsuite:     Microchip PIC32 C-Compiler
Toolsuite Contents:   MPLAB C-32 C-Compiler
Location:             (Folder where the compiler is installed)
```

```
/*********************************************************************

                    PIC32 STARTER KIT LED CONTROL
                    =============================

This is a very simple MPLAB PIC32 program which shows how to use the PIC32 Starter Kit and
the steps to compile and run a program. The LED connected to PORT D pin RD0 is turned ON
when switch RD6 is pressed, otherwise the LED is tuned OFF when the button is released.

The PIC32 Starter Kit is operated from 8 MHz crystal. With the PLL, the actual clock frequency
of the microcontroller is 80 MHz.




Author:       Dogan Ibrahim
Date:         September, 2012
File:         LED-SWITCH.C
*********************************************************************/
//
// Adds support for PIC32 Peripheral library functions and macros
//
#include <plib.h>

//
// Microcontroller clock
//
#define SYS_FREQ            (80000000)




//
// Start of main program
//
int main(void)
{

//
// Configure RD0 as output
//
    PORTSetPinsDigitalOut(IOPORT_D, BIT_0);

// Turn OFF the LED to start with
//
    PORTClearBits(IOPORT_D, BIT_0);

//
// Configure RD6 as input
//
    PORTSetPinsDigitalIn(IOPORT_D, BIT_6);

    for(;;)

    {
        if(PORTDbits.RD6 == 0)                        // If switch is pressed
            PORTSetBits(IOPORT_D, BIT_0);             // Turn ON LED
        else
            PORTClearBits(IOPORT_D, BIT_0);           // Turn OFF LED
    }

}
```

Figure A.3
Program listing.

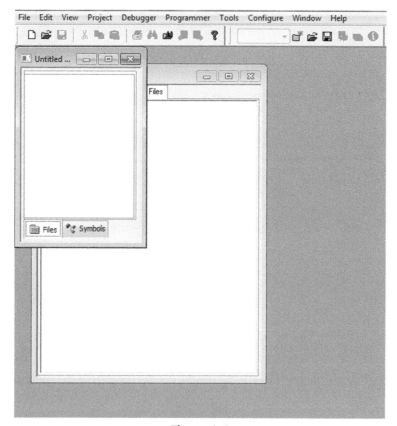

Figure A.4
MPLAB IDE startup window.

Step 4: Click *Next* and choose a new project name. In Figure A.7, project name *PICSTART-LED* is chosen in folder called *ALEVS*.

Step 5: Click *Next*. Add file shown in Figure A.2 to the project (filename LED-SWITCH.C) by highlighting filename *LED-SWITCH* on the left window, and then click Add≫ (Figure A.8).

Step 6: Click *Finish* to finish as shown in Figure A9. A summary of the project files will be displayed.

Step 7: As shown in Figure A.10, you should see a list of project folders and the source files you have added to the project.

Double-click source file LED-SWITCH.C, maximize the window, and the program listing will be displayed as shown in Figure A.11

Step 8: Check the Configuration Bits. Click *Configure* → *Configuration Bits* from the top drop-down menu of the IDE. Unclick the box "Configuration bits set in code" so

Figure A.5
Select the device type.

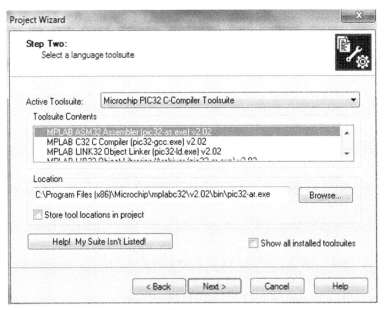

Figure A.6
Selecting the compiler.

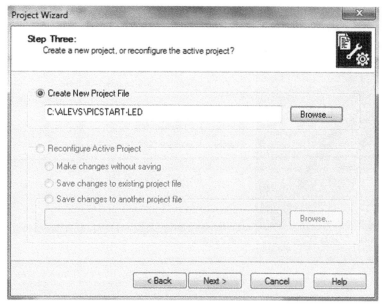

Figure A.7
Choose a new project name.

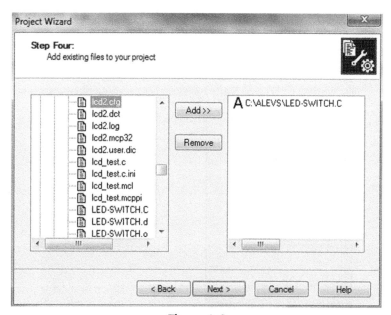

Figure A.8
Add source file to your project.

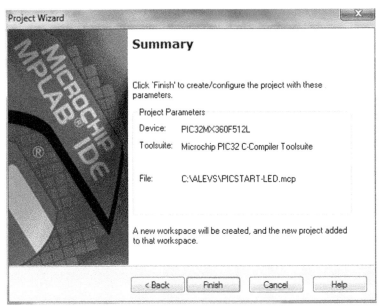

Figure A.9
Click finish to complete the project wizard.

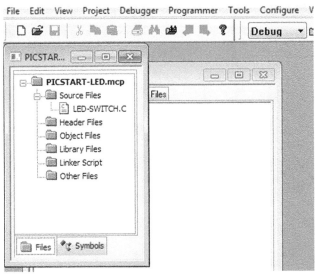

Figure A.10
List of project files.

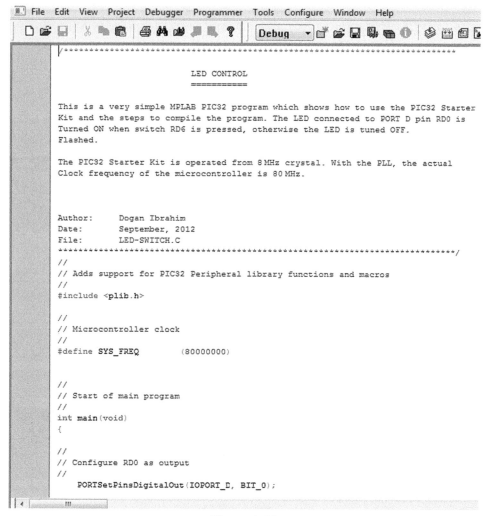

Figure A.11
File LED-SWITCH.C.

that the values given here take effect. Change the following configuration bits by clicking on the bit to be changed:

PLL Input Divider:	2
PLL Multiplier:	20
System PLL Output Clock Divider:	1
Oscillator Selection Bits:	Primary osc w/PLL (XT+,HS+,EC+PLL)
Primary Oscillator Configuration:	XT osc mode
Peripheral Clock Divisor:	Pb_Clk is Sys_Clk/1
Watchdog Timer Enable:	WDT Disabled

Address	Value	Field	Category	Setting
1FC0_2FF0	FFFFFFFF	USERID		
1FC0_2FF4	FFF8FFD9	FPLLIDIV	PLL Input Divider	2x Divider
		FPLLMUL	PLL Multiplier	20x Multiplier
		FPLLODIV	System PLL Output Clock Divider	PLL Divide by 1
1FC0_2FF8	FF7FCDFB	FNOSC	Oscillator Selection Bits	Primary Osc w/PLL (XT+,HS+,EC+PLL)
		FSOSCEN	Secondary Oscillator Enable	Enabled
		IESO	Internal/External Switch Over	Enabled
		POSCMOD	Primary Oscillator Configuration	XT osc mode
		OSCIOFNC	CLKO Output Signal Active on the	Enabled
		FPBDIV	Peripheral Clock Divisor	Pb_Clk is Sys_Clk/1
		FCKSM	Clock Switching and Monitor Sele	Clock Switch Disable, FSCM Disabled
		WDTPS	Watchdog Timer Postscaler	1:1048576
		FWDTEN	Watchdog Timer Enable	WDT Disabled (SWDTEN Bit Controls)
1FC0_2FFC	7FFFFFFF	DEBUG	Background Debugger Enable	Debugger is disabled
		ICESEL	ICE/ICD Comm Channel Select	ICE EMUC2/EMUD2 pins shared with PGC2/PGD2
		PWP	Program Flash Write Protect	Disable
		BWP	Boot Flash Write Protect bit	Protection Disabled
		CP	Code Protect	Protection Disabled

Figure A.12
List of Configuration bits.

```
Debug build of project `C:\ALEVS\PICSTART-LED.mcp' started.
Language tool versions: pic32-ar.exe v2.02, pic32-gcc.exe v2.02, pic32-ld.exe v2.02, pic32-ar.exe v2.02
Preprocessor symbol `__DEBUG' is defined.
Target debug platform is `__MPLAB_DEBUGGER_PIC32MXSK=1'.
Wed Oct 24 12:33:32 2012
```

```
Clean: Deleting intermediary and output files.
Clean: Deleted file "C:\ALEVS\LED-SWITCH.o".
Clean: Deleted file "C:\ALEVS\PICSTART-LED.elf".
Clean: Deleted file "C:\ALEVS\PICSTART-LED.hex".
Clean: Done.
Executing: "C:\Program Files (x86)\Microchip\mplabc32\v2.02\bin\pic32-gcc.exe" -mprocessor=32MX36C
Executing: "C:\Program Files (x86)\Microchip\mplabc32\v2.02\bin\pic32-gcc.exe" -mprocessor=32MX36C
Executing: "C:\Program Files (x86)\Microchip\mplabc32\v2.02\bin\pic32-bin2hex.exe" "C:\ALEVS\PICST/
Loaded C:\ALEVS\PICSTART-LED.elf.
```

```
Debug build of project `C:\ALEVS\PICSTART-LED.mcp' succeeded.
Language tool versions: pic32-ar.exe v2.02, pic32-gcc.exe v2.02, pic32-ld.exe v2.02, pic32-ar.exe v2.02
Preprocessor symbol `__DEBUG' is defined.
Target debug platform is `__MPLAB_DEBUGGER_PIC32MXSK=1'.
Wed Oct 24 12:33:33 2012
```

BUILD SUCCEEDED

Figure A.13
Compiling the program.

A list of the Configuration Bit settings for this project is shown in Figure A.12. You can set all the configuration bits in software by the following statements:

```
#pragma config FNOSC = PRIPLL          // Oscillator Selection
#pragma config FPLLIDIV = DIV_2        // PLL Input Divider
#pragma config FPLLMUL = MUL_20        // PLL Multiplier
#pragma config FPLLODIV = DIV_1        // PLL Output Divider
#pragma config FPBDIV = DIV_1          // Peripheral Clock divisor
```

```
#pragma config FWDTEN = OFF          // Watchdog Timer
#pragma config WDTPS = PS1           // Watchdog Timer Postscale
#pragma config FCKSM = CSDCMD        // Clock Switching & Fail Safe Clock Monitor
#pragma config OSCIOFNC = OFF        // CLKO Enable
#pragma config POSCMOD = XT          // Primary Oscillator
#pragma config IESO = OFF            // Internal/External Switch-over
#pragma config FSOSCEN = OFF         // Secondary Oscillator Enable
#pragma config CP = OFF              // Code Protect
#pragma config BWP = OFF             // Boot Flash Write Protect
#pragma config PWP = OFF             // Program Flash Write Protect
#pragma config ICESEL = ICS_PGx2     // ICE/ICD Comm Channel Select
#pragma config DEBUG = OFF           // Debugger Disabled
```

Step 9: Select Debug from the top drop-down menu. Then Click *Build All* from the top drop-down menu to compile the program. You should have no errors as shown in Figure A.13.

Step 10: Connect the PIC32 Starter Kit to the USB port of your PC. Click *Debugger → Select Tool → PIC32 Starter Kit* from the top drop-down menu.

Step 11: Click *Debugger → Run* to start the program. You should see a green progress bar at the bottom-left corner of the screen. Pressing switch SW1 on the PIC32 Starter Kit will turn ON the LED next to it.

Step 12: Stop the program by clicking *Debugger → Halt*.

Index

Note: Page numbers followed by "f" denote figures; "t" tables.

Printed and bound by CPI Group (UK) Ltd, Croydon, CR0 4YY

03/10/2024

01040315-0005